Collins

CORAL REEF GUIDE

RED SEA TO GULF OF ADEN, SOUTH OMAN

EWALD LIESKE • ROBERT F. MYERS

In collaboration with Klaus E. Fiedler

DEDICATION

It is with great pleasure that we dedicate this book to our friend and kindred spirit John E. Randall. 'Jack', as he is known among friends and colleagues, was born in 1924 and grew up near the ocean in southern California. He made his first dive with primitive military surplus scuba equipment in 1947. In 1950 he graduated from UCLA then sailed his self-built 12-m ketch 'Nani' to Hawaii to pursue graduate studies in ichthyology. Jack has done more to advance our knowledge of Indo-Pacific fishes than any other living ichthyologist. Over a span of 55 years he has made over 17,000 dives in nearly every tropical marine faunal region of the world. This and his keen ability to notice subtle differences among species has given him a rare insight into the identities of, and relationships among, coral reef fishes. He has participated in 10 expeditions to the Red Sea and several others to Oman and the Arabian Gulf. Either alone or with colleagues, he has discovered and described 555 species of fishes, including 29 from the Red Sea. He has also documented another 29 species previously unknown from the Red Sea, and reviewed several Red Sea fish families including the moray eels, cardinalfishes, goatfishes, damselfishes and wrasses. His great knowledge of the taxonomy and systematics of Indo-Pacific coral reef fishes has enabled him to make insightful comparisons of Red Sea populations to those outside the Red Sea, a crucial step in determining their status.

HarperCollins*Publishers* Ltd.
1 London Bridge Street
London SE1 9GF

HarperCollins *Publishers*
Macken House, 39/40 Mayor Street Upper,
Dublin 1, D01 C9W8, Ireland

The Collins website address is:
www.collins.co.uk

Collins is a registered trademark of
HarperCollins*Publishers* Ltd.

First published in 2004

Text © 2004 Robert F. Myers and Ewald Lieske
Artwork © 2004 Ewald Lieske
Photographs © as indicated in the captions and acknowledgements. Cover photo by Klaus Hilgert (Egypt).
Editor: Helen Brocklehurst

23

13

ISBN: 978 0 00 715986 4

Designed by Robert F. Myers and Ewald Lieske

Edited by D & N Publishing, Hungerford, Berkshire

Printed and bound in India by Replika Press Pvt. Ltd.

MIX
Paper from
responsible sources
FSC™ C007454

This book is produced from independently certified
FSC™ paper to ensure responsible forest management.

CONTENTS

FOREWORD

Whoever has the great privilege of transforming himself into an aquatic being, diving into the wonder-world of Red Sea coral reefs, or as a snorkeller, becomes aware of these magnificent abysses and forgets all the daily problems of life on land. He becomes immersed in a community of fishes which are not just aesthetic wonders, but which awaken an interest in knowing more about them and all the details of their behaviour. I am very pleased that this excellent book will help you as a companion and a competent guide. When in 1949, I was the first human to invade these still virgin and haunted waters, I had no idea that so many would follow.

I would like to use this occasion to pay homage to my colleague, the eminent ichthyologist John Randall, who during endless hours of keen observation, has explored the private lives of the many marine creatures in and around the reefs and captured them on film.

All life stems from the sea, including mankind. Nothing seems to me more important today than to bring our technological achievements into harmony with all the variety of living creatures. The advance of man into the sea and close contact with it as divers may help bring this about.

Hans Hass
Vienna, Austria
January 2003

Painting by Stephanie Naglschmid

▶ *Xarifa*, the research platform for Hans Hass' pioneering expeditions to document marine life.
Photo by G. Scheer

INTRODUCTION

The Red Sea is unique. Its lush reefs stand in stark contrast to the surrounding desert landscape. It is the only deep ocean that remains warm to great depths, yet its surface waters remain pleasantly cool. While biologically part of the great Indo-Pacific region, it has a character all its own. Many of its species are endemic, that is they are found nowhere else. The eastern portion of the adjacent Gulf of Aden is influenced by seasonal upwelling of cooler, nutrient-rich water. Here, stands of kelp compete with corals in a unique mix of endemic species and Indian Ocean species that do not occur in the Red Sea. This guide covers the coral reefs of the Red Sea and Gulf of Aden. It has been prepared in order to enable the visitor to identify nearly all species of marine life likely to be encountered from the intertidal to 40 m, the lower limit of safe sport diving. The metric system is used throughout. The following conversions can be used to obtain English equivalents: 1 cm = 0.394 in; 1 m = 3.28 ft or 39.4 in; 1 km = 0.6214 miles; 1 kg = 2.205 lbs; °C = °F − 32 × (5/9).

GEOLOGY AND OCEANOGRAPHY
BIRTH OF AN OCEAN

The Red Sea is a young ocean with origins dating to the Eocene. Forty million years ago, a split developed in the continental crust of northeastern Africa and what is now the Arabian Peninsula started drifting east. Through a process called sea floor spreading, this rift widens at the rate of 1 to 2 cm per year. At its centre there is considerable geological activity with hot brines of mineral-rich mud forming precious knolls of iron, manganese, zinc and silver. The rift splits east into the Indian Ocean through the Gulf of Aden and south through eastern Africa as the Great Rift Valley. The floor of the Red Sea remained dry until the Miocene, 20 million years ago, when water from the Mediterranean region of the Tethys Sea poured in from the north. This connection was tenuous with alternating periods of flooding and drying for both seas. Large salt deposits formed during periods of drying. A dramatic change occurred during the Pliocene, 5 million years ago, when the southern end of the Red Sea lowered as the Sinai area uplifted. This separated the Red Sea from the Mediterranean and connected it to the Indian Ocean, bringing in an Indo-Pacific fauna and setting the stage for coral reef development. Subsequent isolation from the Indo-Pacific set the stage for the evolution of the region's many endemic species. For most of the Pleistocene (the Ice Ages), from 2 million to 10,000 years ago, long periods of glaciation lowered the sea level as much as 145 m. This created land bridges at Bab Al Mandeb and the Straits of Tiran, isolating the Gulf of Aqaba from the Red Sea and both from the Indian Ocean. During these periods, hypersaline conditions lethal to all life may have prevailed while adjacent areas served as refuges. When the Red Sea reopened near the end of the Ice Ages 15,000 years ago, a new invasion of Indo-Pacific and regional endemic faunal elements occurred. For those species presently found only in the Red Sea, conditions in their Pleistocene refuges outside the Red Sea must have subsequently deteriorated, leaving them where they are now. Today, Red Sea coral reefs are rich and unique with over 1,000 species of fishes and about 230 species of hard corals. At least 15 per cent of the fishes and a similar proportion of other marine species are endemic, found nowhere else in the world.

The Red Sea reached its present sea level about 5,000 years ago. Today's reefs developed on older reefs that grew during earlier interglacial periods. Above many shores, there are fossilized reefs that formed at times of higher sea level. At Bab Al Mandeb (Arabic for 'gate of lamentations'), the straits are narrow (29 km wide) and only 130 m deep so there is little inflow from the Indian Ocean. With a length of 2,270 km, width of 100 to 350 km and maximum depth of 3,040 m, the Red Sea is like a large lake lying between barren deserts and mountains. Along the coast of the Sinai Peninsula, fringing reefs extending only a few metres from shore may drop precipitously to a depth of 500 m. In the southern Red Sea, there are two major archipelagos of offshore platform reefs and a single atoll, Sanganeb Reef.

HYDROGRAPHY AND CLIMATE

THE GREAT WATER-BATTERY

During the summer, air temperatures in the region typically exceed 40°C, warming Red Sea surface waters to 30°C. In winter, northern Red Sea surface waters cool to slightly below 20°C. Coral reefs and mangroves are thus able to grow under favourable conditions in the far northern parts of the Gulfs of Aqaba and Suez. The predominant north wind (shamal; up to 7 m/s during the winter) drives water circulation throughout the year bringing a unique year-round constant temperature of 20°C to depths of 1,000 m. Strong evaporation creates high surface salinity which ranges from 42‰ in the north to 37‰ in the south. In the Gulf of Aden it is 35‰ and in the Arabian Gulf, 46‰. Evaporation is balanced by surface water flowing in from the Indian Ocean. The area gets only 6 to 60 mm rain annually resulting in insignificant and usually episodic inflow from rivers. With such little sedimentation, visibility in most areas is good, ranging from 15 to 50 m. In the shallow Gulf of Suez (20–50 m) and the shelf areas around the Farasan and Dahlak Islands, visibility is usually poor. Tidal range is minimal, 0.5 m in the north and 1.5 m in the south.

Sea surface temperatures and winds (in Bft scale) over the Red Sea through the seasons. *(changed after H. Schuhmacker)*

TROPICAL KELP

The waters of the eastern Gulf of Aden and southern Oman are subject to seasonal and episodic periods of cold, nutrient-rich upwelling, bringing waters as cool as 18°C to the surface. This favours the growth of brown algae including such temperate species as the southern kelp, *Ecklonia radiata*, to thrive while inhibiting the growth of corals. The reef fauna here is relatively impoverished with a high proportion of endemic species as well as several warm-temperate species tolerant of these conditions. The Gulf of Aden thus has a unique mix of Red Sea endemics, Arabian Sea endemics, and Indian Ocean species that do not occur in the Red Sea.

STRUCTURE AND ZONATION OF A FRINGING REEF

A Inner lagoon: this region consists of a shallow channel between the intertidal beachrock zone and an extended area of seagrass (*Thalassia hemprichi*). It is often covered with *Ulva* and *Enteromorpha* algae. Other seagrasses (*Halodule uninervis*, *H. stipulacea*) and algae (*Halimeda* sp.) are common in areas of coral rock, dead corals and mud-filled troughs. There are few living corals (primarily *Stylopora pistillata*). The sea cucumber *Holothuria atra*, brittle star *Ophiocoma scolopendrina* and sea urchin *Echinometra mathaei* inhabit the troughs.

B Eroded coral heads: an area of massive eroded coral heads subject to a slight longshore current and moderate wave action. The coral heads may be covered with smaller epizoic corals (*Favia pallida*, *Goniastrea pectinata* and *Stylophora pistillata*) and soft corals (*Litophyton arboreum* and *Sinularia* spp.). The brown algae *Sargassum dentifolium* occupies large areas. The sea urchins *Diadema setosum* and *Echinometra mathaei*, starfish *Linkia multifora* and brittle star *Ophiocoma scolopendrina* occupy crevices. Snapping shrimps (*Alpheus* spp.) and their symbiotic gobies inhabit burrows in sand or mud.

C Microatoll zone: an area of large flat-topped coral heads subject to a slight longshore current and moderate wave action. Here, the massive stony corals (*Porites*, *Favia*, *Goniastrea* and *Platygyra lamellina*), some 2 m high, grow outward from dead centres. Less common are bushy yellow soft corals *Litophyton arboreum*, the delicate *Tubipora musica*, and mustard-hued fire corals *Millepora dichotoma* and *Millepora platyphylla*. The crabs *Dardanus lagopodes* and *Ashtoret lunaris* shelter between corals and the gastropod *Murex ramosus* inhabits sandy pockets.

D Reef platform and back-reef: here, wind-driven water displaced by the longshore current flows back through barren channels, small tunnels and canyons in the reef flat. Massive corals of the microatoll zone coalesce to form the back edge of the reef flat where rubble, dead coral and sediments are cemented together by the calcareus red algae *Lithothamnion*. *Fungia* corals occur in tidepools. Closer to the reef edge, mucus nets of the worm snail *Dendropoma maxima* cover the corals in which they live and the colourful giant clam *Tridacna maxima* is common. The slate-pencil urchin *Heterocentrotus mammillatus* and brittle star *Ophiocoma scolopendrina* hide in crevices.

E Reef margin and upper fore-reef slope: the seaward edge of the reef flat is subject to wave-induced turbulence and rip-currents. Here the reef margin is dissected into alternating ridges and deeply notched canyons and tunnels termed the 'spur and groove zone'. Where the wave assault is strong the predominant coral, the branched fire coral *Millepora dichotoma*, grows against the prevailing current. In calmer areas, the predominant corals are the branched *Acropora humilis* and *A. variabilis*. In large canyons exposed to strong surge and currents, a few corals such as *Acropora variabilis*, *Favia* sp. and *Seriatophora hystrix* may be found. Enlarged side channels may have depressions and chimneys with the same coral community as the back-reef (*Platygyra*, *Favia* and *Porites*). Massive *Porites* corals hundreds of years old occur on the upper slope as well as large tabular *Acropora* corals. The gorgonian *Acabaria pulchra*, soft coral *Litophyton arboreum* and cup coral *Tubastrea micrantha* occur in the shadows of overhangs. The hermit crab *Dardanus lagopodes*, gastropod *Murex ramosus*, bivalve *Pteria aegyptiaca*, starfish *Fromia ghardaqana*, crinoid *Lamprometra klunzingeri*, giant clam *Tridacna maxima* and Christmas tree worm *Spirobranchus giganteus* are common.

F Lower fore-reef and pinnacles: the base of the seaward reef slope consists primarily of coral rubble and sand. Among the few corals growing here are *Acropora maryae*, *Sinularia leptoclados* and the octocoral *Heteroxenia fuscescens*. Large coral pinnacles may rise from the base to the surface. The top may be eroded by wind and water but is often covered by massive *Porites* and *Platygyra* corals. The northern margin typically faces the wind-induced current and is covered with abundant coral growth while the southern margin is relatively barren and covered by the coralline red algae *Lithothamnion* sp. Rich coral growth typically covers the walls. In areas exposed to current below 25 m there may be large black coral trees (*Antipathes* spp.) and large fans (*Annella mollis*).

Fringing Reef (20–200 m)

Seaward Reef (Fore-reef)

Depth: 5 m, 10 m, 15 m, 20 m, 25 m, 30 m

Lagoon — A, B, C

Reef Platform — Back-reef, Reef Flat — D

Outer Reef Slope — E, F

Pinnacle

beach — seagrass — sand — eroded coral heads — living corals — microatolls — sand — coral rock — bedrock — algal ridge — reef margin — surge channels — submarine terrace — cave — living coral

A
Among rocks:
Gymnothorax pictus
Abudefduf sordidus
Istiblennius edentulus
Bathygobius spp.
Over sand:
Gerres spp.
Terapon jarbua
In seagrasses:
Cheilio inermis
Leptoscarus vaigiensis
Siganus rivulatus

B
Myrichthys colubrinus
Synanceia verrucosa
Dactylus arnanus
Amblygobius
albimaculatus
Cryptocentrus spp.
Lactoria cornuta

C
Echidna nebulosa
Chaetodon auriga
Chromis viridis
Chrysiptera unimacula
Dascyllus trimaculatus
Hipposcarus harid
Cirripectes castaneus
Canthigaster margaritata

D
Chaetodon auriga
Acanthurus sohal
Acanthurus nigrofuscus
Cheilinus lunulatus
Thalassoma rueppellii
Oxymonacanthus halli

E & F shallow (above 20 m):
Gymnothorax javanicus
Pseudanthias squamipinnis
Pseudochromis fridmani
Cephalopholis hemistiktos
Caesio striata, C. suevica
Chaetodon fasciatus
C. paucifasciatus
C. semilarvatus
Pomacanthus imperator
P. maculosus
Amphiprion bicinctus
Cetoscarus bicolor
Scarus ferrugineus
Zebrasoma desjardinii
Z. xanthurum
Pseudobalistes fuscus
Rhinecanthus assasi

E & F deep (below 20 m):
Sphyrna lewini (>40 m)
Gymnothorax nudivomer
Pseudanthias taeniatus
P. heemstrai (>40 m)
Genicanthus caudovittatus (>20 m)
Chromis pembae (>20 m)
C. pelloura (>35 m)
Bodianus opercularis (>30 m)
Paracheilinus octotaenia
Odonus niger

E & F all depths:
Carcharhinus amblyrhynchos
Triaenodon obesus
Lutjanus bohar
Cheilinus undulatus

On sand, most zones:
Taeniura lymma
Papilloculiceps longiceps

Structure and zonation of a fringing reef with characteristic marine life. Fishes are listed by zone.

REEFS UNDER ASSAULT

Coral reefs face many dangers, both natural and man-made. On a geological timescale, entire seas have come and gone while coral reefs have persisted. Reefs with a suite of organisms recognizable to us arose during the Eocene, a time of renaissance following the great extinctions that ended the Cretaceous. Through changing sea levels, drifting continents and shifting climates, the organisms that made these reefs have migrated, adapted and evolved. Like the forest fires of the land, plagues of coral-eating starfish and bleaching events have swept through coral reefs for as long as they have existed. As long as they are infrequent and localized, such events may be overcome. However, man has upset the balance on an unprecedented scale. Rare natural events have become frequent unnatural events with the survival of coral reefs hanging in the balance.

Coral bleaching is probably the single greatest threat facing coral reefs today. Bleaching is the loss of the symbiotic zooxanthellae that give reef-building corals their colour. Zooxanthellae are single-celled algae that live within the tissues of corals and other animals including many species of anemones and other Cnidaria, *Tridacna* clams, and some flatworms and tunicates. The algae aid its host by producing nutrients through photosynthesis, the process that converts CO_2 and metabolic wastes into sugars and oxygen. Certain stresses, particularly persistent high temperature, cause corals to expel their zooxanthellae. The white skeletons showing through the clear coral tissue gives them a bleached appearance. Different strains of zooxanthellae show different responses to stresses, so the expulsion may allow the coral to find new strains of zooxanthellae better suited to the offending stress. In some cases, bleached corals regain their zooxanthellae and return to normal. But if this condition persists, the coral will die. The single greatest cause of coral bleaching is prolonged periods of anomalously high temperatures. This phenomenon has increased in frequency during the past 20 years and was nearly global in scope in 1997–98. During that event, some areas such as Palau experienced unusually low temperatures followed by prolonged high temperatures resulting in coral mortality reaching as high as 90 per cent. At Bali, bleached anemones were able to survive longer and regain their zooxanthellae, while many corals died. Staghorn and tabletop *Acropora* corals were the hardest hit. El Niños and coral bleaching are all indications of global warming. Other causes of bleaching are increased exposure to UV radiation, exposure to air or rain during extreme low tides, and exposure to pollutants.

Pollution adversely affects coral reefs. Many chemicals, particularly petroleum products, gather at the very surface of the water. This is particularly dangerous to coral gametes which rise to the surface after spawning. Most corals synchronize their spawning in a single annual event. Surface pollution at that time can wipe out an entire year's reproduction. Most invertebrates are extremely sensitive to heavy metals, much more so than fishes. Reproduction and embryonic development are the most sensitive stages. Even minute concentrations can prevent coral embryos from settling. High concentrations of certain nutrients such as ammonia, nitrates and phosphates promote the growth of algae which may choke out or overgrow newly settled corals. In the Caribbean, recent plagues of black band disease, an insidious bacterial infection that rots living corals, have been traced to spores blown across the Atlantic from the Sahara. This has dramatically increased with desertification, the process in which grasslands and scrub are turned into desert by overgrazing and destructive agricultural practices.

Destructive fishing practices, particularly the use of explosives and cyanide, have destroyed vast areas of reef throughout much of the underdeveloped world. Both methods are a convenient way to get a lot of fish quickly. Explosives are used to stun and kill fishes. Those that float to the surface are harvested, those that sink are often left to rot. The corals closest to the blast are pulverized. Sodium cyanide is used to collect fishes for the live reef fishery. This practice started in the Philippines with the aquarium trade, then spread throughout the Asia-Pacific region, most recently driven by the Chinese restaurant industry. Cyanide is squirted into the reef to immobilize the target species which are caught, then revived in clean water and kept alive. Everything else is left to die. The fishes exposed to cyanide suffer liver damage which eventually kills them. In the aquarium trade, most of these fishes die long before they are sold, those that live long enough die later in the hobbyist's aquarium. In the restaurant trade, the most valuable species, groupers (p. 66) and the Napoleon wrasse (p.150), are kept in pens, then shipped in live wells to restaurants, primarily in Singapore and Hong Kong, and more recently throughout China. In the restaurant they are kept in aquariums where the customer can choose the fish destined for his plate. Rich Chinese are willing to pay exorbitant prices for such fish. In the same deadly theme that has cursed endangered species throughout the world, the false belief that a particular animal or animal part is an aphrodisiac is a driving force behind the live reef fishery. A live Napoleon wrasse may fetch up to US$180 per kg ($82 per lb). A dish of the lips, considered an aphrodisiac, may cost as much as $300. The Stonefish (p. 64) is also considered an aphrodisiac with live ones fetching a high price. The giant Bumphead parrotfish (p. 170) is also in great danger. It is a prime target of cyanide fishers as well as native spearfishermen who seek them as they sleep at night.

Overfishing, even if by non-destructive fishing practices, is harmful if stocks are exploited beyond their sustainable yield. Refrigeration and the shift from subsistence to cash economies has made it possible for coastal villagers to obtain short-term profits by taking more than their immediate local needs. Exploding human populations and the ability to get frozen seafood to distant markets have driven populations of many coral reef animals to near extinction. The best solution is to establish strictly enforced preserves. These must be large enough to protect enough of a species' population for it to reproduce and sustain itself. At least 20 per cent of appropriate habitat is required to meet this need. It is critical that particularly sensitive areas such as spawning sites are included. Spawning sites outside the preserves should also be protected at critical spawning times. It is important that the largest individuals of a species be preserved. Fecundity is exponential to size: a 2 kg fish may produce 10 times as many eggs as a 1 kg fish of the same species. Preserves are also the best way of attracting the tourists that sustain many economies. Here fishes can grow to a large size and become tame as they grow accustomed to the presence of divers who pose no threat. A Napoleon wrasse worth $1,000 to a fisherman is worth much more alive as a star attraction in a national park that draws thousands of visitors a year. Preserves also serve to educate local peoples by providing a place where species can be observed in their natural condition, fostering an appreciation for nature that is often lost to the urban lives of the masses.

Crown-of-thorns starfish infestations are a major cause of damage to coral reefs. This starfish (*Acanthaster planci*, p. 341) feeds on live coral, particularly species of *Acropora*. Under normal conditions, *Acanthaster* is an uncommon nocturnal predator. When its population explodes, it may form an unbroken front several starfish deep that advances across the reef 24 hours a day, leaving in its wake a scene of white bleached corals. These are soon covered with filamentous algae, then begin to decay and crumble. Many species that depend on the living coral for food or shelter soon disappear, while certain herbivores may temporarily increase in number. Ultimately the reduction in shelter results in a reduction in the population of all the species that depend on that shelter. One possible cause of *Acanthaster* explosions is reduced populations of its natural enemies. Triton's trumpet (*Charonia tritonis*, p. 290), Harlequin shrimp (*Hymenocera elegans*, p. 324) and Napoleon wrasse feed on *Acanthaster*. Triton's trumpet is a highly prized shell that has been over-collected in many areas. The spectacular Harlequin shrimp is sought after for the aquarium trade, but is difficult to find, is too small, and feeds on its prey too slowly to have much of an impact. The threatened Napoleon wrasse feeds on many other prey species and its importance as a predator on *Acanthaster* is poorly known. Taken together, the reduced populations of predators may help give *Acanthaster* an edge when conditions are right for its population to explode. A possible trigger for such explosions is a torrential rain event that washes nutrients onto the reef that favour the growth and development of larval *Acanthaster*. These develop deep beneath coral rubble before emerging as young adults. Studies of reef cores taken on the Great Barrier Reef have shown that *Acanthaster* plagues have occurred for millenia at intervals averaging 400 years, allowing plenty of time for recovery. However, in places where coastal agriculture and development have contributed to massive runoff, the frequency has increased dramatically, leaving very little time for recovery.

CONSERVATION

In the Red Sea the pressure on coral reefs near urban areas is high. In the early 1980s, Egypt recognized the damage occurring on the reefs of the Sinai and responded by creating the Egyptian Environmental Affairs Agency (EEAA). Ras Mohammed National Park was established in 1983, followed by the Nabq and Abu Galum Nature Reserves in 1992. Today, 52 per cent of the Egyptian Sinai coastline is protected. Coral reefs of intenational importance as well as unique coastal and desert habitats have been saved.

The situation around Hurghada and Safagha remains problematic. Hotels and settlements, often poorly planned, have severely damaged most of the area's fringing reefs. Near Jeddah, Saudi Arabia, similar urban developments have damaged a 100 km stretch of coastal reefs. The situation in Egypt south of Quseir is critical: a 300 km stretch of coast down to Ras Banas has been sold by the government to developers to boost tourism. As an underdeveloped country, Egypt needs tourism. Unfortunately, many new hotels have been built in innapropriate places. In other areas this has resulted in ruined investments and empty hotels. Along this stretch of coast, there are only 34 natural bays (marsas) offering good access for swimming, diving or boats. Hotels built on broad reef flats without swimming beaches or access to boats face a difficult time staying in the diving market. Long piers are a poor solution since they are often pounded by dangerously high surf generated by strong northerly winds.

The continental shelf of much of the Red Sea is narrow. Exceptions are the Straits of Tiran at the entrance to the Gulf of Aqaba and the Farasan and Dahlak Archipelagos in the south. The Egyptian coast south of Hurghada has few offshore reefs and islands, hence a limited area to support diving or fishing. The area north of Ras Banas (Siyul, Fury and Lahami Reefs) has about 60 small detached reefs. About half of them are still in good condition and should be declared a national park. The Gebel Elba area to the south also has great potential for diving.

A newly opened private international airport near Marsa Alam is poised to bring in thousands of tourists. Many new hotels have been built or are planned for the surrounding area. The demand for fresh water and food, particularly reef fishes will increase dramatically. Divers who love to eat fish as well as see them underwater will be a problem. Marine preserves are urgently needed to save the local fish populations before they are over-exploited.

Parts of the northern Red Sea have been severely affected by the Crown-of-thorns starfish. Two years after the great flash-flood of 1996, a Crown-of-thorns infestation spread from Hurghada north to Ras Mohammed and south to Ras Banas. Reefs in current-poor nearshore shelf areas were hardest hit while those exposed to waves and currents faired well. Ras Mohammed National Park was saved by the removal of nearly 180,000 of the starfishes.

What is the situation in other countries? Saudi Arabia has only a few protected areas: Asir National Park (4,500 km^2) and the Farasan and Dawat/Qamari islands (2,100 km^2). Most of the reefs away from urban areas are still in relatively good condition. In 1990, Sudan declared the Red Sea's only atoll, Sanganeb Reef, a national park.

Shark populations in the southern Red Sea have been severely impacted by the Yemeni shark-fin trade. Finless Hammerheads have been seen piled high on the Hanish islands. With only the fins in demand, the despicable practice of throwing away the rest of the animal to die has decimated shark populations throughout the world. The future of sharks worldwide is bleak with fins bringing prices of up to several hundred dollars per kg in Asian markets. The newest threats to Red Sea coral reefs are the aquarium fish and live reef fish fisheries out of Jeddah operated by hired Philippinos. The area's butterflyfishes and angelfishes are already becoming scarce. Live groupers, particularly *Plectropomus pessuliferus* (p. 66) are highly sought for Chinese restaurants and their populations are rapidly declining. Throughout the Red Sea, even in well-protected areas, a major attaction to divers, the Napoleon wrasse, *Cheilinus undulatus*, has nearly disappeared.

LIFE ON THE CORAL REEF

ECOLOGY AND BEHAVIOUR

Coral reefs contain many habitats, each with its own set of conditions and each species of plant and animal has its own set of requirements for life and procreation. Some are specialists that occur only in specific habitats, while others occur in a wide variety of places and habitats. Some may be inconsistently present due to seemingly random reasons, such as differential recruitment of juveniles or occasional disturbances such as storms.

At the base of the coral reef's food web are marine plants. These include unicellular forms (diatoms, phytoplankton and zooxanthellae), benthic algae (seaweeds) and seagrasses. Next are the consumers, in three trophic levels: herbivores that feed on plants, omnivores that feed on plants and animals, and carnivores that feed on animals. Consumers may be further subdivided according to their diet: planktivores feed on plankton, corallivores feed on corals and detritivores feed on detritus (decomposing plant and animal particles). Corals and other animals that contain zooxanthellae are unique because they are consumers that host their own primary producers, the zooxanthellae. Most major groups of marine animals include herbivores, omnivores and carnivores.

Coral reef organisms exhibit virtually every known form of behaviour and lifestyle. Even seemingly inanimate animals such as corals have surprisingly complex behaviours: most actively prey on drifting plankton and some engage in chemical warfare by producing toxins that keep them from being overgrown. Behaviour is more easily seen in mobile free-living animals such as fishes. Virtually all behaviour is related to eating, avoiding being eaten, or reproducing. These three requirements of survival are accomplished through a broad spectrum of lifestyles that determine whether an animal is solitary or gregarious, site-attached or free-ranging, territorial or peaceful, active by day (diurnal) or night (nocturnal) and conspicuous or cryptic.

PROTECTIVE RESEMBLANCE AND MIMICRY

Protective resemblance and mimicry are ploys used by many species to capture prey or escape predators. Camouflage is used by many animals to blend in with their surroundings. Scorpionfishes and frogfishes have fleshy tassels or warts that resemble fronds of algae and some even have algae growing on them. Many groupers become virtually invisible when overlain with a series of darker bands that break up their outline. Juvenile Rockmover wrasses (*Novaculichthys taeniourus*, p. 163) look and swim like a clump of detached seaweed.

Protective resemblance is known as mimicry when one organism, the mimic, resembles another, the model, that is protected from predation by virtue of distastefulness, toxicity, or some other characteristic. There are four basic forms of mimicry: Batesian mimicry when an otherwise unprotected animal resembles a protected one, Müllerian mimicry when two or more protected animals closely resemble one another, aggressive mimicry when a mimic uses its disguise for predation or other hostile motives, and social mimicry when two or more species resemble and interact with one another in order to gain protection through numbers.

Among coral reef animals, Batesian mimicry is the most common. Both model and mimic share a distinctive conspicuous colour pattern (aposematic colouration) that predators learn to avoid. The model is usually more common than the mimic. A typical example is that of the slow-swimming noxious Red Sea toby (*Canthigaster margaritata*, p. 222) and the edible Spotted toby mimic (*Paraluteres arqat*, p. 217). In aggressive mimicry, the mimic uses the guise of a harmless or beneficial model to take advantage of another animal. This is well-developed among blennies (p. 177). In a classic example, the Cleaner mimic (*Aspidontus taeniatus*, p. 182) has evolved a colour pattern identical to that of the Bluestreak cleaner wrasse (*Labroides dimidiatus*, p. 162). The wrasse services other fishes by removing parasites and pieces of damaged tissue. Predatory fishes recognize the cleaner's service and will not attempt to eat it. The blenny tricks its victim into posing to be cleaned, then darts in to make a meal of a piece of fin or scale! Experienced fishes usually learn to distinguish the two and avoid the mimic. Social mimicry is used by schooling or aggregating species to gain protection through increased numbers. A predator has a more difficult time singling out prey from a large group of similar looking individuals than from a smaller group. Different species can increase their group size by resembling one another. Females of the anthiases *Pseudanthias squamipinnis* and *P. taeniatus* and the blenny *Ecsenius midas* have similar colour patterns and mingle together.

Benefactor or nemesis? The cleaner wrasse (left) provides a valuable service to other fishes. The mimic, a blenny, takes advantage of this in order to approach closely unsuspecting prey and take a bite.

SYMBIOSIS

Symbiosis is the close association of two dissimilar organisms. There are three basic forms: mutualism when both organisms depend on each other, commensalism when one organism depends upon the other without harming it, and parasitism when one benefits to the detriment of the other. The coral reef itself is built upon the symbiotic relationship between corals and zooxanthellae (p. 241). A classic example of mutualism is the relationship between shrimpgobies and burrowing shrimps (pp. 187, 196, 326). Perhaps the best-known example of commensalism is the relationship among anemonefishes and their host anemones (p. 148). Virtually every coral reef animal harbours parasites. Most are internal and unseen, but a few are external; certain species of copepods and isopods live attached to the skin of fishes and some species of pearlfishes (p. 47) may feed on the respiratory trees of their host echinoderms.

REPRODUCTION AND DEVELOPMENT

Reproduction among coral reef animals is highly varied. Among invertebrates, many primitive forms reproduce asexually as well as sexually. Flatworms may simply divide into two while sponges and cnidarians may bud new organisms and produce vast colonies of identical clones. In this way colonial organisms often have growth forms and patterns that resemble those of plants. Vertebrates reproduce sexually by combining sperm and eggs from separate individuals to create offspring. This may be in ways that seem bizarre. Hermaphroditism is quite common among marine animals. Some species are simultaneously male and female. They may exchange gametes in both directions, or take turns playing the roles of female and male. Others are sequential hermaphrodites by starting life as one sex then switching to the other. Fertilization may be internal in which the male delivers sperm through a penis or similar organ, or external in which the gametes are shed into the water.

Most marine animals produce enormous numbers of widely dispersed expendable young. Mortality is high, but just enough survive through maturity to maintain the species. A few of the larger species, such as sharks, rays and marine mammals, produce a few fully developed young that stand a good chance of surviving through reproduction. Marine animals that produce large numbers of eggs distribute them by either depositing them on the bottom (nesting) or by spewing them into the water (broadcast spawning). Most molluscs and some fishes (dottybacks, damselfishes, blennies, gobies, triggerfishes, filefishes, trunkfishes and puffers) are nesters. Jawfishes and cardinalfishes are mouthbrooders, with the male keeping the eggs in his mouth until hatching. Most fishes and macroinvertebrates are broadcast spawners.

Early development of marine animals occurs in two basic ways: by direct development into juveniles or through a larval stage. Direct development is uncommon among coral reef animals. It occurs in some molluscs and all sharks and rays, but in only three species of bony fishes, none from this region. Direct developers tend to produce relatively few young and have limited distributions, usually along continental coasts. Most marine animals have a lengthy pelagic larval stage and are more broadly distributed. Larvae start out as tadpole-like creatures, some with an external yolk sack to nourish them until their gut develops. They then drift with the currents and feed in succession on phytoplankton then larger zooplankton as they grow. The open sea is a much safer place for planktonic eggs and larvae than the reef with its surface of filter feeders and hoards of planktivores. Some larvae actively swim, guided by environmental cues that may help them find a suitable settling site.

Reproduction among coral reef animals usually occurs in cycles, most often around either the new or full moon when the tidal range is highest and currents are their strongest. This offers the best chance for eggs and larvae to disperse. Many species spawn near the edge of the reef at sunset which gives the eggs a chance to drift away from the reef when most planktivorous fishes are asleep. Many species have reproductive peaks in the spring, so that settlement and growth of juveniles occurs when water temperature is the warmest. Some highly mobile animals such as large groupers and snappers migrate once a year to favoured sites where they gather in large numbers to spawn. These sites are usually at the entrances of major channels or promontories where currents can quickly carry the eggs out to sea. Many corals spawn only once or twice a year, usually shortly after dark during a midsummer full moon.

Growth rates and lifespans vary greatly among coral reef animals. Some small species may reach maturity within a few months and live less than a year. Many familiar reef fishes (groupers, snappers, emperors, butterflyfishes, angelfishes, damselfishes and surgeonfishes) have surprisingly long lifespans of 10 to 20 years, while larger species may take several years to mature and live as long as 80 years or more. Some colonial clonal organisms, such as corals, could be thousands of years old.

Reproductive strategy and maximum lifespan are crucial to a species' ability to withstand human exploitation. Species that grow and mature fast and reproduce in enormous numbers tend to be resilient to over-exploitation by having the ability to recover quickly when offered protection. Those that grow and mature slowly and have few young tend to be highly vulnerable. They are among the first to disappear, and in some cases may face extinction.

DANGEROUS MARINE LIFE

VERTEBRATES

Fishes pose a threat to humans by either biting, stinging or being poisonous to eat. Among the biters, only a few species of **sharks** have fatally attacked humans. Among these, only the Tiger shark *Galeocerdo cuvier* (p. 27) regularly visits coral reefs. During the day it usually occurs in deep water along steep seaward slopes, but at night it is more likely to visit shallow areas. The Oceanic whitetip shark *Carcharhinus longimanus* (p. 24) is an open sea species that occasionally visits steep offshore pinnacles. The dreaded Great white shark is a cooler water species that very rarely enters shallow tropical waters, and the Bull shark *Carcharhinus leucus* (p. 28) is a coastal species that enters estuaries and rivers. The Grey reef shark *Carcharhinus amblyrhynchos* (p. 25) is a territorial and occasionally aggressive species commonly encountered by divers. It may view man as a competitor and has been involved in numerous non-fatal attacks. Before an attack it engages in ritualized agonistic behaviour called 'threat-posturing', by swimming with exaggerated movements. If any shark gets overly curious, it is advisable to back off or get out of the water. Sharks should never be intentionally attracted by speared or hooked fishes. Other potentially dangerous biters are barracudas, moray eels and large triggerfishes. Most attacks by **barracudas** (p. 207) occur in murky water as a result of mistaken identity. Most bites from **moray eels** (p. 34) result from thrusting a hand in a hole, or carrying speared fish, but occasionally large morays may be aggressive. Large **triggerfishes**, usually either *Balistoides viridescens* or *Pseudobalistes flavimarginatus* (p. 212), aggressively guard their nests if eggs are present. Their fierce attacks have caused severe injuries requiring stitches. In one case, a dive master struck on the head nearly lost consciousness and had to be rescued and in another, a man required over 30 stitches to close a wound on his cheek.

    Scorpionfishes and **rabbitfishes** have venomous spines. The stout dorsal spines of the **Stonefish** *Synanceia verrucosa* (p. 64) can penetrate tennis shoes. Its excruciatingly painful stings may be fatal without adequate medical care. The highly conspicuous **lionfishes** (p. 60) make little effort to avoid the diver. Their stings are extremely painful but rarely fatal. Stings by rabbitfishes (p. 200) may be nearly as painful as those of the stonefish, but have not caused fatalities. **Surgeonfishes** (p. 202) have either a pair of razor-sharp movable sheathed blades or two pairs of sharp fixed bucklers at the base of the tail. **Squirrelfishes** (p. 50) have prominent cheek spines. **Stingrays** (p. 31) have one or more sharp detachable spines at the base of a long tail which are capable of causing severe wounds. Many other species of reef fishes can cause minor wounds with sharp spines, scales, or teeth, and should be handled cautiously.

    Some reef fishes are poisonous to eat. Toxicity may be due to substances either made by the fish itself, or ingested with other organisms. Many **puffers** (p. 220) and **boxfishes** (p. 218) have highly toxic skin and viscera. The toxin, **tetradotoxin**, is among the most powerful neurotoxins known and has caused many fatalities. **Soapfishes** (p. 77) and possibly juvenile **sweetlips** (p. 110) secrete distasteful toxins through the skin that discourage predators. Ciguatera fish poisoning may be present in certain fishes. It reaches its highest concentrations in the piscivores at the top of the food chain. It does not affect the fishes themselves but can cause extreme lingering illness and even death in humans. Symptoms vary from a tingling of the lips and extremities to reversal of the sensations of hot and cold, muscular weakness, vomiting, diarrhoea, shortness of breath and cardiac arrest. There is no known cure, but treatment by intravenous injection of manitol (sugar) often reduces the symptoms. The greatest danger of ciguatera is its unpredictability of occurrence which varies from place to place and through time. The ciguatera toxin is produced by a small dinoflagellate *Gambierdiscus toxicus* that colonizes bare surfaces. It is eaten along with filamentous algae by herbivorous fishes which in turn are eaten by predators. Since the toxin is not metabolized, it accumulates in the flesh and particularly the liver and reproductive organs. Each time a predator eats a smaller fish, it accumulates its victim's lifetime accumulation of ciguatera. Consequently, the highest concentrations occur in large predators. In the Indo-Pacific, the **Twinspot snapper** *Lutjanus bohar* (p. 104) and **Giant moray** *Gymnothorax javanicus* (p. 36) are the most frequently toxic species. Any unusually large individuals of groupers, snappers, emperors, jacks, barracudas and triggerfishes should be treated with caution and local knowledge is often helpful.

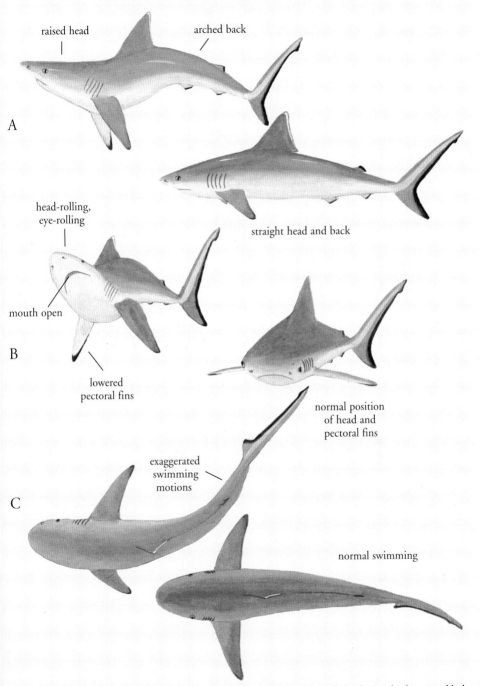

raised head

arched back

A

head-rolling,
eye-rolling

straight head and back

mouth open

B

lowered
pectoral fins

normal position
of head and
pectoral fins

exaggerated
swimming
motions

C

normal swimming

Agonistic behaviour compared with normal swimming in the **Grey reef shark** *Carcharhinus amblyrhynchos* showing A) raised head and arched back, B) head- and eye-rolling and lowered pectoral fins and C) exaggerated swimming motions.

Sea snakes (p. 225): none of the 54 known species of sea snakes occur in the Red Sea but at least one occurs in the Gulf of Aden and several occur in the Arabian Gulf and Gulf of Oman. Although docile, their neurotoxic venom is highly potent: theoretically, 1.5 mg can cause death. The **Banded snake eel** *Myrichthys colubrinus* (p. 41) is often misidentified as a snake in the Red Sea!

INVERTEBRATES

Sponges (p. 234): many species have minute silica spicules or bio-active chemicals that irritate the skin or cause painful dermatitis.

Hydrozoans: all cnidarians (fire corals, hydroids, hydrozoans, jellyfishes, anemones and corals; p. 240) have nematocysts, specialized venomous cells containing a small harpoon launched by a spring-like trigger. Fire corals, *Millepora* spp. and hydroids (p. 243) inflict very painful stings. Fire corals are abundant along the upper edges of fringing reefs where they are easily brushed by snorkellers. They are easily recognized by their mustard colour and small hair-like surface structures.

Portuguese man-of-war (*Physalia physalis*, p. 244): this floating hydrozoan is often mistaken for a jellyfish. Its retractile tentacles may extend several metres and produce extremely painful stings. They are often washed ashore.

Jellyfishes (p. 246): all jellyfishes may irritate the skin. A few small species of sea wasps (Cubomedusae) in the area can cause painful stings, but none is as virulent as the Australasian *Chironex fleckeri* which can cause death within minutes.

Sea anemones (p. 254): the small (2 cm) *Triactis producta* may inflict an extremely painful sting and the more common *Alicia pretiosa* may cause a severe inflammatory rash. Most other species can cause inflammation to sensitive areas of skin.

Corals: most hard corals (p. 258) have sharp skeletons which may cause severe lacerations that heal slowly and easily become infected. Coral wounds must be thoroughly cleaned and may require antibiotics.

Cone shells (*Conus* spp., p. 291): these molluscs have venomous harpoon-like darts that are thrust from the tip of a long worm-like proboscis which can reach all regions of the shell (this is not the hollow siphon used to breathe). The venom causes paralysis and a few species (*Conus geographus*, *C. striatus*, *C. textile*) have caused death. All species should be handled with great care and the soft parts never touched. Treatment is to remove the venom by suction and seek immediate medical attention.

Cephalopods (p. 318): all octopus species have a venomous beak located in the centre of their underside where the tentacles meet. None of the local species are dangerous, but they may give a painful bite if mishandled. The bite of the tiny Australasian Blue-ringed octopus (*Hapalochlaena lunulata*) can cause death within an hour. Its neurotoxic venom paralyses the respiratory muscles. The only treatment is to be put on a respirator until symptoms pass.

Crown-of-thorns starfish (*Acanthaster planci*, p. 340): this coral predator's spines have a venomous sheath and may cause highly painful wounds or even paralysis, particularly if broken off. The venom may be destroyed with hot water (more than 50°C).

Sea urchins (p. 348): many species have extremely sharp and delicate spines that easily pierce the skin and desintegrate. While painful, such wounds usually heal within a few days and the unremoved particles dissolve. The globular tips of the spines of the fire urchin *Asthenosoma marisrubri* contain a venom that causes intense pain. Spines of long-spined *Diadema* species have a mild toxin that may affect breathing in sensitive people.

INTERESTING DIVE SITES IN EGYPT, ISRAEL AND JORDAN

There is a map of the region on pages 382–3 showing where the best dive sites are.

1. Elat and Aqaba Heavily impacted by sewage, overbuilding, and potash sediments from the ports of Elat and Aqaba. Some good areas with rich coral growth remain 10 km south of Aqaba (Arabian Border, Cedar Pride, Wrack, Canyon). An astonishing variety of deep-dwelling fishes can be found in only 15 m: Heemstra's anthias, Red Sea angelfishes, and rare morays. The Royal Diving Center founded by King Hussein is the best starting point.

2. Nuweiba A small bedouin village in a large sandy wadi. Litter from ferries and many new hotels is a problem, mosquitoes are rampant. Good coral growth (60–80 per cent coverage) in the north and south. Few large fishes compared to other places, perhaps due to overfishing. Up to about 100 species of fishes observed per dive. Bottlenose dolphin and rare morays. Former Helnan Hotel now under Italian management has dive centre and campsite with interesting small house reef. Boat and jeep tours available.

3. Abu Galum Nature Reserve (400 km²) 20 km south of Nuweiba. Access by coastal dirt road skirting steep mountains and large wadis. Ranger station in the north. The bedouin village of Galum in the south hosts camel/dive treks from Dahab. Ras Mamlah and Galum boast rich coral growth and high diversity of fishes. Dropoffs feature Rainbow runner, tunas, rare anthiases (Heemstra's and Red-striped), Dusky chromis, Red Sea angelfishes and frogfishes. Jackals, Red foxes, Dorca gazelles and 165 species of plants, 44 of them endemic, inhabit the preserve.

4. Dahab This large village located in a vast wadi still has its own special atmosphere with many quaint hotels, pensions and cafés. There are about 50 dive operators, some with good house reefs (Inmo, Laguna and Daniella Village). Jeep tours available to famous dive sites to the north which are often overcrowded (Blue Hole, Canyon) or pounded by northerly winds (The Bells). Sheltered dive sites are to the south (Blocks, Three Pools and Um Sid). Many problems due to overbuilding; Lighthouse Reef is littered by trash. Still recommended to beginners. Camel tours to Abu Galum 10 km north are usually expensive. Notable fishes include rare sea horses and pipefishes, Heemstra's anthias, Dusky chromis and many wrasses and triggerfishes.

5. Nabq Nature Reserve (600 km²) 35 km north of Sharm El Sheik. Difficult access by coastal dirt road, best route is from south, jeep tours available from Sharm El Sheik. Wadi Kid features high dunes, Dorca gazelles and Nubian bighorns. Shoreline at Al Arkana has large area of mangroves and a shrimp farm. Fine reef crest 200 m offshore shelters a labyrinth of *Porites* bommies and silty valleys with extensive seagrass beds. Seaward reef has rich coral gardens and thick seagrass beds in crystal clear water.

6. Ras Mohammed National Park 12 km south of Sharm El Sheik at tip of Sinai Peninsula and Straits of Tiran. Many famous dive sites with large tame fishes: Marsa Bareka (a large bay with high fish diversity), Shark Reef (sharks, large snappers and schools of jacks), Jolanda wreck, Shark Observatory (dropoffs with gorgonians and soft corals), Anemone City and Eel Garden (Garden eels). The land features 2 million-year-old fossil reefs and is an important stopover for thousands of migrating storks and eagles. Marsa Bareka has many campsites available through park headquarters. Boat dives available through many dive operators in Sharm, including the oldest, Sinai Divers at Naama Bay.

7. Hurghada A large city with international airport. Rampant development during the 1980s and Crown-of-thorns infestation of 1998 destroyed most fringing reefs near town. Many offshore islands and reefs with several wrecks, some with good variety of fishes. Many hotels and dive operators from budget to high-end. Some hotels on the road south to Safaga still have good house reefs.

8. Safaga A busy port and major terminus for potash export. Many hotels and pensions in all categories with some fine hotels north of town (Shams Hotel, Menaville). Fringing reefs near town destroyed by pollution, overfishing and Crown-of-thorns. A few offshore patch reefs (ergs) and wrecks still offer good diving with lush dropoffs and pinnacles. Many operators dive the ferry Salem Express wrecked in 1991 while carrying 690 pilgrims from Jeddah.

9. Mövenpick (Quadim Bay) 6 km north of Quseir, 0.5 km², 65 m deep, somewhat exposed. Very good house reef with over 200 species of fishes (flashlightfishes, large groupers, velvetfishes and stargazers) and Bottle-nose dolphins, and fine coral growth, especially at entrance. Subex runs excellent dive operation with access by pier for shore- and boat-diving. Jeep tours available. Mövenpick Hotel is efficiently managed and the reef well-cared for. High-end, but highly recommended for snorkellers and divers.

10. Mangrove Bay 40 km south of Quseir, large sheltered bay 1.5 km². Very good house reef with rich coral growth and over 180 species of fishes (Zebra angelfishes, Yellowbar angelfishes, devilfishes, stonefishes, schools of mackerel and fusiliers, many butterflyfishes, wrasses, gobies). South side of bay with excellent coral growth. Extensive seagrass beds outside the bay occasionally visited by Dugongs. Arabian-German dive operation with pier access for shore- and boat-diving. Very good for snorkelling and UW photography. Mangrove Hotel close to bay.

11. Shagra Bay A small protected bay with a fine house reef and extremely tame fishes. Fringing reefs outside of bay have Bottlenose and Spinner dolphins and good diversity of fishes. Crown-of-thorns has affected a few offshore patch reefs (ergs) still good for seeing sharks. Scalloped hammerheads and Oceanic whitetips are consistently present at Elphinstone Reef 9 km offshore. Efficient dive operation run by Red Sea Diving Safari offers shore and boat access to bay, and boat and truck access to distant sites regularly visited by sharks, dugongs, turtles and dolphins. Shagra Ecolodge offers small bungalows and safari tents for divers of all budgets. Highly recommended for UW photographers. El Nabaa, a small bay 10 km north with a long fringing reef, features Spinner dolphins and turtles, but the fishes are shy. Equinox Hotel nearby.

12. Marsa Alam Small fishing village, Coral Cove Camp 14 km north. Fine fringing reefs and sandy bays with good sea life. Access over reef flat sometimes difficult. Dugongs and Spinner dolphins visit regularly, cold water spring at 50 m. Quality of camp has deteriorated.

13. Shams Alam Reefs Fine beaches, small house reef and very good fringing reef to the north. Some damage by Crown-of-thorns. Shallow shelf (15–20 m) west of Wadi Gimal Island with 40 offshore patch reefs, about 18 still with good coral growth. About 180 species of fishes including *Chaetodon larvatus*, many frogfishes and triggerfishes. Platform reefs with dropoffs to 100 m at Dolphin and Shab Sharm offer turtles, pods of dolphin, and Oceanic whitetip sharks. Primarily boat diving by Wadi Gimal dive operation. Shams Alam Hotel nearby.

14. Hamata/15. Lahami Reefs (Siyul–Lahami–Fury Reefs) Small fishing village with important harbour for safari boats. Some wrecks, 60 offshore reefs with 30 offering among the best and most versatile diving in Egypt. Some damage by Crown-of-thorns, but still moderate to good coral growth. Few large fishes due to overfishing. In summer the outer ergs have large tunas, mantas, whale sharks, reef and hammerhead sharks, turtles and dolphins. This area should be declared a National Park. Lahami Ecolodge (Red Sea Safari Diving) and Zabargad Hotel (Orca Diving) offer the best access. Lahami Ecolodge, a small camp 10 km south of Hamata, offers 4–5 boat dives a day to outlying reefs. Zabargad Dive Hotel has excellent fringing house reef with over 150 species of fishes. Access by pier or boat diving through Orca.

16. Zabargad and Rocky Island Exposed islands 100 km south of Ras Banas. Among the best dive sites in the Red Sea, access only by Safari boats. Eastern sides have magnificent pinnacles, caves, and terraced dropoffs to 600 m. Hilly Zabargad Island (235 m) was mined by Romans for olivine. Its lagoon has abundant fish life and excellent coral growth. Rocky island offers large fishes like Napoleons, tunas and sharks (hammerheads, Reef whitetip, Oceanic whitetip and Silky) throughout the year. The northern part is barren, southern part has fine coral growth. Area is threatened by Crown-of-thorns which appeared at Ras Banas in 2002.

PART I: FISHES, REPTILES AND MAMMALS

This section covers the vertebrates, those animals with a spinal cord protected by a skeleton. The vast majority are fishes, which are covered first, followed by reptiles and mammals.

Living fishes are divisible into three major groups: jawless fishes (lampreys and hagfishes), elasmobranchs (sharks and rays: class Chondrichthyes) and bony fishes (class Osteichthyes). The primitive jawless fishes live in cold waters while the elasmobranchs and bony fishes inhabit all seas. Elasmobranchs are most diverse in continental shelf and slope waters with many species resident or transient on coral reefs, while bony fishes reach their greatest diversity on coral reefs. Elasmobranchs have a skeleton of soft flexible cartilage while bony fishes have a skeleton of harder calcified bone. Externally, elasmobranchs and bony fishes share the same basic anatomy with respect to jaw, gill and fin location. These and other features useful for identification in the field are shown here.

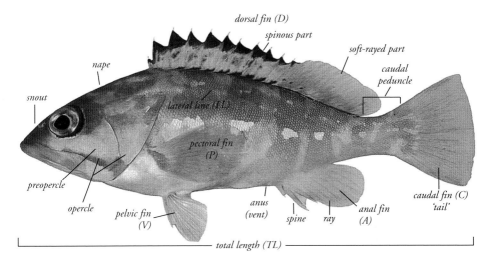

External anatomy of a bony fish.

Major external differences include the number of gill slits (5–7 in elasmobranchs, 1 in fishes) and skin surface (placoid scales with the same structure as teeth in elasmobranchs, and usually overlapping plates in bony fishes). Male elasmobranchs also have a pair of penis-like structures called claspers to facilitate internal fertilization. Internally and physiologically, there are many differences. Elasmobranchs have a spiral-valve intestine, a large oil-filled liver to aid in buoyancy and high levels of urea in the blood. Bony fishes have a tubular intestine, gas-filled swim bladder (in most) and blood with lower levels of urea and other electrolites. In most bony fishes the dorsal, anal and pelvic fins are supported by spines in front and flexible segmented rays behind. In fishes with two dorsal fins, the first consists of spines, the second of flexible rays behind a single spine. Numbers of spines, rays, rows of scales and other counts and measurements are useful for distinguishing similar species of fishes but are not practical as a visual aid and not used in this guide.

SHARKS

WHALE SHARK

RHINCODONTIDAE

◀ **Whale shark** *Rincodon typus*   12.2 m (possibly 14 m), 20,400 kg
Upper body with pale lines and spots; back with prominent ridges; mouth terminal without teeth. **Biology:** a slow moving giant often accompanied by small jacks and tunas and occasionally by pilot whales, dolphins, or sharks. Harmless, but unprovoked attacks on boats have been reported. Often curious towards divers. Whale sharks are active by day and night and feed on plankton and small pelagic animals including squids, crustaceans and fishes which are sieved from the water by spongy tissue between the gill arches. They can engulf 6 tonnes of water in a single gulp! Like all sharks, males have a pair of claspers between the pelvic fins. Copulation is probably belly-to-belly. Unlike most sharks, females may contain over 300 embryos in thin egg cases that hatch internally at a size of 70 cm. Whale sharks are exclusively pelagic, but enter deep lagoons and visit outer reef slopes. Subadults may aggregate in small groups, adults are solitary. At Ningaloo Reef, W. Australia, up to 100 individuals aggregate annually at the time of synchronous coral-spawning. Severely overharvested in many areas. **Range:** circumglobal in all warm seas. Generally rare, but consistently present seasonally at certain places including Cocos Is., Christmas Is. (Indian Ocean), Seychelles, Papua New Guinea, W. Thailand, Ryukyus, Sea of Cortez and NW. Australia; nearly extirpated from the Philippines. *(Photo: WT.)*

ZEBRA SHARK
STEGOSTOMATIDAE

Zebra shark   2.8 m
Stegostoma varium
Tail very long and without lower lobe, back with prominent ridges, mouth small with pebble-like teeth; juv. black with white bars. **Biology:** inhabits sand, rubble, or coral bottoms of lagoons and channels, 5 to 65 m. Feeds at night on molluscs, crustaceans and small fishes. Uncommon in most areas. Oviparous, the young hatch from egg cases at a size of 20–26 cm. Harmless. **Range:** Red Sea to Samoa, n. to S. Japan, s. to S. Africa.

Thailand, WP

NURSE SHARKS
GINGLYMOSTOMATIDAE

Tawny nurse shark   3.2 m
Nebrius ferrugineus
Tail without lower lobe; mouth small with bicuspid teeth. **Biology:** inhabits lagoons, channels and seaward reefs, 1 to 70 m. Typically rests on sand near coral heads or in caves in areas exposed to currents. More active at night than during the day. Feeds on cephalopods, crustaceans, fishes and even sea urchins and sea snakes which are sucked in, then crushed. Up to eight fully developed 40 cm young born per litter. Generally harmless, but unprovoked attacks known. **Range:** Red Sea to Tuamotus, n. to S. Japan, s. to S. Africa.

Palau, RM

REQUIEM SHARKS   CARCHARHINIDAE

Sharks have changed little in 300 million years. They have highly effective auditory (vibration reception), olfactory (smell) and electroreceptory systems, and vision sensitive to movement and dim light. They can detect vibrations and blood from great distances, and electrical impulses of hidden prey nearby. Most sharks are voracious feeders of fishes, cephalopods, sea turtles, and even smaller sharks and seabirds. Sharks lack a swim bladder and must swim continuously or will sink. Most reef sharks are members of this large family of 49 species. Many reef sharks are territorial and patrol a home range. Some species (particularly the Grey reef shark) exhibit 'threat-posturing', an exaggerated sinuous movement with head arched up and pectoral fins depressed (p. 17). This territorial display may be followed by a single quick bite. Females often have mating scars caused by the bites of males during courtship and copulation. During copulation, one of the paired claspers is inserted at a time. In the ovoviviparous Tiger shark, embryos develop within an egg and hatch internally. All other species of requiem sharks are viviparous, with a yolk-sac placenta. About 30 shark attacks are reported annually, averaging one on divers. Few are fatal. Attacks are usually stimulated by speared fishes or result from a combination of poor visibility, awkward movements or nearby prey. The Bull shark and Tiger shark are the most dangerous species in warm waters, the Great white shark (family Lamnidae) prefers waters cooler than 20°C (68°F). All large species are potentially dangerous and divers should be cautious if approached.

▲ **Oceanic whitetip shark**
Carcharhinus longimanus   3.5 m
Fins rounded with white tips, P fins very large. **Biology:** pelagic, near surface to at least 150 m over deep water, rarely near shore. In the Red Sea, regularly visits dive boats moored at offshore islands and reefs. Almost always accompanied by pilot fish, often accompanies pilot whales. Potentially dangerous, may persistently circle divers. Few attacks reported. Litter size: 2 to 15.
Range: circumtropical, usually in seas over 20°C. *(Photo: Elphinstone Reef, WT.)*

Silvertip shark   2.7 m
Carcharhinus albimarginatus
Tips and rear margins of D, P and
tail fins white. **Biology:** usually along
deep dropoffs and offshore banks,
occasionally in deep lagoons or
channels, 2 to 400 m. Solitary or
in groups. Cautious and rarely
approaches divers, but has chased
divers out of the water. Juveniles
more inquisitive and usually
shallower. Litter size: up to 11
after 12 months' gestation.
Range: Red Sea to Galapagos,
n. to S. Japan, s. to S. Africa.

Burma Banks, WT

Grey reef shark   1.8 m
Carcharhinus amblyrhynchos
Tail margin dusky. **Biology:** inhabits
lagoons, seaward reefs and channels,
1 to 274 m. Usually around channels
and steep reef slopes. Attacks on
humans are territorial and usually
as a single, non-fatal bite following
threat-posturing (see p. 17). May
be stimulated into feeding frenzy by
speared fish. Indian Ocean population
considered less aggressive than Pacific.
Litter size: 1 to 6; lifespan: 25 years.
Range: Red Sea to Easter Is., n. to
Hawaii, s. to S. Africa.

Maldives, RM

Silky shark   3.3 m
Carcharhinus falciformis
Slender, first D fin small.
Biology: pelagic near surface over
deep water, 3 to 500 m. Rarely visits
steep outer reef slopes, offshore banks
and seamounts. Usually indifferent
towards divers, but may get aggressive
if attracted by speared fish. Swift
swimming, feeds primarily on
fishes. Litter size: 2 to 14.
Range: circumtropical.

Ras Mohammed, MH

Blacktip reef shark   1.8 m
Carcharhinus melanopterus
Fins and tail with black tips, sides with light lateral streak. **Biology:** inhabits reef flats and shallow lagoon and seaward reefs, 0 to 75 m. Fins and back often exposed when in shallows. A fast-swimming predator of reef fishes and cephalopods. Timid and difficult to approach, but known to bite waders mistaken for prey. Litter size: 2 to 4. **Range:** Red Sea to Fr. Polynesia, n. to S. Japan, s. to S. Africa.

Bora Bora, SM

Sandbar shark   2.4 m
Carcharhinus plumbeus
D fin very large. **Biology:** primarily inshore but to 280 m. Feeds primarily on fishes, also on cephalopods and crustaceans. Generally inoffensive. Migrates to temperate seas in summer. Young live in shallow nursery areas. Litter size: 1 to 14. Matures in 13 years, at 1.3–1.8 m, lives up to 30 years. **Range:** continental shelves and around large islands of tropical and warm-temperate seas.

Turkey, HL

Whitetip reef shark   1.7 m
Triaenodon obesus
Tips of D and tail fins white; nasal flaps. **Biology:** inhabits lagoon and seaward reefs, 1 to 122 m. Often rests on sand in caves, channels or under ledges during the day. Feeds on reef fishes and octopuses, primarily at night. Returns to habitual daytime resting sites. Solitary or in small groups. Inoffensive unless provoked. Litter size: 1 to 5; growth rate: 2 to 4 cm annually. **Range:** Red Sea to Panama, n. to S. Japan and Hawaii, s. to S. Africa.

Maldives, RM

Tiger shark 5.5 m
Galeocerdo cuvier
Head broad, snout nearly square;
upper sides with dusky bars; teeth
with distinct notch. **Biology:** inhabits
deep lagoons, bays, outer reef slopes
and offshore banks, 1 to 300 m.
Usually deep during the day, shallower
at night. Feeds on a wide range of
animals including sea turtles,
porpoises, seabirds, sharks, rays and
bony fishes. The most dangerous
tropical shark. Litter size: up to 80;
♂ matures at 2.9 m, ♀ at 3.4 m.
Range: all tropical seas, seasonally
in temperate seas.

Coral Sea, CR

Sicklefin lemon shark 3.1 m
Negaprion acutidens
D fins nearly the same size; colour with
yellow hue. **Biology:** in shallow water
habitats from turbid inshore lagoons
and estuaries to exposed seaward reefs.
May rest on bottom. Has small home
range. Feeds on bottom-dwelling fishes
and rays. Shy, but easily enraged and
potentially dangerous. Never provoke
this species! Litter size: 1 to 11.
Range: Red Sea to Fr. Polynesia; n. to
Taiwan, s. to S. Africa.

Society Islands, SM

HAMMERHEAD SHARKS
SPHYRNIDAE

Scalloped hammerhead 4.0 m
Sphyrna lewini
Biology: inhabits steep slopes of
offshore islands and banks, usually
below thermocline, also pelagic, 5 to
275 m. Feeds primarily on fishes,
including rays which may be rammed
into the bottom with head. Non-
aggressive unless stimulated.
Occasionally displays threat posture
before retreating. Litter size: 15 to 31.
Range: circumtropical. **Similar:** *S.
mokarran* has huge falcate D fin and
smoother snout profile. ▼

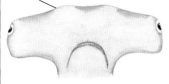

Elphinstone Reef, KH

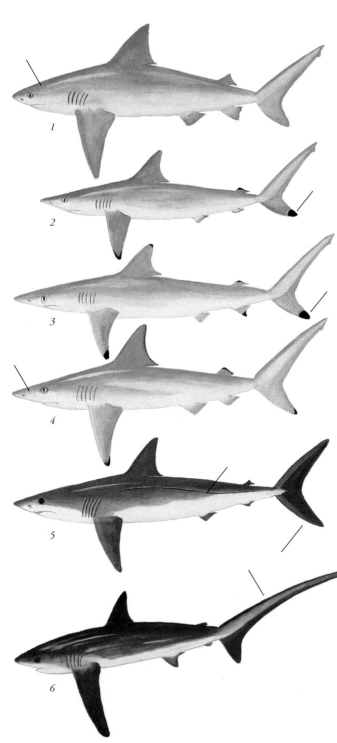

CARCHARHINIDAE

1. Bull shark   3.4 m
Carcharhinus leucas
Biology: fresh to coastal waters
from the lower reaches of rivers
to 150 m. Extremely dangerous.
Range: circumtropical; Oman to
Arabian Gulf, not in Red Sea.

2. Spot-tail shark   1.6 m
Carcharhinus sorrah
Biology: continental shelves,
intertidal to 80 m. Usually near
surface over mud bottoms. Feeds on
fishes, cephalopods and crustaceans.
Range: Red Sea to Solomon Is., n. to
E. China, s. to S. Africa, N. Australia.

3. Spinner shark   2.8 m
Carcharhinus brevipinna
Biology: coastal waters to 75 m.
Migrates to temperate seas during
summer to feed and breed. Feeds on
schooling fishes and squids, may leap
from water. **Range:** Indo-Pacific,
Atlantic, Mediterranean.

4. Blacktip shark   2.5 m
Carcharhinus limbatus
Biology: young coastal, adults
primarily pelagic. Feeds on fishes.
Range: circumtropical.

MACKEREL SHARKS
LAMNIDAE

5. Shortfin mako   4.0 m
Isurus oxyrinchus
Biology: pelagic. The fastest
shark. Feeds primarily on fishes.
Dangerous, has attacked swimmers.
Range: circumglobal in seas over
16°C.

THRESHER SHARKS
ALOPIIDAE

6. Pelagic thresher   3.3 m
Alopias pelagicus
Biology: pelagic, surface to 152 m.
Occasionally visits cleaning stations
along steep seaward slopes. Feeds on
small schooling fishes. Uses tail to
concentrate, then stun prey.
Range: Red Sea to Galapagos, n. to S.
Japan.

BATOID FISHES: RAYS

The rays evolved from sharks 100 million years ago and share the same cartilaginous skeleton. Their body is flattened and disc-like with confluent pectoral fins. Mantas and eagle rays 'fly' through the water with graceful flapping movements, while stingrays undulate the edges of their discs and wedgefishes, guitarfishes and torpedo rays swim primarily with their tails. The bottom-dwelling species have white bellies and often lie buried in sand or mud with only the eyes and respiratory openings showing. Rays respire by drawing water through a pair of small holes located behind the eyes called spiracles and expelling it through gill slits located under the disc. They are usually solitary while manta and devil rays often occur in groups. Most rays feed primarily on molluscs, worms, crustaceans and fishes, leaving a groove in the sand. Mantas feed on zooplankton and small schooling fishes. Copulation is belly-to-belly while the male inserts a clasper into the female. All the species featured here are ovoviviparous. The embryo feeds initially on yolk and later on albuminous fluid secreted by the uterine wall. The related skates lay horny egg cases. Stingrays have an effective defence in the form of one or more long serrated venomous bony spines along the base of the tail. When stepped on, the ray whips its tail, driving the spine into the offender. The wound causes excruciating pain and may be followed by vomiting, loss of blood pressure, sweating, muscular paralysis or death. About 1,500 incidents are reported annually in the USA.

TORPEDO RAYS
TORPEDINIDAE

Small rays with nearly square disc, round dorsal and tail fins and small eyes; electric organs located behind eyes used for stunning prey or predators can give a powerful jolt.

Leopard torpedo ray 45 cm
Torpedo panthera
Tan with tiny white stellate spots.
Biology: on sand or mud near coral reefs, 0.5 to 55 m. Usually buried. Not shy. Moves slowly away when disturbed. Common. Moves to deep water during winter. **Range:** Red Sea and Arabian Gulf only.

Mövenpick Bay, RM

Marbled torpedo ray 1.3 m
Torpedo sinuspersici
Brown with pale reticulations and spots. **Biology:** on sand or rubble bottoms of inshore coral or rocky reefs, 2 to 200 m. Often buried. Feeds on fishes, crustaceans and molluscs. Usually solitary. Aggregates during mating season. Litter: 9 to 22 pups. Gives birth in shallows. **Range:** Red Sea and Arabian Gulf to Sri Lanka, s. to S. Africa and Mauritius.

Maldives, GG

Jeddah, HS

WEDGEFISHES
RHINIDAE

Bowmouth wedgefish   2.7 m
Rhina ancylostoma
Snout round, body with spiny
medial ridge, lesser ridges on sides.
Biology: inhabits areas of coarse sand
or coral rubble among mangroves or
on coral reefs, 3 to 90 m. Usually close
to bottom, but also swims above
bottom. Feeds on shrimps, crabs and
shellfishes. **Range:** Red Sea to
Australia, n. to S. Japan, s. to S. Africa.

GUITARFISHES
RHINOBATIDAE

Similar to wedgefish but with rear of
P fins overlapping with beginning of
V fins.

Whitespotted guitarfish
Rhynchobatis djiddensis   3.0 m
Snout pointed, P fins fully in front
of V fins; tan with scattered white
spots, pair of ocelli above P fins.
Biology: inhabits shallow sandy
bottoms of estuaries and coastal
reefs, reef flats to 50 m. Feeds on
crabs, squids and small fishes.
Uncommon and shy. Litter size:
up to 10. **Range:** Red Sea to
w. Indian Ocean; similar species e. to
Fiji, n. to S. Japan, s. to S. Africa.

Halavi guitarfish   1.7 m
Rhinobatus halavi
Snout pointed, P fins overlap with
origin of V fins; tan with scattered
white spots. **Biology:** inhabits sandy
slopes and seagrass beds, 1 to 45 m.
Feeds primarily on crustaceans.
Common in N. Red Sea where
females give birth in shallow water
from May to October. Litter size of up
to ten, 29 cm young. **Range:** Red Sea
and Gulf of Oman to China; a recent
migrant to Mediterranean. **Similar:**
Arabian guitarfish *R. punctifer* has
white spots (Red Sea to Gulf of Oman
only; 81 cm). ▼

Marsa Alam, MM

Marsa Alam, BE

STINGRAYS
DASYATIDAE

2 fins confluent with body forming disc, mouth small with pebble-like teeth, base of tail with venomous barbs. Dangerous if stepped on.

Darkspotted stingray 3.2 m;
Himantura uarnak 1.5 m width
Close-set dark spots; disc rhomboid.
Biology: sandy or muddy areas from estuaries to clear lagoons and seaward reefs, 0.5 to 50 m. Often partially buried. Feeds on fishes, shrimps, molluscs and jellyfishes. Litter size: 1 to 5. **Range:** Red Sea to Fr. Polynesia, n. to S. Japan, s. to S. Africa; recent immigrant to E. Mediterranean.

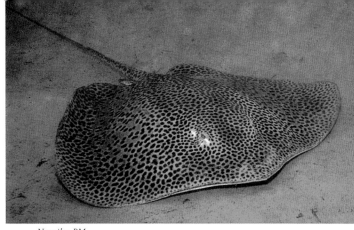
Nuweiba, PM

Bluespotted stingray 90 cm
Taeniura lymma
Tan with bright blue spots; disc round, somewhat oblong.
Biology: inhabits sandy areas of coral reefs, often under ledges or in caves, 2 to 30 m. Sometimes buried in sand. Active by day and night. Feeds on molluscs, worms and shrimps. Often visits cleaning stations. Litter size: 1 to 7. **Range:** Red Sea to Solomon Is., n. to Oman, s. to E. Africa and Maldives.

Mövenpick Bay, RM

Blackblotched stingray 3.0 m;
Taeniura meyeni 1.64 m width
Grey with irregular small black blotches; disc round. **Biology:** inhabits coral and rocky reefs with patches of sand, 3 to 500 m. Not aggressive, but has fatally wounded divers attempting to ride it. Feeds on benthic fishes and invertebrates. Uncommon in Red Sea. Litter size: 1 to 7. **Range:** Red Sea to Galapagos, n. to S. Japan, s. to E. Africa.

Egypt, GG

Nuweiba, MT

Fantail stingray 3.0 m;
Pastinachus sephen 1.8 m width
Uniformly brown; disc rhomboid, underside of tail with broad cutaneous fold. **Biology:** inhabits sand and mud, from brackish and estuarine areas to seaward reef slopes, 1 to 60 m. Feeds on fishes, crustaceans, molluscs and worms. **Range:** Red Sea and Arabian Gulf to Melanesia, n. to Ryukyus, s. to S. Africa and SE. Australia.

Marshall Islands, JR

Thorny ray 1.0 m width
Urogymnus africanus
Grey; disc round, covered with thornlike tubercles, no venomous spine. **Biology:** inhabits lagoons and seaward reefs, 1 to 100 m. Rests on sand or rubble bottoms of seagrass beds and coral reefs, or in caves. Feeds on sipunculids, polychaetes and crustaceans. Uncommon. Easily approached. Litter size: up to 12. **Range:** Red Sea and Oman to Fiji, s. to E. Africa.

EAGLE RAYS
MYLIOBATIDAE

Disc wide and wing-like with pointed tips; snout protruding, jaws powerful with plate-like crushing teeth; 2–6 barbs at base of tail. Fully developed at birth.

Spotted eagle ray 2.3 m width
Aetobatis narinari
Black with white spots or rings. **Biology:** inhabits sandy areas of reef flats, outer reef slopes and open ocean, 1 to 80 m. Solitary, paired or in groups of up to 200. Digs for molluscs and crustaceans. Preyed upon by sharks. May leap from water. Copulation belly-to-belly, mating time short, to 1.5 min. Timid. **Range:** circumtropical. **Similar:** Cownose rays, *Rhinoptera* spp. (Rhinopteridae) in Gulf of Aden and Arabian Gulf have bilobed snout which resembles a cow's nose and lack spots.

Mövenpick Bay, KF

MANTA AND DEVIL RAYS

MOBULIDAE

Mantas are enormous rays with a wide wing-like disc, horn-like flaps on the sides of a wide mouth containing minute teeth on the lower jaw only, and filter plates inside the gill slits. They feed by straining zooplankton and tiny fishes from the water.

(Previous page) **Manta** *Manta birostris*   6.7 m width; 1,400 kg
Cephalic flaps flexible. **Biology:** inhabits surface or mid-waters of lagoons and seaward reefs, particularly near channels during outgoing tides, 1 to 50 m. Often visits cleaning stations. Solitary or in groups of up to 50. A harmless giant which performs somersaults and other graceful manoeuvres. **Range:** circumtropical. **Similar:** devil or box rays, *Mobula* spp. are smaller and differ by having inflexible cephalic flaps. *(Photo: Maldives, PH.)*

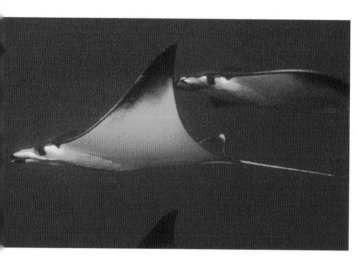

◄ Thurston's devil ray
Mobula thurstoni   1.8 m width
Cephalic flaps stiff, smaller. **Biology:** occurs in groups in coastal waters near reefs as well as offshore waters. Identification of species tentative, almost impossible without a specimen. Two other similar species reported from Red Sea (*M. ergoodootenkee, M. tarapacana*) and one from Gulf of Oman (*M. kuhli*). **Range:** circumtropical. *(Photo: Eritrea, JR.)*

BONY FISHES

MORAY EELS   MURAENIDAE

Moray eels are a large and diverse group characterized by a large mouth containing numerous teeth, restricted gill openings and no pectoral or pelvic fins. Most have distinct dorsal and anal fins, but a few (*Gymnomuraena, Scuticaria, Uropterygius*) are snakelike with undeveloped fins. Most species are secretive and remain hidden among rocks or corals. Many hunt in the open at night. All are predators with a keen sense of smell, most feed on fishes, cephalopods and crustaceans. Those with fang-like teeth feed primarily on fishes, while those with conical or nodular teeth feed on crustaceans, or even molluscs or urchins. Morays are normally docile, but most species will bite if provoked. Bites from large morays may be severe and require medical attention or result in permanent injury. Although widely used as food, some large species may be ciquatoxic. Sexuality in morays varies greatly: some species are gonochorists, that is they are permanently male or female; others are hermaphrodites, either starting life as females then changing to males (protogynous), starting life as males then changing to females (protandrous), or simultaneously both male and female. Morays have long larval-stages and most species are widely distributed. Thirty-two species have been reported from the Red Sea, 15 from Oman.

Yellowmouth moray   1.2 m
Gymnothorax nudivomer ►
Brown with small white spots becoming larger and ocellated towards rear; inside of mouth yellow. **Biology:** inhabits deep lagoon and seaward reefs, 1 to 165 m. Solitary or in pairs, in holes by day. Common in shallow water in the Red Sea. Preys on fishes at night. Exhibits a warning display by showing the bright yellow inside of mouth. Its skin mucus is toxic, a possible defence against parasites or predation. **Range:** Red Sea to Fr. Polynesia and Hawaii, n. to the Ryukyus, s. to E. Africa. *(Photo: Gulf of Aqaba, EL.)*

Mövenpick Bay, RM

Giant moray 2.3 m
Gymnothorax javanicus
Brown with black flecks, those on body and tail forming leopard-like spots. The largest moray.
Biology: inhabits lagoon and seaward reefs, 1 to 46 m. Common. Feeds primarily on fishes, occasionally on crustaceans. Normally docile and occasionally 'tamed' by dive guides, but has been implicated in unprovoked attacks and caused severe injury. Often ciguatoxic. **Range:** Red Sea to Marquesas and Pitcairn, n. to Ryukyus and Hawaiian Is., s. to S. Africa; absent from Arabian Gulf and Oman.

Mövenpick Bay, KF

Yellowmargin moray 1.2 m
Gymnothorax flavimarginatus
Brown with fine yellow mottling, orange eye; fins of juv. with yellow-green margin. **Biology:** inhabits reef flats and lagoon and seaward coral and rocky reefs, 0.3 to 150 m. Often with only head protruding. Feeds on small fishes and crustaceans. Common in Red Sea. A protogynous hermaphrodite. **Range:** Red Sea to Panama, n. to Ryukyus, s. to S. Africa and New Caledonia.

Gulf of Aqaba, AK

Undulated moray 1.1 m
Gymnothorax undulatus
Dark brown with undulating light lines and speckles; head olive-green.
Biology: inhabits reef flats and lagoon and seaward reefs, 1 to at least 50 m, juveniles in tidepools. Among rubble, rocks and debris. Common. Hunts at night for fishes, octopuses and crustaceans. May be aggressive towards divers. **Range:** Red Sea to Panama, n. to S. Japan, s. to S. Africa and GBR.

Lidspot moray   50 cm
Gymnothorax chilospilus
Lower lip with white spot; body
with dendritic brown blotches.
Biology: inhabits lagoon and seaward
reefs, 0.3 to 45 m. Usually on seaward
reef slopes in less than 5 m. In Oman,
in rubbly areas of algae-rich rocky
reefs. **Range:** Oman to Fr. Polynesia,
n. to S. Japan, s. to SE. Australia.

Salalah, KF

Whitemouth moray   1.2 m
Gymnothorax meleagris
Dark brown with small round
white spots, inside of mouth white.
Biology: inhabits coral-rich areas
of lagoon and seaward reefs, 0.3 to
36 m. Active by day and night, feeds
primarily on fishes and crabs. Comet
(*Calloplesiops altivelis*, p. 78) may be
a mimic. A gonochorist with sex
predetermined. Common in
Mauritius and Hawaii; uncommon
in S. Red Sea. **Range:** S. Red Sea
to the Galapagos, n. to Ryukyu and
Hawaiian Is., s. to E. Africa and SE.
Australia.

Mauritius, EL

Yellow-speckled moray   90 cm
Gymnothorax punctatus
Brown with irregular fine yellow
speckles. **Biology:** inhabits coastal
rocky and coral reefs and offshore
banks. Poorly known. **Range:** Red Sea
to Sri Lanka and E. India.

Gulf of Aqaba, EL

Salalah, KF

Honeycomb moray 2.2 m
Gymnothorax favagineus
Close-set black spots forming honeycomb pattern on white; juv. white with large curved black blotches. **Biology:** inhabits reef flats, lagoons and fringing reefs, 1 to 50 m. Primarily coastal reefs. Feeds on fishes and octopuses by day and night. Occasionally in the open in seagrass. Not timid, occasionally aggressive. **Range:** S. Red Sea to Samoa, n. to Taiwan, s. to S. Africa and GBR.

Marsa Shagra, RM

Yellowhead moray 80 cm
Gymnothorax rueppelliae
Head yellowish, body with dark bands. **Biology:** inhabits reef flats, clear lagoon and seaward reefs, 0.3 to 55 m. In crevices during day, hunts at night on crabs, fishes and octopuses. A nervous species that will readily bite. **Range:** Red Sea to Hawaii, n. to Ryukyus, s. to S. Africa and GBR.

Lahami, KF

Grey moray 65 cm
Gymnothorax griseus
Off-white with black spots in pattern on head; snout blunt. **Biology:** inhabits coastal rocky and coral reefs, 1 to 30 m. Common, often seen in the open among seagrass and rubble. Hunts for small fishes and crustaceans, sometimes in the company of goatfishes, groupers, or scorpionfishes. A simultaneous hermaphrodite. Releases up to 12,000 eggs per spawn. **Range:** Red Sea and Oman, s. to S. Africa and Mauritius.

Peppered moray   1.2 m
Gymnothorax pictus
Off-white, peppered with tiny black
dots; snout blunt. **Biology:** inhabits
reef flats and shallow protected areas,
intertidal to at least 20 m. Common
along shorelines of fringing reefs,
sometimes in groups of up to 5.
Hunts primarily crabs and fishes, may
leave water to capture grapsid crabs. A
simultaneous hermaphrodite.
Range: Red Sea to Galapagos, n. to
Ryukyu and Hawaiian Is., s. to S.
Africa and GBR. **Similar:**
G. monochrous, without black dots
(Red Sea to Oman; 56 cm).

Kenya, RM

Zebra moray   1.5 m
Gymnomuraena zebra
Dark brown to black with narrow
white bars; body snake-like, teeth
nodular. **Biology:** inhabits lagoon and
seaward reefs, 1 to 50 m. Feeds
primarily on crabs, molluscs and sea
urchins, probing prey before striking.
Secretive, rarely ventures into the
open. Docile. A protogynous
hermaphrodite. **Range:** Red Sea to
Panama, n. to Ryukyus, s. to S. Africa
and Fr. Polynesia.

Thailand, KF

Snowflake moray   75 cm
Echidna nebulosa
White with yellow spots in black
dendritic blotches; teeth conical.
Biology: inhabits reef flats and
shallow lagoon and seaward reefs,
intertidal to 12 m. Juv. often in
tidepools on the reef flat. Primarily
nocturnal, may leave water to hunt
grapsid crabs on beach rock. Feeds
primarily on crabs and mantis
shrimps, occasionally on small fishes
and cephalopods. A protogynous
hermaphrodite. Common. **Range:** Red
Sea to Panama, n. to Ryukyu and
Hawaiian Is., s. to S. Africa and Fr.
Polynesia.

Arabian Gulf, AK

Guam, RM

Barred moray   60 cm
Echidna polyzona
Pale with dark bands that fade with age; teeth conical. **Biology:** inhabits reef flats, clear lagoons and coastal reefs, intertidal to 15 m. Secretive; hides in crevices by day, hunts crabs and shrimps by night. Probes prey before striking. Non-aggressive. **Range:** Red Sea to Hawaii and Fr. Polynesia, n. to Ryukyus, s. to S. Africa and GBR.

s. Japan, RM

Dragon moray   80 cm
Enchelycore pardalis
Orange-brown with dark-edged rounded white marks; rear nostrils in long tubes, jaws hooked. **Biology:** inhabits rocky and coral reefs, 15 to 50 m. Solitary and secretive. Feeds on fishes and octopuses. Gives impressive threat display with gaping jaws and erect D fin. Rare in tropics, common in S. Japan. **Range:** primarily anti-tropical; Oman to Hawaii and Fr. Polynesia, n. to S. Korea, s. to Réunion and New Caledonia.

Hawaii, RM

Tiger snake moray   1.2 m
Scuticaria tigrina
Pinkish-tan with scattered large dark spots; snake-like, fins greatly reduced. **Biology:** inhabits lagoons and reef slopes, 0.2 to 24 m. Secretive and rarely seen. Uncommon. **Range:** Oman and E. Africa to Panama, n. to Ryukyu and Hawaiian Is., s. to S. Africa, New Caledonia and Fr. Polynesia.

SNAKE EELS
OPHICHTHIDAE

Snake-like with reduced D, A and C fins, most have pointed snouts with downturned nostrils. Usually remain buried in sand.

Marbled snake eel   87 cm
Callechelys marmorata
White with dark spots; nostrils overhang mouth. **Biology:** inhabits sand patches of clear seaward reefs, to 25 m. Usually buried in sand with only cone-like head protruding. Feeds on small fishes and crustaceans. **Range:** Red Sea to Polynesia, s. to Mozambique.

Dahab, AK

Stargazer snake eel   1.59 m
Brachysomophis cirrocheilos
Tan with small dark specks; eyes far forward; mouth large, lips fringed. **Biology:** inhabits silty sand of coastal reefs, 0.5 to 38 m. Nearly always completely buried with only eyes exposed. An ambushing predator of small fishes and cephalopods, more active at night, sometimes with head exposed. Will attempt to bite if harassed. **Range:** Red Sea to PNG, n. to Arabian Gulf and S. Japan, s. to Mozambique. **Similar:** *B. crocodilinus* and *B. henshawi* have eyes closer to tip of snout (Indo-Pacific).

Indonesia, SM

Banded snake eel   88 cm
Myrichthys colubrinus
Off-white with widely-spaced black bars, a few replaced by spots. **Biology:** inhabits sand flats and seagrass beds, 0.3 to 25 m. Usually buried, occasionally hunts in the open on fishes and crustaceans, primarily at night. Common. Often mistaken for the venomous Banded sea-snake (*Laticauda colubrina*) which does not occur in the Red Sea. Mimicry is unlikely due to broader distribution of the eel. Seems to have poor eyesight, often ignores divers. **Range:** Red Sea to Fr. Polynesia, n. to Ryukyus, s. to S. Africa and GBR.

Guam, RM

Spotted snake eel 1.0 m
Myrichthys maculosus
Tan with large round dark spots.
Biology: inhabits reef flats, lagoons
and seaward reefs, 0.3 to 262 m.
Feeds on fishes and crustaceans by day
or night which are detected by smell.
The tip of the tail is hard for
facilitating rapid backward burrowing.
Common in shallow silty areas.
Range: Red Sea to Fr. Polynesia, n. to
Ryukyus, s. to S. Africa and SE.
Australia.

Dahab, AK

CONGER AND GARDEN EELS
CONGRIDAE

Conger eels stout bodied with
conspicuous P fins; garden eels
extremely elongate with short blunt
heads and large eyes.

Moustache conger 1.3 m
Conger cinereus
Brownish-grey with dusky upper lip;
broad dark bars at night. **Biology:**
inhabits reef flats and lagoon and
seaward reefs, 1 to 80 m. A nocturnal
predator of fishes and crustaceans,
rarely seen during the day. **Range:**
Red Sea to Easter Is., n. to S. Japan
and Hawaii, s. to SE. Australia.

New Caledonia, RM

Red Sea garden eel 96 cm
Gorgasia sillneri
Olive with small close-set dark spots.
Biology: inhabits broad sand slopes of
seaward reefs and clear lagoons
exposed to current, 3 to 46 m. Lives
in colonies in tube-like burrows, with
heads facing the current to feed on
plankton. Juveniles extremely secretive
among corals.
Range: N. Red Sea
s. to Jeddah only;
many similar species
elsewhere. **Similar:**
Heteroconger balteatus
(Jeddah area only;
33 cm) white with
larger widely-spaced
orange spots. ▶

Marsa Alam, BE

EEL CATFISHES
PLOTOSIDAE

Striped eel catfish   32 cm
Plotosus lineatus
P and 1st D fins with stout venomous
spine, 2 pairs of barbels; body eel-like.
Stripes of adults pale. **Biology:** inhabits
estuaries, seagrass beds, and lagoon and
coastal reefs, 1 to 60 m. Juveniles
form dense ball-shaped schools. Adults
solitary and under ledges
by day. Venomous P and D spines
dangerous, even fatal. **Range:** Red Sea
to Samoa, n. to S. Korea, s. to S. Africa
and SE. Australia. **Similar:** *P. limbatus*
(Oman to Kenya) lacks stripes.

Kenya, KF

CUSK EELS
OPHIDIIDAE

Reef cusk eel   43 cm
Brotula multibarbata
Single D fin confluent with A fin;
3 pairs of barbels, body eel-like.
Biology: inhabits lagoon and seaward
reefs, 1 to 220 m. Common but
nocturnal and secretive, seen only at
night when it retreats into crevices
when illuminated. Feeds on small
invertebrates and fishes. Livebearing.
Male has penis which is inserted into
genital pore. Newborns are 5 to 12 mm
long. **Range:** Red Sea to the Hawaiian
and Pitcairn Is., n. to S. Japan, s. to SE.
Australia.

Marsa Shagra, PM

TOADFISHES
BATRACHOIDIDAE

Red Sea toadfish   34 cm
Thalassothia cirrhosus
Bottom-dwellers with large somewhat
flattened head and mouth. **Biology:**
in crevices or under ledges, 5 to 200
m. Shallow off Jordan. Toadfishes feed
on crustaceans and fishes. They nest
in cavities, ♂ guards eggs. Secretive.
Range: Red Sea only. **Similar:**
colourful *Bifax lacinia* (Oman; 33 cm)
inhabits rocky bottoms, 5 to 10 m. ▼

Jordan, JN

Guam, RM

LIZARDFISHES
SYNODONTIDAE

Slender lizardfish   32 cm
Saurida gracilis
Sides of jaws with tiny teeth.
Biology: on sand or rubble of silty
lagoon and protected seaward reefs,
1 to 135 m. Common along the base
of reefs, often partly buried. An
ambushing predator of small fishes,
capable of amazing bursts of speed.
Changes ambush sites regularly.
Range: Red Sea to Fr. Polynesia, n. to
Ryukyu and Hawaiian Is., s. to S.
Africa and GBR.

Nabq, RM

Clearfin lizardfish   23 cm
Synodus dermatogenys
Mid-lateral series of 8 to 9 light-
centred dark blotches; bluish lateral
stripe. **Biology:** inhabits sand patches
of lagoon and seaward reefs, 1 to 70
m. Often buries itself leaving only
eyes and nostrils exposed. Up to 7
males form leks before sunset to
attract females. Spawns about 3 times
a night. **Range:** Red Sea to Fr.
Polynesia, n. to Ryukyu and Hawaiian
Is., s. to S. Africa and SE. Australia.

Marsa Shagra, EL

Reef lizardfish   28 cm
Synodus variegatus
Mid-lateral series of rectangular
blotches often form dark lateral stripe;
colour highly variable from brown to
red. **Biology:** inhabits lagoon and
seaward reefs, 5 to 70 m. Frequently
in pairs, more often on hard bottoms
than other lizardfishes, rarely buries
in sand. Reported to change position
an average of once every 4 minutes
and attack prey every 35 minutes.
Range: Red Sea to Fr. Polynesia, n.
to Ryukyu and Hawaiian Is., s. to S.
Africa and SE. Australia.

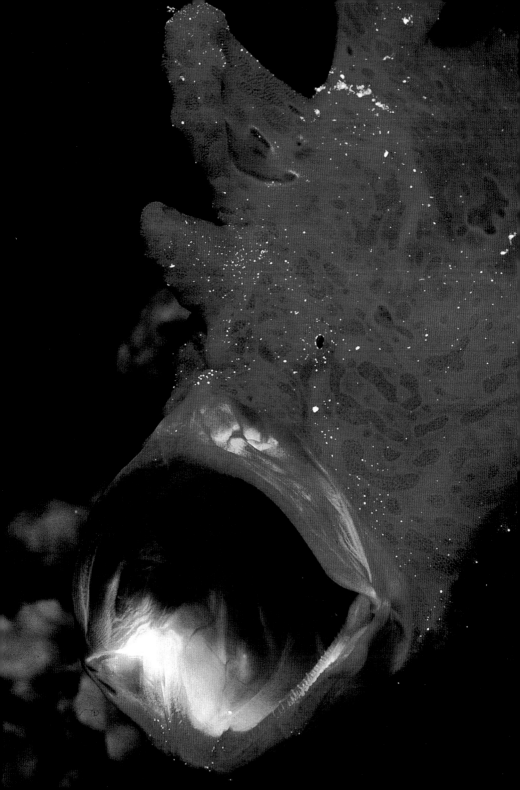

Painted frogfish (previous page)
Antennarius pictus   21 cm
Colour variable from white to yellow, red, brown or black, usually with small dark ocelli. **Biology:** inhabits lagoon and seaward reefs, 1 to 75 m. Usually associated with colourful sponges, rubble, algae, or corals. Juveniles may mimic toxic flatworms as well as sponges. The small ocelli may represent the openings of sponges. Has been known to feed on lionfishes. **Range:** Red Sea to Fr. Polynesia, n. to S. Japan and Hawaii, s. to S. Africa and GBR. *(Photo: Marsa Alam, BE.)*

FROGFISHES

ANTENNARIIDAE

Frogfishes have bulbous bodies, elbow-like pectoral fins that are used like arms, and a large upturned mouth. The first dorsal spine is modified into a fishing pole (termed *illicium*) tipped with a lure (*esca*) which may look like a small worm, fish or shrimp. The esca is flicked enticingly above the mouth to attract prey that is sucked up in only 6 milliseconds with an explosive sound. The prey can be of equal size or even venomous (e.g. lionfish). Frogfishes have loose prickly skin which may be adorned with fleshy or filamentous appendages. They may be brightly coloured or drab and mimic sponges and other bottom features by means of their cryptic coloration. This combined with long periods of stillness makes them almost invisible to predators and divers as well as prey. At intervals of 3 to 4 days, reproductive females may lay up to 280,000 tiny eggs in a long buoyant gelatinous raft of more than 2 metres in length. Adults occasionally fill their stomach with water or air.

Marsa Bareka, MH

Warty frogfish   11 cm
Antennarius maculatus
Numerous wart-like protuberances; esca large, mimics tiny fish. Colour variable, from cream to yellow, brown, or black with scattered dark circular spots. Heavily spotted ones resemble *A. pictus*. **Biology:** inhabits protected reefs at depths of 1 to 15 m. Juveniles may resemble a toxic nudibranch. Not previously known from the Red Sea. Identification based on photos is tentative, confirmation by collection of specimens is needed. **Range:** Red Sea and Mauritius to the Solomon and Mariana Is.

Papua New Guinea, JR

Giant frogfish   36 cm
Antennarius commersoni
Colour highly variable, including yellow, orange, green, brown and black; blotched or uniformly coloured. The largest frogfish. **Biology:** inhabits coral reefs at depths of 1 to 70 m. More common on protected coastal reefs than exposed reefs. Usually near sponges. **Range:** Red Sea to Panama, n. to S. Japan and the Hawaiian Is., s. to S. Africa, SE. Australia and Society Is. **Similar:** *A. nummifer* has shorter 'fishing pole' and usually a single basidorsal spot (Red Sea to C. America; 10 cm).

Sargassumfish   19 cm
Histrio histrio
Mottled brown, yellow or olive with white spots and lines and weed-like dermal flaps. **Biology:** pelagic, in rafts of floating *Sargassum* seaweed which it closely resembles. Occasionally drifts to reefs with detached weeds. Feeds on fishes, polychaetes and crustaceans, even cannibalizing its own species. Can inflate by pumping water in its stomach. Rarely seen unless searched for in seaweed.
Range: all tropical and temperate seas except E. Pacific.

Hawaii, RM

PEARLFISHES
CARAPIDAE

Partially transparent eel-like fishes lacking scales and pelvic fins. Many live commensally in the body cavities of echinoderms, clams, or tunicates.

Pincushion star pearlfish
Carapus mourlani   17 cm
Biology: known from depths of 3 to 116 m. Prefers pincushion starfishes *Culcita* spp. for a host but also occurs in certain sea cucumbers (*Bohadschia argus*, *Stichopus* sp.) and the Crown-of-thorns starfish *Acanthaster planci*.
Range: Red Sea to Fr. Polynesia and Hawaiian Is., s. to S. Africa. **Similar:** Silver pearlfish *Encheliophis homei* lacks melanophores on body.

Nuweiba, MH

MILKFISH
CHANIDAE

Milkfish   1.8 m
Chanos chanos
Silvery with large forked tail and small mouth. In the distance, may resemble a shark. **Biology:** inhabits lagoons and coastal reefs, 0 to 30 m. Common in estuaries and among mangroves. An important foodfish, the basis of an important aquaculture industry throughout S. Asia.
Range: Red Sea to Panama, n. to S. Japan and Hawaii, s. to S. Africa. **Similar:** Bonefishes *Albula* spp. have larger mouth under conical protruding snout.

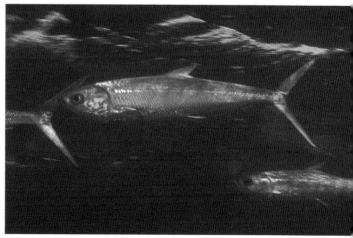
Hurghada, HG

Aqaba, EL

SILVERSIDES
ATHERINIDAE

Hardyhead silverside　　14 cm
Atherinomorus lacunosus
Biology: in large schools along sandy shorelines and reef margins, in calm clear water, 1 to 15 m. Feeds on zooplankton. Important prey for jacks. **Range:** Red Sea to Hawaii and Rapa, n. to S. Japan, s. to SE. Australia. **Similar:** *Hyperatherina* spp. more slender. Similar families with single D fin, abdominal V fins: herrings and sprats (Clupeidae) with similar mouth, and anchovies (Engraulidae) with jaws under snout.

NEEDLEFISHES
BELONIDAE

Elongate fishes with greatly elongate jaws full of needle-like teeth. Surface-dwelling predators of small fishes. May leap from water towards light at night and have fatally impaled fishermen.

Marsa Shagra, RM

Red Sea needlefish　　1.2 m
Tylosurus choram
Biology: coastal waters, usually along upper edge of reef margin. Often visits cleaning stations. The most common needlefish in the Red Sea. **Range:** Red Sea to Gulf of Oman only. Most similar to Houndfish; positive identification requires examination of specimen.

Houndfish　　1.5 m
Tylosurus crocodilus
Biology: coastal waters, usually along upper edge of reef margin. Often visits cleaning stations. Has fatally impaled fishermen by leap from water towards light at night. **Range:** circumtropical, different ssp. in Atlantic, E. Pacific. **Similar:** nearly identical to Red Sea needlefish and Agujon *T. acus* (Indo-Pacific; 1.0 m); Banded needlefish *Strongylura leirua* smaller with longer jaws (Indo-Pacific; 80 cm); Keeled needlefish *Platybelone argalus platyura* (Indo-Pacific; 38 cm) with flattened tail base, commonly in groups.

Sulawesi, RM

HALFBEAKS
HEMIRHAMPHIDAE

Elongate silvery fishes with short upper jaw and elongate spike-like lower jaw. Omnivores of algae, zooplankton and fishes.

Spotted halfbeak   40 cm
Hemirhamphus far
Biology: in small groups at surface above shallow reefs. **Range:** Red Sea to Samoa. **Similar:** several species that lack spots, Gambarur halfbeak *Hyporhamphus gambarur* is the most common (Red Sea and Gulf of Aden only; 20 cm).

Papua New Guinea, JR

FLASHLIGHTFISHES
ANOMALOPIDAE

Small black fishes with bony heads, small rough scales and a light organ beneath each eye.

Indian Ocean flashlightfish
Photoblepharon steinitzi   11 cm
Biology: seaward reefs with caves, 1 to 25 m. Normally in the far reaches of caves, they emerge only during moonless periods of the night. They get highly agitated by diver's lights and seek cover or freeze. The soft green light is produced by symbiotic bacteria and is shut on or off by an eyelid-like flap. It is used to communicate and attract zooplankton prey. **Range:** Red Sea to Maldives, n. to Oman, s. to Comoros Is.

Mövenpick Bay, EL

PINEAPPLEFISHES
MONOCENTRIDAE

Japanese pineapplefish   17 cm
Monocentris japonicus
A small armoured fish that resembles a pineapple. **Biology:** in caves or under ledges of rocky or coral reefs, 15 to 200 m, juveniles as shallow as 3 m. At night they disperse. A luminescent organ near tip of lower jaw contains symbiotic bacteria that produce a cold blue-green light which can be turned on and off and is used to locate food and probably for communication. Aggregations of up to 100 reported in Japan. **Range:** primarily anti-tropical; Red Sea to Fr. Polynesia, n. to S. Japan, s. to E. Africa and SE. Australia.

Al Mukallah, HGr

SQUIRRELFISHES AND SOLDIERFISHES

HOLOCENTRIDAE

Squirrelfishes and Soldierfishes are medium-sized fishes with bony heads, stout fin spines, large spiny scales and large eyes. Most species are nocturnal and predominately red, a colour that shows up black in low light. Soldierfishes (subfamily Myripristinae) have blunt snouts and large scales. During the day they hover in or near caves and crevices. At night they swim well above the bottom to feed on large zooplankton, primarily late larval crustaceans. Squirrelfishes (subfamily Holocentrinae) have longer snouts, a prominent preopercular spine on each side of the head, and smaller scales. A number of species are striped. Extreme pain associated with wounds from the spine of the Longjawed squirrelfish indicate that it may be venomous, but no definitive studies have been made. Squirrelfishes also hover in or near shelter during the day, but roam the reef at night to feed on benthic invertebrates and small fishes. All members of this family are important subsistence food fishes in many areas.

▲ **Longjawed squirrelfish**
Sargocentron spiniferum   45 cm
Preopercular spine extremely long and possibly venomous; spinous D fin uniformly red. **Biology:** inhabits reef flats, lagoon and seaward reefs, 1 to 122 m. Common, hovers in caves or under ledges or tabular corals by day, returning to the same spot each morning. At night it forages singly or in small groups for small fishes, crabs and shrimps. **Range:** Red Sea to Fr. Polynesia and Hawaii, n. to S. Japan, s. to S. Africa and SE. Australia.
(Photo: Marsa Shagra, RM.)

Tailspot squirrelfish   25 cm
Sargocentron caudimaculatum
Red, base of tail pale with large white
spot. **Biology:** inhabits deep lagoon
and seaward reefs with rich coral
growth, 2 to 50 m. Singly or in small
groups in the vicinity of holes or
ledges by day. The white tailspot gets
reddish during night when it hunts
for small fishes and crabs. **Range:** Red
Sea and S. Oman to Fr. Polynesia, n.
to S. Japan, s. to S. Africa and GBR.

Marsa Shagra, RM

Crown squirrelfish   17 cm
Sargocentron diadema
Red with narrow white stripes;
D fin mostly black with curved white
stripe. **Biology:** inhabits subtidal reef
flats, and lagoon and seaward reefs,
2 to 60 m. Typically near crevices,
caves, or under ledges by day. Roams
over open sand by night to feed on
gastropods, polychaetes and small
crustaceans. Solitary or in groups. Not
shy. Has invaded the Mediterranean.
Range: Red Sea and Oman to Fr.
Polynesia, n. to Ryukyu and Hawaiian
Is., s. to S. Africa and SE. Australia.

Marsa Alam, KH

Redcoat squirrelfish   27 cm
Sargocentron rubrum
Red with thick white stripes.
Biology: among rocks or corals of
protected coastal reefs, 3 to 50 m.
Absent from oceanic islands.
Range: Red Sea to New Caledonia,
n. to S. Japan, s. to E. Africa.
Similar: Yellowfin squirrelfish
S. seychellense (Oman to Seychelles)
has yellow fin rays and D spine
membranes. ▼

Bali, RM

Spotfin squirrelfish   32 cm
Neoniphon sammara
Silvery with narrow white stripes;
front of D fin with large reddish-black
spot. **Biology:** inhabits reef flats,
protected lagoons and seaward reefs,
2 to 45 m. Common and less secretive
than other squirrelfishes. Hovers in
small groups near branching corals,
rocks, or ledges by day. Disperses at
night to feed on small fishes and
crabs. Not shy. **Range:** Red Sea to Fr.
Polynesia, n. to S. Japan, s. to S.
Africa and GBR.

Mövenpick Bay, EL

Blotcheye soldierfish   25 cm
Myripristis murdjan
Dirty red; soft D, A and tail fins dark,
usually with narrow white margins.
Biology: inhabits lagoon
and seaward reefs, 1 to 50 m.
Hovers in caves or under ledges by
day, disperses at night to hunt for
small crabs, polychaetes and shrimp
larvae in the water above the reef.
Returns to shelter about 30 minutes
before sunrise. **Range:** Red Sea to
Samoa, n. to Ryukyus, s. to S. Africa
and GBR.

Mangrove Bay, RM

Yellowtip soldierfish   20 cm
Myripristis xanthacra
Red; tips of soft D, A and tail
fins yellow. **Biology:** inhabits silty
lagoon and seaward reefs, 1 to 18 m.
Usually in groups in caves and under
overhangs. In S. Yemen and Gulf of
Aden, common on algae-covered
rocky reefs with plenty of holes. Not
shy. **Range:** Central Red Sea to Gulf
of Aden only.

Sudan, JR

CORNETFISHES
FISTULARIIDAE

Smooth cornetfish   1.07 m
Fistularia commersonii
Extremely long, cylindrical, slightly
depressed body with long tubular
mouth and filamentous tail.
Biology: in most habitats, except
areas of heavy surge, 1 to 128 m.
Feeds on small fishes and crustaceans.
Swims close above bottom, singly
or in small groups. Often in small
aggregations 1 to 2 m above
protected slopes during the day.
Range: Red Sea to Panama, n. to
Ryukyu and Hawaiian Is., s. to S.
Africa and N. New Zealand.

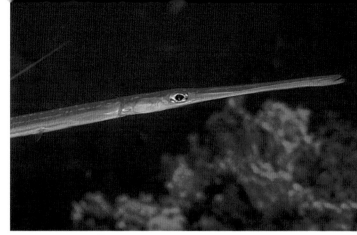

Mangrove Bay, KF

SEA MOTHS
PEGASIDAE

Little dragonfish   8 cm
Eurypegasus draconis
Body encased in bony plates; snout
proboscis-like, P fins wing-like.
Biology: inhabits protected sand,
mud, gravel and rubble bottoms, 1 to
90 m. Territorial and in pairs. ♂
protects ♀ against intruding males.
Feeds on minute invertebrates. Rarely
seen, but probably more common
than thought. **Range:** Red Sea to Fr.
Polynesia, n. to S. Japan, s. to S.
Africa and SE. Australia.

Dahab, AK

SHRIMPFISHES
CENTRISCIDAE

Spotted shrimpfish   15 cm
Aeoliscus punctulatus
Thin elongate body encased in bony
plates with ventral keel and D fin
modified into spike-like 'tail'; this
species pale with tan stripe and tiny
black dots; D spine with joint.
Biology: inhabits protected lagoons
and bays, often among seagrasses, 1 to
17 m. Solitary or in groups of up to
100 that swim synchronously in a
head-down posture. Feeds on minute
invertebrates. **Range:** Red Sea to
S. Africa, e. to Seychelles.

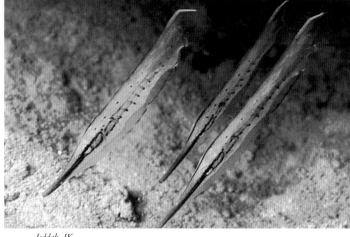

Jeddah, JK

◄ Ornate ghost pipefish 11 cm
Solenostomus paradoxus
Base of tail narrow, fins sculpted,
numerous tassels; colour variable:
white or black with red, yellow or
orange markings. **Biology:** inhabits
rocky or coral reefs, 3 to 30 m.
Associates with gorgonians, soft corals,
black corals and crinoids. In pairs or
temporary small groups. Often
overlooked because of its perfect
camouflage. Male displays by
swimming back and forth and circling
the female. **Range:** Red Sea to Fiji,
n. to S. Japan, s. to Mauritius and SE.
Australia. *(Photo: Abu Dabab, AK.)*

GHOST PIPEFISHES SOLENOSTOMIDAE

Ghost pipefishes resemble outstretched seahorses with thin bodies and
large fins. Females are larger than males and have modified pelvic fins
that form a brood pouch. Up to 350 eggs are laid into the pouch. The
embryos are ventilated by contractions and hatch after 10 to 20 days.
The translucent larvae settle near crinoids, soft corals, gorgonians,
sponges, seagrasses and algae. Adults usually occur in pairs, occasionally
in small groups. They typically swim in a head-down posture and blend
in with their surroundings. They feed primarily on minute crustaceans
including shrimps, isopods and amphipods which are sucked into their
tubular mouths. There are only 3 or 4 validly described species and several undescribed ones. One species mimics green *Halimeda* algae when
young then grows a thick coat of red hair-like tassels as it ages.

Robust ghost pipefish 17 cm
Solenostomus cyanopterus
Base of tail wide; colour variable:
green, grey or dark brown.
Biology: inhabits shallow sheltered
coastal reefs, 0.2 to 20 m. In the Red
Sea, normally among algae or
seagrasses (*Halimeda*, *Halophila* and
Sargassum). Normally in head-down
posture but may adopt a horizontal
posture when swimming rapidly.
Often seen in pairs. **Range:** Red Sea
to Fiji, n. to S. Japan, s. to S. Africa
and SE. Australia.

Dahab, AK

Slender ghost pipefish 8 cm
Solenostomus leptosomus
Base of tail wide; fins sculpted,
numerous tassels. **Biology:** inhabits
reef edges near sand, 2 to 15 m. Rare,
primarily pelagic, settling at full adult
size to breed. Usually seen at night.
Recently photographed in the Red Sea
for the first time by T. M. Hackenberg.
Range: Red Sea and Mauritius to
SE. Australia, n. to S. Japan.

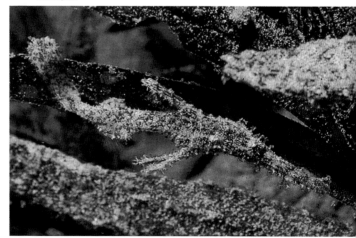

Marsa Shagra, BE

Dahab, AK

Egyptian seahorse *Hippocampus suezensis*   25 cm
Colour variable; corners of plates with small knobs.
Biology: estuaries to seaward reefs, 1 to 68 m, usually
below 20 m. Clings to seagrasses, sponges, gorgonians, or
occasionally to *Sargassum* seaweed. **Range:** Red Sea to
Pakistan. **Similar:** possibly identical to *H. kelloggi* (S. Japan
to China) and other species outside our area.

Jayakar's seahorse *Hippocampus jayakarai*   15 cm
Straw yellow; corners of plates with black thorns.
Biology: inhabits shallow sheltered reefs, 1 to 82 m. Often
attached to clumps of algae or seagrasses, particularly
Halophila spp. Adults form pairs. **Range:** Red Sea to
Pakistan. **Similar:** *H. hystrix* (E. Africa to Fr. Polynesia and
Hawaii, n. to S. Japan, s. to S. Africa).

SEAHORSES AND PIPEFISHES
SYNGNATHIDAE

Members of this large family of over 220 species typically have long
tubular snouts and elongate bodies encased in rings of bony plates.
Parental care is unusual: male possesses a ventral brood pouch embedded in the skin in which the eggs (up to 150) are deposited by the female,
then fertilized and incubated for about 3 weeks. The pouch becomes vas-
cularized to facilitate nourishment of the eggs as well as provide oxygen
and protection. In some pipefishes the eggs are attached directly to the
skin. Courtship may take as long as 3 days and parents are often monogamous. Before mating each morning they exhibit changes in colour and
engage in ritualized dancing or entwining to synchronize reproduction
readiness. Most species prefer protected habitats and live in weedy or seagrass areas close to reefs, others inhabit crevices or overhangs. They feed
on tiny crustaceans (copepods, larval shrimps, amphipods) that are
sucked in by the tubular mouth. A flourishing trade threatens seahorses
on a global scale. About 20 million seahorses (60 tonnes) are harvested
annually, primarily in Vietnam, the Philippines, China and Thailand.
Three quarters are dried for the Chinese herbal medicine market while
the remainder (4 to 5 million) are kept alive for the aquarium market.
This has resulted in an estimated worldwide reduction of natural populations by 50 to 70 per cent.

▲ **Soft coral seahorse**   2.5 cm
Hippocampus sp.
Highly cryptic with long soft spines.
Biology: only among branches of the
soft coral *Dendronephthya* sp. (p. 250)
on seaward reef slopes below 20 m.
Range: Red Sea only. **Similar:** *H.
lichtensteini* is a larger species without
long spines.

56   **SEAHORSES AND PIPEFISHES**

Network pipefish   11.5 cm
Corythoichthys flavofasciatus
Snout short; irregular black and yellow
blotches and lines forming network.
Biology: among algae-covered rocks
and living corals of subtidal lagoon
and seaward reefs, 1 to 25 m. Also
on rubble and coarse sand. Common,
often in small groups or in pairs.
Feeds on copepods, isopods and
ostracods. Sexually mature at about
7 cm. **Range:** Red Sea to Maldives,
s. to Mauritius; populations from
Pacific are most likely a different
species, *C. conspicillatus* (Ryukyus and
PNG to Fr. Polynesia, s. to GBR).

Marsa Shagra, RM

Guilded pipefish   16 cm
Corythoichthys cf. *schultzi*
Snout long; black and yellow horizontal
lines and dashes. **Biology:** usually
included with *C. schultzi*, possibly
a distinct species. In the Red Sea,
in sheltered bays and harbours, usually
3 to 10 m. On sand, rubble, corals,
algae and gorgonians. In pairs or small
groups. Monogamous, often seen
engaging in courtship ritual. Matures
at about 9 cm. **Range:** possibly Red
Sea only; *C. schultzi* e. to Fr. Polynesia,
n. to Ryukyus, s. to E. Africa and
Tonga.

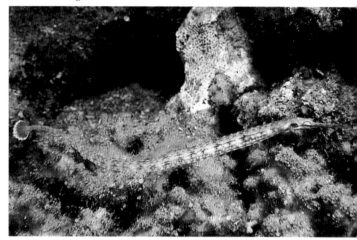
Abu Galum, RM

Red Sea pipefish   12 cm
Corythoichthys sp.
Curved red lines on head, large ochre
patch on cheek. **Biology:** on rubble
or corals of protected reef slopes, often
in caves, 2 to 30 m. Often in pairs.
Range: Red Sea only. **Similar:** *C.
insularis* (W. Indian Ocean) and *C.
nigripectus* (W. Pacific).

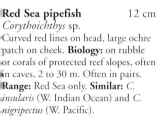

Marsa Shagra, RM

Bluestripe pipefish   7 cm
Doryramphus excisus abbreviatus
Orange with blue lateral stripe.
Biology: inhabits lagoon and seaward
reefs, 1 to 45 m. In crevices or caves,
occasionally among spines of long-
spined *Diadema* and *Echinothrix* sea
urchins. Often swims upside down.
A cleaner that picks parasites from
other fishes, particularly morays.
Like the cleaning wrasses, they
advertise their trade with a bobbing
dance. They occur in pairs and are
territorial. One male's brood pouch
contained 137 eggs. **Range:** Red Sea
only; other subspecies from E. Africa
to Mexico.

Jeddah, JK

Broad-banded pipefish   16 cm
Dunckerocampus boylei
Alternating broad orange-brown and
slightly narrower white bars.
Biology: in crevices and caves, usually
below 25 m. **Range:** Red Sea to Bali,
s. to Mauritius and S. Africa; replaced
by a very similar undescribed species
in W. Pacific. Previously confused with
Banded pipefish, *D. dactyliophorus*
(W. Pacific).

Eritrea, AK

Multibar pipefish   16 cm
Dunckerocampus multiannulatus
Reddish-brown with about 60 thin
white bars. **Biology:** free-swimming
in crevices, caves, under overhangs, 1
to 75 m. May hover upside down near
ceilings of caves, often in pairs. May
clean other fishes. Male carries eggs in
a ventral skin flap. **Range:** Red Sea to
Sumatra, s. to Mauritius and
S. Africa. Replaced by the similar
Yellow-banded pipefish, *D.
pessuliferus*, in W. Pacific.

Marsa Shagra, KH

Short-bodied pipefish   6.5 cm
Choeroichthys brachysoma
Snout long, trunk deep; black to brown
with while flecks becoming bars on
tail. **Biology:** shallow protected reefs
among algae and seagrasses, 1 to
42 m. **Range:** Red Sea to Fr. olynesia,
n. to Philippines, s. to NE. Australia;
Red Sea–Indian Ocean population
possibly distinct from *C. brachysoma*.

Jeddah, JK

Double-ended pipefish   40 cm
Trachyrhamphus bicoarctatus
Black to brown with tiny black spots
and while flecks, sometimes with
broad light bars; resembles a thin bent
stick. Rudimentary tail fin often lost,
can not be regenerated. **Biology:** on
rubble, sand, mud, or among algae
and seagrasses in lagoons and coastal
bays, 1 to 42 m. Often seen lying on
sand with raised head or drifting like
debris. Slow moving, captures drifting
zooplankton. Matures at about 26 cm.
Range: Red Sea to New Caledonia, n.
to S. Japan, s. to E. Africa and SE.
Australia.

Marsa Alam, BE

Hairy pygmy pipehorse   5.5 cm
Acentoneura tentaculata
Cryptic with short snout, prehensile
tail and numerous finely branched
tassels; resembles outstretched seahorse.
Biology: shallow sheltered areas among
algae and seagrasses. **Range:** Red Sea
to Comore Islands. **Similar:** Alligator
pipefish, *Syngnathoides biaculeatus*
much larger (28 cm; Red Sea to
Samoa) with shorter branched tassels,
in protected coastal shallows among
algae and seagrasses and in rafts of
floating *Sargassum* seaweed.

Nuweiba, MT

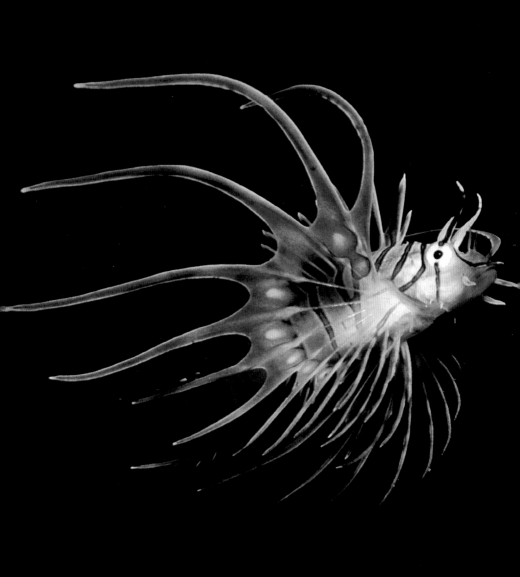

SCORPIONFISHES

SCORPAENIDAE

Scorpionfishes are typically stout-bodied bottom-dwelling carnivores characterized by venomous dorsal, anal and pelvic fin spines and small spines on the head. Most species are well camouflaged with numerous flaps and tassels and a highly variable cryptic coloration that may be altered to match their surroundings, and are often overlooked by divers. They rely on this concealment to ambush their prey of fishes, crustaceans and cephalopods. Their feeding movement may be incredibly rapid, under 15 milliseconds. When disturbed most species raise their venomous dorsal spines then dart a short distance away. Some also have brightly coloured undersides to their pectoral fins which are flashed in a startling display that may deter predation by serving as a warning or by buying time to escape. Many species are seen only at night when they emerge to hunt. Little is known of their reproduction. The 1 mm eggs may be embedded in a floating gelatinous mass which is carried away by currents. Lionfishes and stonefishes (Synanceiidae) have potent venoms which can cause excruciating pain, vomiting, swelling, difficulty in breathing, and fever. The venom of stonefishes has caused fatalities. Treatment is by immersion in hot water (45°C for 40 min.) or blowing with hairdryer which helps denature the protein. The pain usually dissipates within a few days, but the side effects of a stonefish sting may last for weeks. Stonefish antivenom is seldom available, since it must be refrigerated. The stonefishes (Synanceiidae) are treated here as a separate family.

Common lionfish   38 cm
Pterois miles
Red to black with narrow white bars; P fin rays mostly free and feathery. **Biology:** inhabits lagoons, bays and seaward reefs, 0.2 to 60 m. Hovers near edges, caves and in wrecks by day. At dusk and at night, roams the reef hunting for fishes, crabs and shrimps. Their feathery pectoral fins are used to corral prey to within striking distance. Often attracted by divers' lights which distract prey. May charge divers with dorsal spines pointed forward. Stings extremely painful. **Range:** Red Sea to Sumatra, s. to S. Africa. Replaced by *P. volitans* in Pacific. **Similar:** *P. russelli*, with wider interspaces between bars and no spots on soft D, A and tail fins, inhabits deeper mud habitats (Red Sea to Australia).
◀ *Pterois miles* juv., 3 cm *(Jeddah, JK.)*

Marsa Abu Dabab, RM

Clearfin lionfish   24 cm
Pterois radiata
P rays white and mostly free and filamentous; tail base with two thin white stripes. **Biology:** inhabits reef flats, lagoons and seaward reefs, to 25 m. Solitary or in small groups hovering in crevices, caves, or under ledges by day. Feeds primarily on crabs and shrimps. Common. In Yemen, often in algae-covered areas with poor coral growth. **Range:** Red Sea to Fr. Polynesia, n. to Ryukyus, s. to S. Africa and New Caledonia.

Mangrove Bay, RM

Kenya, LS

Mombasa lionfish   19 cm
Pterois mombasae
P fin rays mostly attached, filamentous at ends, membrane with dark spots. **Biology:** on hard bottoms of deep seaward reefs, 20 to at least 60 m, in areas rich in soft corals and sponges. **Range:** Red Sea to Papua New Guinea, n. to Sri Lanka, s. to S. Africa and NW. Australia. **Similar:** *P. antennata* (Gulf of Aden to Pacific; to 20 cm) with longer P rays. ▼

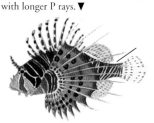

Jordan, EL

Shortfin dwarf lionfish   17 cm
Dendrochirus brachypterus
P fin fan-like, rays fully attached by membrane; inner surface with concentric bands. **Biology:** inhabits reef flats and shallow lagoon and coastal reefs, 2 to 80 m. Typically on isolated coral heads or algae- or sponge-covered rocks in silty inshore areas. Often rests at base of rocks or corals or hangs upside-down under ledges. Solitary or in small harems (of up to 10) headed by a dominant male. **Range:** Red Sea to Samoa, n. to Ryukyus, s. to S. Africa and SE. Australia.

Salalah, KF

Zebra lionfish   20 cm
Dendrochirus zebra
P fin fan-like, rays fully attached by membrane and banded, membranes dusky. **Biology:** inhabits sheltered coastal reefs, from reef flats to 73 m. Among rubble, corals and debris. Territorial male aggressively defends territory containing several females. Starts to hunt 3 hours before sunset on crabs, shrimps and small fishes. Absent in N. Red Sea, uncommon s. of Port Sudan. **Range:** C. and S. Red Sea and E. Africa to Samoa, n. to Ryukyus, s. to S. Africa and SE. Australia.

Smallscale scorpionfish 36 cm
Scorpaenopsis oxycephala
Snout much longer than eye diameter;
second D fin spine as long as third.
Biology: inhabits lagoons, sheltered
bays and seaward reefs, 1 to at least 60
m. Prefers coral-rich reefs in clear
water. Rests motionless on hard
bottoms, living corals, rubble, coralline
algae, rocks or between soft corals or
sponges. **Range:** Red Sea to Moluccas,
n. to Taiwan, s. to Natal, GBR.
Similar: Spinycrown scorpionfish *S.
rossi* (Red Sea to Pitcairn; 25 cm) has
shorter snout, fewer P rays (17 vs 20).

Salalah, KF

Devil scorpionfish 30 cm
Scorpaenopsis diabolus
Back highly arched; inner surface of
P fin bright yellow. **Biology:** inhabits
reef flats and lagoon and seaward
reefs, 1 to 70 m. Lies motionless on
rubble, sand, algae, or corals. Feeds
on small fishes. Often misidentified
as the stonefish, but flashes colourful
underside of P fins. **Range:** Red Sea
to Hawaii and Pitcairn Is., n. to
Ryukyus, s. to S. Africa and GBR.
Underside of P fin. ▼

Marsa Shagra, RM

Bearded scorpionfish 25 cm
Scorpaenopsis barbatus
Snout short, nearly equal to eye
diameter. **Biology:** inhabits rocky and
coral reefs, 3 to 30 m. In areas of
mixed sand and corals, common on
rocky reefs of Yemen and Oman.
Often in pairs. Lies motionless on
bottom to ambush prey of fishes and
crabs. **Range:** Red Sea to Arabian
Gulf and N. Somalia. **Similar:** many
smaller cryptic species; *Scorpaenodes
parvipinnis* (Red Sea to Fr. Polynesia;
13 cm) usually seen only at night. ▼

Nuweiba, PM

Mövenpick Bay, KF

Yellow-spotted scorpionfish
Sebastapistes cyanostigma 8.5 cm
Brown with tiny yellow specks and
larger yellow blotches. **Biology:** in
Pocillipora branching corals of exposed
seaward reefs, 0.3 to 29 m. Secretive,
remains among branches of home
corals. Feeds on small fishes and
crustaceans. **Range:** Red Sea to
C. Pacific, s. to S. Africa.

STONEFISHES
SYNANCEIIDAE

Stonefish 38 cm
Synanceia verrucosa
Bulbous body with vertical mouth,
tiny eyes and large P fins; D spines
hidden beneath warty skin.
Biology: usually on sand or rubble of
reef flats, lagoons and seaward reefs,
0.3 to 45 m. Often partially buried or
under rocks, may stay on site for
months. Feeds on small fishes and
crustaceans. Stings extremely painful,
occasionally fatal without medical
attention. **Range:** Red Sea to Fr.
Polynesia, n. to S. Japan, s. to S.
Africa. **Similar:** *S. nana* (Red Sea
to Oman; 13.5 cm). ▼

Ras Mamlah, RM

Filamented devilfish 25 cm
Inimicus filamentosus
D spines long and free; inner surface
of P fin orange-brown, the lower
rays free. **Biology:** on mud, sand, or
rubble of lagoons and sheltered bays,
3 to 55 m. Often buried. Uses free
pectoral rays to 'walk' on bottom,
flashes bright underside as warning.
Spines highly venomous. **Range:** Red
Sea to Maldives, s. to Mauritius.
Similar: *Choridactylus multibarbus* has
blunt snout, D spines connected by
membrane (Red
Sea to Philippines;
12 cm). Underside
of P fin of *C.
multibarbus.* ►

Mövenpick Bay, KF

VELVETFISHES
APLOACTINIDAE

Crested velvetfish　　　10 cm
Ptarmus gallus
Tail with yellow eye-spots.
Biology: on rubble and sand
of sheltered lagoons and bays,
10 to 30 m. Solitary or in pairs,
lies motionless. **Range:** Red Sea only.
Similar: Günther's waspfish *Snyderina
guentheri* (Tetrarogidae; Gulf of Aden
to India; 20 cm). ▼

Jeddah, JK

FLATHEADS
PLATYCEPHALIDAE

Indian Ocean crocodilefish
Papilloculiceps longiceps　　1.0 m
Elongate depressed fish with large
head and mouth. **Biology:** common in
sheltered bays and lagoons, 1 to 40 m.
Usually lies on sand or rubble close to
reefs, often partly buried. Highly
cryptic, an ambushing predator of
crabs, shrimps and fishes. Top of pupil
with expandable branched sunshade,
called an iris lappet. Easily pproached.
Range: Red Sea to Oman, s. to S.
Africa. **Similar:** Long-snout flathead,
Thysanophrys chiltonae smaller with
eyes closer together (Red Sea to Fr.
Polynesia; 25 cm).

Mövenpick Bay, RM

HELMET GURNARDS
DACTYLOPTERIDAE

Common helmet gurnard
Dactyloptena orientalis　　38 cm
Large armoured head, rough scales,
and large wing-like P fins.
Biology: inhabits sandy flats and
slopes of lagoon and seaward reefs,
1 to 100 m. Feeds primarily on
invertebrates, occasionally on small
fishes. When alarmed it spreads
its colourful pectorals and speeds
away. Can produce sounds with its
swim-bladder. **Range:** Red Sea to
Hawaii and Fr. Polynesia, s. to S.
Africa and N. New Zealand.

Guam, RM

GROUPERS, ANTHIASES AND SOAPFISHES SERRANIDAE
Subfamily Epinephelinae

This large and diverse family includes groupers, anthias and soapfishes. All have a single dorsal fin with well-developed spines, continuous lateral line, 2 or 3 flattened opercular spines, and small ctenoid scales.

Groupers are medium to large robust-bodied bottom dwellers that live in a wide range of habitats from the shoreline to depths of over 200 m. They are voracious carnivores of crustaceans, fishes and cephalopods. Species of *Cephalopholis* and *Epinephelus* are sequential hermaphrodites that mature first as females, then change to males. Small species may mature in as little as a year, large ones may take several years and live for several decades. Spawning is typically seasonal and synchronized by moon phase. Some species migrate to aggregate at favoured spawning sites. Eggs are small (under 1 mm) and pelagic and the planktonic larval stage lasts up to several weeks. Some of the large species, particularly the piscivores, may be ciguatoxic in certain areas. Groupers readily take a baited hook and large species that take years to mature are highly vulnerable to overfishing. If the larger individuals are harvested, leaving only small or immature females, the reproductive potential of a population may be lost and new stocks must depend on recruitment or migration from less exploited areas. Large groupers that are conspicuous in protected areas are usually the first species to disappear from fished areas. In many areas of the world, the discovery of seasonal spawning aggregations has resulted in one or two years of great catches followed by rapid population collapse.

▲ **Red Sea coralgrouper** 1.1 m
Plectropomus pessuliferus
Brown to red with blue spots becoming vertically elongate below, occasionally with pale saddles; emarginate tail, conspicuous canine teeth. **Biology:** inhabits lagoon and seaward reef slopes, 3 to 50 m. A roaming predator that feeds primarily on fishes, including garden eels. **Range:** subspecies *marisrubri* in Red Sea only; subspecies *pessuliferus*: Indian Ocean and Fiji. *(Photo: Mangrove Bay, KF.)*

Redmouth grouper 60 cm
Aethaloperca rogaa
Deep body; inside of mouth red.
Biology: inhabits areas of rich coral
growth of bays and protected seaward
reefs, 1 to 54 m. Usually near caves,
crevices and wrecks. Feeds on fishes
and crustaceans. Juveniles mimic
the angelfish *Centropyge multispinus*
(p. 136) in order to approach prey
closely. **Range:** Red Sea to Fiji, n. to
Ryukyus, s. to Mozambique, GBR.

juv.

Bali, RM

Slender grouper 52 cm
Anyperodon leucogrammicus
Round tail; greenish with red spots.
Biology: inhabits areas of rich coral
cover and clear water of lagoon and
seaward reefs, 1 to 50 m. Fairly
secretive, but approachable. Preys
on small fishes. Juveniles mimic
the wrasse *Halichoeres cosmetus* in
the Indian Ocean and other related
wrasses in the W. Pacific. **Range:** Red
Sea to Samoa, n. to Ryukyus, s. to E.
Africa, GBR.

juv.

Kenya, KF

Ocelot grouper 85 cm
Dermatolepis striolata
Deep body; brown with irregular pale
blotches and dark specks.
Biology: inhabits rocky and coral
reefs, 5 to 70 m, usually in turbid
areas. Shy, usually found lurking
near caves or crevices. Common in
S. Oman. **Range:** Gulf of Oman
s. to S. Africa, e. to Seychelles;
absent from Red Sea.

Salalah, KF

GROUPERS 67

Marsa Shagra, RM

Peacock grouper   55 cm
Cephalopholis argus
Dark brown with blue spots,
sometimes up to 6 pale bars on lower
body; tail round. **Biology:** inhabits
shallow lagoon and seaward reefs
with rich coral growth and clear
water, 1 to 50 m. Often under ledges
or on reef margins. Feeds primarily on
fishes. Juveniles in protected coral
thickets, adults often in pairs or small
groups. Not shy. **Range:** Red Sea
to Marquesas and Pitcairn Group,
n. to S. Japan, s. to S. Africa and SE.
Australia.

Coral hind   41 cm
Cephalopholis miniata
Red with blue spots, ground colour
may be mottled; tail round.
Biology: inhabits shallow lagoon and
seaward reefs, 1 to 150 m. Common
in coral-rich areas with clear water.
Solitary or in groups. Not shy. **Range:**
Red Sea to Line Is., n. to S. Japan, s.
to S. Africa and SE. Australia.

juv.

Mangrove Bay, EL

Halfspotted hind   35 cm
Cephalopholis hemistiktos
Brown to red with blue spots on head
and lower body. **Biology:** inhabits
shallow coral-rich areas of bays,
lagoons and seaward reefs, 4 to 55 m.
Common around coral heads of open
patch reefs with caves and crevices.
Feeds primarily on fishes and
crustaceans. Often at cleaning
stations, remains close to bottom.
Range: Red Sea to Pakistan, n. to
Arabian Gulf, s. to Somalia.

Mövenpick Bay, RM

Sixspot grouper 50 cm
Cephalopholis sexmaculata
Orange-red with small blue spots,
and broad bifurcated dark bars.
Biology: inhabits steep seaward
reefs with caves and holes, 6 to
150 m. Usually in caves, upside
down in crevices, or vertically along
walls, rarely in open water. Frequently
at *Periclimenes* shrimp cleaning
stations. Solitary or in small groups.
Range: Red Sea to Fr. Polynesia,
n. to S. Japan, s. to S. Africa and
SE. Australia.

Palau, RM

Vermilion hind 30 cm
Cephalopholis oligosticta
Orange-red with widely-spaced bright
blue spots. **Biology:** inhabits barren
patch reefs in silty areas of lagoons
and fringing reefs, 15 to 48 m.
Generally uncommon, but in some
areas, particularly Hurghada, up to
10 individuals per 250 m³. Shy.
Range: Red Sea only, from Gulf
of Aqaba s. to Farasan Is.

Hurghada, DE

Tomato grouper 57 cm
Cephalopholis sonnerati
Juv. dark brown, adults brown to
orange-red sometimes with scattered
large white flecks, head with small
closely packed red spots; body
moderately deep. **Biology:** inhabits
rocky or coral reefs, 5 to 70 m.
More common below 30 m,
typically around isolated patch reefs.
Range: Gulf of Aden to Line Is. and
Samoa, n. to S. Japan, s. to S. Africa
and New Caledonia.

juv.

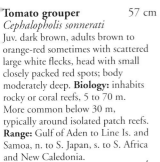

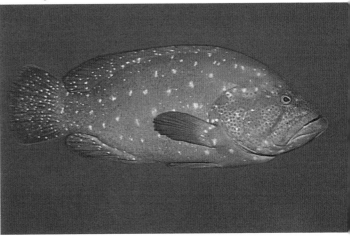

Mauritius, EL

Marsa Shagra, EL

Blacktip grouper 40 cm
Epinephelus fasciatus
Tips of dorsal fin spines black; often a dark red cap on head. **Biology:** a common inhabitant of lagoons, bays and seaward reefs, 0.5 to 160 m. In southern Egypt, very common on isolated dead coral blocks in seagrass beds. An ambushing predator of fishes and crustaceans. **Range:** Red Sea to Marquesas and Pitcairn Group, n. to S. Japan, s. to S. Africa and SE. Australia.

Safaga, KF

Summana grouper 52 cm
Epinephelus summana
Dark brown with white blotches and spots; tail round. **Biology:** inhabits shallow coral-rich areas of sheltered lagoons and seaward reef slopes, 1 to 30 m. May enter brackish water. Often in pairs or small groups. Not shy. Uncommon in northern Red Sea and Yemen. **Range:** Red Sea and Gulf of Aden only; replaced by the very similar *E. caeruleopunctatus* elsewhere (Oman and Arabian Gulf to W. Pacific; 76 cm); *E. multinotatus* larger (1 m) with truncate tail (Gulf of Aden to W. Australia). ▼

Blue-and-yellow grouper 90 cm
Epinephelus flavocaeruleus
Juv. with bright yellow extending from D fin onto upper rear of body. **Biology:** inhabits rocky and coral reefs, 10 to 150 m. Juveniles shallow, adults usually deeper. Feeds primarily on fishes, occasionally on crustaceans and cephalopods. **Range:** Gulf of Aden and Arabian Gulf to Andaman Sea, s. to E. Africa and Mauritius.

Thailand, RM

Brown-marbled grouper 90 cm
Epinephelus fuscoguttatus
Head profile slightly concave
behind eyes which may bulge slightly.
Biology: inhabits lagoon pinnacles,
sandy channels and outer reef slopes
usually in clear coral-rich areas,
1 to at least 60 m. May be ciguatoxic
in some areas. Wary and difficult to
approach. Feeds on fishes, crabs
and cephalopods. Uncommon.
Range: Red Sea to Samoa, n. to
Ryukyus, s. to Mauritius and
New Caledonia.

GBR, RM

Malabar grouper 1.2 m
Epinephelus malabaricus
Small black spots and pale blotches;
irregular bifurcated dark bars.
Biology: inhabits estuaries, bays and
protected seaward reefs, 2 to 100 m.
Often rests on sand or rubble of deep
channels or at the base of coral slopes.
Wary and difficult to approach.
Solitary. Feeds on fishes, crustaceans
and octopuses. A highly esteemed
foodfish, often confused with
E. coioides. **Range:** Red Sea to Tonga,
n. to S. Japan, s. to S. Africa and SE.
Australia. **Similar:** Orange-spotted
grouper *E. coioides* has red and black
spots and lacks white blotches in the
dusky bars (Red Sea and Arabian Gulf
to Fiji; 95 cm).

Eritrea, AK

Marbled grouper 75 cm
Epinephelus polyphekadion
Head profile convex, black blotch
on caudal peduncle. **Biology:** inhabits
bays, lagoons and seaward slopes with
rich coral growth, 1 to 46 m. Often
along the base of fringing reefs close
to caves or crevices. An ambushing
predator, primarily of small fishes.
Usually easy to approach. May bump
heads with rivals during courtship.
Range: Red Sea to Fr. Polynesia, n. to
S. Japan, s. to Mauritius and SE.
Australia.

Chuuk (Truk), RM

Mövenpick Bay, EL

Greasy grouper 70 cm
Epinephelus tauvina
Small red-brown spots; background
often with dusky diagonal bars.
Biology: inhabits bays, lagoons
and seaward reef slopes, 1 to 46 m.
Common in clear, coral-rich areas.
Usually rests on bottom while waiting
to ambush its prey. Feeds primarily
on fishes. May be ciguatoxic. Easily
approached. Most accounts of this
species in fisheries and aquaculture
are based on misidentification of
E. coioides and *E. malabaricus*.
Range: Red Sea to Fr. Polynesia,
n. to S. Japan, s. to S. Africa and
GBR.

Salalah, KF

Epaulet grouper 38 cm
Epinephelus stoliczkae
Front with spots, rear with 3 dark
bars. **Biology:** sandy areas of shallow
rocky reefs, 1 to 25 m. Often near
isolated coral heads or rocks, rare on
well-developed coral reefs. Not shy.
Range: Red Sea to Pakistan, n. to
Gulf of Oman. **Similar:** *E. rivulatus*
(Yemen to New Caledonia; 39 cm). ▼

Brownspotted grouper 75 cm
Epinephelus chlorostigma
Hexagonal dark spots small and
close-set; tail slightly emarginate.
Biology: inhabits coastal and deep
seaward reefs, 4 to 280 m. More
common in rubble and seagrass
habitats than well-developed coral
reefs. **Range:** Red Sea and Gulf
of Aden to Samoa, n. to S. Japan,
s. to S. Africa and New Caledonia.

Hurghada, DE

Areolate grouper 40 cm
Epinephelus areolatus
Light brown polygonal spots; rear margin of tail with white edge.
Biology: inhabits coastal reefs, 3 to 200 m. Occurs around isolated coral heads or rock outcrops on silty sand or in seagrass beds, usually in turbid areas. Solitary or in pairs. Not shy. Uncommon. **Range:** Red Sea to Fiji, n. to S. Japan, s. to S. Africa and GBR.

Nuweiba, AK

Gabriella's grouper 52 cm
Epinephelus gabriellae
Close-set orange-brown spots, median fins with white edge; may display light saddles. **Biology:** inhabits rocky and poorly developed coral reefs, 1 to at least 40 m. Common on inshore reefs of S. Oman and S. Yemen. Solitary or in small groups. Hovers well above the bottom. Not shy. **Range:** Oman and Yemen to Somalia.

Salalah, KF

Giant grouper 3 m; 400 kg
Epinephelus lanceolatus
Bold juv. pattern breaks up with growth. The largest bony reef fish.
Biology: inhabits deep estuaries, lagoons, channels and seaward reefs, 3 to 100 m. Usually has a home cave or wreck. Solitary and wary. Feeds on fishes, large crustaceans and even small sharks and turtles. Large ones often ciguatoxic. Unconfirmed fatal attacks on humans. Rare, nearly wiped out by spear and line fishing. Juveniles secretive. **Range:** Red Sea to Fr. Polynesia and Hawaii, n. to Ryukyus, s. to S. Africa and GBR.

juv.

Marsa Alam, MM

GROUPERS 73

GBR, RM

Potato grouper 2.0 m
Epinephelus tukula
Light grey with large dark blotches.
Biology: inhabits pinnacles, dropoffs, deep lagoons and wrecks, 3 to 150 m, usually in clear, coral-rich areas. Solitary or in small groups, uncommon in Red Sea. Feeds on fishes, crustaceans and cephalopods. Bold and easily approached, potentially dangerous to inexperienced divers. **Range:** Red Sea to GBR, n. to S. Japan, s. to S. Africa and Mauritius; distribution patchy.

Palau, RM

Squaretail coralgrouper 73 cm
Plectropomus areolatus
Black-edged blue spots; square tail; conspicuous canine teeth.
Biology: inhabits bays, lagoons and seaward reef slopes, 1 to 50 m. Solitary and wary. Often cruises about 1 m above edge of dropoffs. A voracious predator of smaller fishes. Common in some areas of the Red Sea. **Range:** Red Sea to Samoa, n. to Ryukyus, s. to Mauritius and GBR. **Similar:** juv. *P. punctatus* has small whitish elongate pale spots, adult marbled brown (S. Oman to Mauritius; 96 cm).

Mövenpick Bay, RM

Lyretail grouper 81 cm
Variola louti
Lunate tail with yellow margin.
Biology: inhabits lagoons, channels and seaward reefs, 1 to 150 m. A predator of fishes and crustaceans. Juveniles may resemble the goatfishes *Parupeneus forsskali* or *P. macronema* which they shadow as they feed. Common. **Range:** Red Sea to Marquesas and Pitcairn Group, n. to S. Japan, s. to S. Africa and SE. Australia.

juv.

nthiases are relatively small (under 12 cm) colourful inhabitants of reef margins and steep outer reef slopes. lost of those familiar to divers, primarily in the genus *Pseudanthias*, are laterally compressed, sexually dichro- atic, diurnally active planktivores that commonly aggregate a few metres above prominent coral heads or dges. They reach their greatest abundance along current-swept dropoffs of the outer reef slope. In the Red ea, large aggregations of the Lyretail anthias are common in quite shallow water along fringing reef margins id around patch reefs. Aggregations typically consist of one or more brightly coloured territorial males and umerous less colourful females. Males are sex-reversed females, with sex-reversal socially controlled and ter- inal. They spend an enormous amount of time showing off to females and other males. The courtship dis- lay consists of an upside-down U-shaped rush with fins flared. At the approach of danger and at night, ithias shelter within the reef. Other members of this subfamily, the species of *Plectranthias*, more closely semble small hawkfishes with large mouths and thickened lower pectoral rays. Most of them are deep- welling or highly cryptic predators of small benthic fishes and crustaceans.

yretail anthias   15 cm
seudanthias squamipinnis
range with yellow-centred scales;
with elongate dorsal filament and
d patch on P fin. **Biology:** in
gregations near coral outcrops
clear lagoons, patch reefs and
ep slopes, 0.3 to 35 m. Abundant,
metimes in enormous swarms above
e reef. Male is territorial and
aintains a harem of 5 to 10 females.
ange: Red Sea to Fiji, n. to S. Japan,
to S. Africa and SE. Australia; absent
om Oman and Arabian Gulf.

♂ *Mövenpick Bay, RM*

man anthias   16 cm
seudanthias marcia
range, pale below; ♀ with red tips on
il lobes, resembles female *P. taeniatus*
d *P. townsendi*; tail of ♂ with red
argin and filamentous upper lobe.
iology: in aggregations over rugged
cky bottoms, 14 to 30 m. Locally
undant. **Range:** Gulf of Oman to
kha Is. (Gulf of Aden) only.

♂ *Oman, JPH*

Red Sea anthias 13 cm
Pseudanthias taeniatus
♀ orange with pale belly, ♂ pale
with 2 broad red to burgundy stripes.
Biology: inhabits coral-rich areas
of clear lagoon and seaward reefs,
12 to 50 m. Common on isolated
deep patch reefs and outcrops.
Females stay close to bottom while
males swim 1 to 2 m above.
Often aggregates with other anthias.
Harems tend to be larger (8 to 15
individuals) than those of *P.
squamipinnis*. **Range:** Red Sea only;
replaced by the similar *P. townsendi*
from Gulf of Aden to Arabian Gulf.

♂ Dahab, EL

Townsend's anthias 9 cm
Pseudanthias townsendi
♀ orange with pale belly, ♂ pale
yellow with 2 broad red to burgundy
stripes, red D fin and semicircular red
bands on the tail. **Biology:** inhabits
rocky bottoms, 15 to 63 m. More
common in Gulf of Oman than
Arabian Sea, rare at Sikha Island.
Range: Gulf of Aden (Sikha Island)
to Arabian Gulf only; replaced in the
Red Sea by the similar *P. taeniatus*
which has nearly identical ♀ and ♂
with white D fin and mostly white
tail.

♂ Oman, JPH

Heemstra's anthias 13 cm
Pseudanthias heemstrai
♀ orange above, lavender below;
♂ lavender with yellow-edged
red tail and yellow-orange head.
Biology: inhabits seaward reef slopes
and pinnacles, 13 to 67 m, usually
near reef bases below 40 m. In loose
aggregations 1 to 6 m above reef,
sometimes with Red Sea anthias.
Courting males tend to swim
above females in U-shaped
pattern. Shy. **Range:** Red Sea only.
Similar: ♀ *P. marcia*, *P. taeniatus* and
P. townsendi are pale below.

♂, Dahab, AK

Redstripe anthias 20 cm
Pseudanthias fasciatus
♀ orange with broad red stripe; ♂ without red stripe. **Biology:** in caves or under overhangs, 20 to 150 m, always below 40 m in Red Sea. Typically upside down near the ceiling of large caves. Uncommon. **Range:** Red Sea to New Caledonia, n. to S. Japan, s. to GBR. **Similar:** *P. lunulatus* lacks red stripe (S. Red Sea to Indonesia, s. to Somalia, 10 cm). ▼

♀, Ras Mohammed, BF

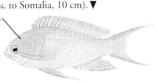

SOAPFISHES

Small members of the grouper family that produce a bitter toxin *grammistin* in mucus which protects them from predation.

Sixstriped soapfish 27 cm
Grammistes sexlineatus
Black with narrow white stripes that break up with age. **Biology:** reef flats, lagoons, channels and seaward reefs, 1 to below 40 m. Subadults common, but secretive in or near holes at base of coral reefs. Adults move to quite deep water. **Range:** Red Sea to Fr. Polynesia, n. to S. Japan, s. to S. Africa and New Caledonia.

Dahab, EL

Red Sea soapfish 14 cm
Diploprion drachi
Grey with black D fin, yellow band through eye. **Biology:** sheltered rocky and coral reefs, 3 to 40 m. Common on steep coral-rich slopes. At sunset they leave crevices to patrol walls and slopes and may hide behind large nonpredatory fishes to ambush their prey of small fishes and crustaceans. Solitary or in pairs. **Range:** Red Sea and Gulf of Aden only. **Similar:** Gold ribbon soapfish *Aulacocephalus temmincki* is deep blue with a broad yellow stripe along the back and through the eye. It lives in relatively deep water (Red Sea to Fr. Polynesia; 30 cm).

Mövenpick Bay, RM

LONGFINS

PLESIOPIDAE

Longfins are small fishes with large mouths, large eyes and long pelvic fins. They feed at night on small benthic invertebrates.

Red-tipped longfin   8 cm▶
Plesiops coeruleolineatus
Biology: Inhabits exposed areas of outer reef flats and seaward reefs to 23 m. Nocturnal. (Red Sea and Gulf of Aden to Samoa, n. to S. Japan.)

Whitespotted longfin   17 cm▶
Plesiops nigricans
Biology: Inhabits rocky and coral reefs, tidepools to 30 m. Nocturnal. (Red Sea to S. Oman only.)

▲ Comet   20 cm
Calloplesiops altivelis
Black with fan-like fins and white spots becoming more numerous with size. **Biology:** inhabits lagoon and seaward reefs, 3 to 45 m. Secretive, in holes and crevices by day, coming out at sunset. May mimic the head of the Whitemouth moray eel *Gymnothorax meleagris* (p. 37). **Range:** Red Sea to Fr. Polynesia, n. to S. Japan, s. to Mozambique, GBR. *(Photo: Dahab, AK.)*

DOTTYBACKS AND SNAKELETS

PSEUDOCHROMIDAE

Pseudochromids are small elongate fishes with a long continuous dorsal fin. Three subfamilies occur around the Arabian Peninsula: the eel-like snakelets (Congrogadinae), and the shorter generally more colourful dottybacks (Pseudochrominae and Pseudoplesiopinae, the latter extremely secretive and unknown to most divers). Many *Pseudochromis* are secretive but the brilliantly coloured ones may be quite conspicuous as they hover near shelter. Dottybacks are carnivores of small crustaceans, polychaete worms and zooplankton. Females produce a spherical mass of eggs which is hidden in a hole and guarded by the male. Dispersal after hatching is limited, resulting in most species having limited distributions. Dottybacks do relatively well in properly maintained reef aquaria and make a stunning display.

Orchid dottyback　　7 cm
Pseudochromis fridmani
Purple with dark streak through eye.
Biology: near crevices and holes of
steep walls and slopes of lagoon and
seaward reefs, 1 to 60 m. Prefers
shallow vertical walls. Always close
to shelter, typically hovers nearby.
Solitary or in groups. Abundant in
N. Red Sea with density of up to
7 per m². **Range:** Red Sea only.

Marsa Shagra, RM

Sunrise dottyback　　7 cm
Pseudochromis flavivertex
♀ yellowish, ♂ blue with yellow back.
Biology: inhabits lagoons and
sheltered bays, 2 to 30 m. Common
at base of coral slopes and patch reefs
on sand or rubble. Females more
secretive than males. Territorial and
shy. Rarely leaves small territory.
Range: Red Sea and Gulf of Aden
only.

♂ Mangrove Bay, RM

Blue-striped dottyback　5.5 cm
Pseudochromis springeri
Black with two brilliant blue stripes
on head. **Biology:** inhabits lagoons
and sheltered reef slopes, 2 to 60 m.
Secretive in small branched corals of
isolated patch reefs and lower reef
slopes. Solitary or in pairs. Territorial
and shy, remains close to home coral
heads of the species *Pocillopora
damicornis*, *Stylophora* spp. and
Acropora spp. **Range:** Red Sea and
Gulf of Aden only.

Marsa Shagra, RM

Lyretail dottyback 9 cm
Pseudochromis dixurus
Pale with 2 yellowish-brown stripes.
Biology: inhabits lagoons and sheltered
bays, 5 to 60 m. Secretive, in holes
and crevices. Very shy, usually seen
only when popping out of a hole.
Uncommon in the N. Red Sea, more
common in the south. Seems to prefer
silty reefs. **Range:** Red Sea only.

Sudan, JR

Olive dottyback 9 cm
Pseudochromis olivaceus
Dark olive with small blue spots.
Biology: inhabits lagoon and seaward
reefs, 2 to 20 m. Secretive, hides
in branching corals or coral rubble.
Shy, quickly retreats when frightened.
Range: Red Sea to Arabian Gulf only.

Marsa Shagra, RM

Arabian dottyback 10 cm
Pseudochromis aldabrensis
Orange with electric blue stripes.
Biology: inhabits rocky and coral
reefs, 7 to 40 m. Among rocks and
coral rubble. Common in Gulf of
Aden and Oman. **Range:** S. Red Sea
and Oman to Sri Lanka, n. to Arabian
Gulf, s. to Aldabra. Replaced in
E. Africa by Bluestriped dottyback
P. dutoiti which has blue P fin base
(8 cm).

Oman, AK

Striped dottyback 7 cm
Pseudochromis sankeyi
Black with 2 broad white stripes.
Biology: occurs among rubble and
dead corals of silty reefs, 2 to 25 m.
Always close to bottom, usually in
aggregations, rarely solitary. In dense
groups of up to 25 per m² in S. Red
Sea. Not shy. Very common in Yemen.
Range: S. Red Sea and Gulf of Aden
only.

Eritrea, AK

Pale dottyback 10 cm
Pseudochromis pesi
Grey with dark back extending to
upper lip. **Biology:** occurs around
small isolated corals or rocks on sand
or rubble, 10 to 45 m. Singly or in
pairs. Leaves territory for extended
periods, occasionally swims well
above the bottom. **Range:** Red Sea
and S. Africa only.

Nuweiba, EL

SNAKELETS
Subfamily Congrogadinae

Spotted snakelet 15.5 cm
Haliophis guttatus
Extremely elongate and eel-like.
Light brown with dark spots on
head, narrow dark bars on body.
Biology: rocky reefs, 1 to 15 m.
Lives in holes among rocks and
corals. **Range:** Red Sea and S. Oman
to Mozambique and Madagascar.
Similar: *Halidesmis coccus* (S. Oman)
and *H. thomaseni* (S. Oman to Bay of
Bengal) have short snout with fleshy
crest; Stars-and-stripes snakelet
Haliophis diadematus is white with
dark stripes and spots (see p. 82).

Eilat, MH

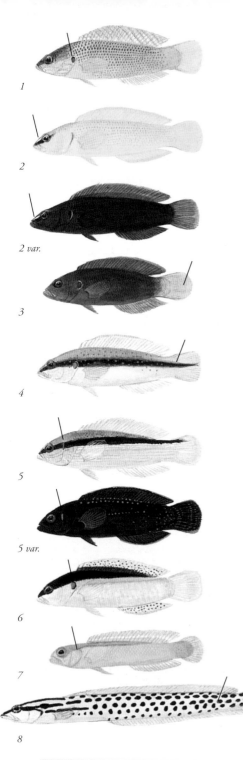

1. Bandtail dottyback   11 cm
Pseudochromis caudalis
Brownish-yellow with blue spot on operculum, large adults with converging dark submarginal bands on tail. **Biology:** on coastal rocky reefs, 12 to 30 m. **Range:** Gulf of Oman to Sri Lanka.

2. Whitelip dottyback   9 cm
Pseudochromis leucorhynchus
Yellow to dark brown with white spot on snout and dark band through eye. **Biology:** common on shallow rocky and coral reefs, 1 to 8 m. A mimic of the blenny *Oman ypsilon*. **Range:** S. Oman to Kenya.

3. Linda's dottyback   11 cm
Pseudochromis linda
Dark olive-brown with yellow tail.
Biology: common on rocky reefs in Gulf of Oman, 1 to 6 m. Secretive and rarely seen.
Range: Gulf of Aden to Pakistan, s. to Somalia.

4. Blackstripe dottyback   9 cm
Pseudochromis nigrovittatus
Dark stripe variable; blue spot on operculum; upper body with tiny blue dots. **Biology:** inhabits rocky coastal reefs, often in tidepools, 0 to at least 12 m. **Range:** Iran s. to Djibouti.

5. Oman dottyback   15 cm
Pseudochromis omanensis
Two phases: brown above dark lateral band, yellow-orange below or dark brown with blue spots. **Biology:** inhabits shallow coastal rocky reefs, 1 to at least 15 m. **Range:** Central and S. Oman, probably also Yemen. **Similar:** Persian dottyback *P. persicus* pale phase is white below, dark phase has dark P fin (Arabian Gulf to Pakistan; 15 cm).

6. Blackback dottyback   10 cm
Pseudochromis punctatus
Dark brown above, white to yellow below; dorsal and anal fin with black dots. **Biology:** inhabits algae-covered rocky reefs, 12 to 65 m. **Range:** Oman to Somalia.

7. Golden dottyback   5 cm
Pseudoplesiops auratus
Yellow-orange with blue-edged orange stripe behind eye. **Biology:** in deep recesses of caves of rocky and coral reefs, 3 to 30 m. **Range:** Gulf of Aqaba, Red Sea and Gulf of Aden.

8. Stars-and-stripes snakelet   10 cm
Haliophis diadematus
White with dark stripes on head followed by dark spots. **Biology:** rocky reefs, 8 to 11 m. May hide among the spines of *Diadema* sea urchins. **Range:** S. Oman only.

JAWFISHES
OPISTHOGNATHIDAE

Blotched jawfish   5 cm
Stalix davidsheni
Jawfishes have a large head and mouth
and narrow tapering body. They live
in burrows they construct in rubble
and sand, leaving only to feed and
search for a mate. Males brood the
eggs in their mouths. **Biology:** lives
in burrows in rubbly sand slopes,
9 to 17 m. Extremely cyptic.
Range: Gulf of Aqaba only.
Similar: *Opisthognathus nigromar-
ginatus* has large ocellus on front part of
D fin (Red Sea to Vietnam; to 19 cm).

Gulf of Aqaba, DS

BIGEYES
PRIACANTHIDAE

Crescent-tail bigeye   40 cm
Priacanthus hamrur
Red, sometimes silvery or barred, with
large eyes and small scales.
Biology: inhabits lagoon and seaward
reefs, 10 to 100 m. Often near caves
or holes of coral-rich slopes during
the day. Becomes semi-pelagic at
night when it feeds on zooplankton.
Common and easy to approach.
Usually solitary, but subadults may
occur in large schools (Oman,
Yemen). **Range:** Red Sea and Oman
to Fr. Polynesia, n. to S. Japan, s. to S.
Africa and SE. Australia.

Sinai, EL

GRUNTERS
TERAPONIDAE

Jarbua grunter   36 cm
Terapon jarbua
Silver with narrow black curved
stripes above. **Biology:** in groups
in shallow sandy areas, particularly
near river mouths, 1 to 30 m.
Juveniles in sandy intertidal areas
close to reefs. Feeds on fishes,
algae, insects and sand-dwelling
invertebrates. **Range:** Red Sea to
Samoa, n. to S. Japan, s. to Mauritius
and SE. Australia.

Philipines, RM

HAWKFISHES

Hawkfishes are small grouper-like fishes with a continuous D fin with numerous short filaments at the tip of each of the 10 spines, thickened and elongate lower P rays and a continuous lateral line. Their common name is derived from their habit of perching on the outermost branches of coral heads or other prominences using their reinforced lower pectoral fin rays to hold them in place. All are carnivores of small crustaceans and fishes that they ambush in a quick burst of speed. Those species studied are protogynous hermaphrodites with territorial males maintaining a harem of females. Courtship and spawning occur at dusk or early night throughout the year in the deep tropics, but only during the warmer months in subtropical areas. Pair spawning occurs at dusk or early night at the apex of a short, rapid ascent. In some species 'sneaker males' posing as females raid a territorial male's harem for a quick spawn. The eggs are pelagic and the larval stage lasts a few weeks. Red Sea members of the family include both widely distributed species that occur as far east as Panama (*Oxycirrhites typus* and *Cirrhitichthys oxycephalus*) and narrowly distributed species (*Cirrhitichthys calliurus*).

Marsa Bareka, EL

Longnose hawkfish   13 cm
Oxycirrhites typus
White with crosshatched red lines; snout long. **Biology:** inhabits steep slopes of deep lagoon and seaward reefs, 12 to 100 m, usually below 30 m. Usually on gorgonians and black corals, occasionally on sponges or hard corals. Feeds on small planktonic crustaceans. Territorial. Often seen in pairs. **Range:** Red Sea to Panama, n. to S. Japan and Hawaii, s. to Mauritius and New Caledonia.

Hurghada, MB

Stocky hawkfish   30 cm
Cirrhitus pinnulatus
Tan with small dark spots and larger pale blotches. **Biology:** inhabits rocky shores, outer reef flats and seaward reef margins subject to surge, 0.3 to 3 m. Feeds primarily on crabs, also on fishes, brittle stars and sea urchins. Common, but somewhat cryptic and quite wary. **Range:** Red Sea and Oman to SE. Polynesia, s. to S. Africa and Rapa.

Freckled hawkfish 22 cm
Paracirrhites forsteri
Variably reddish-brown to olive above, in adults usually becoming black posteriorly and pale below; head with small dark red to black spots. **Biology:** inhabits clear lagoon and seaward reefs, 1 to 40 m. Common. Perches on small heads of *Stylophora*, *Pocillopora* and *Acropora* corals. Feeds primarily on small fishes, occasionally on shrimps. Easily approached. **Range:** Red Sea and S. Oman to Hawaii and Polynesia, n. to S. Japan, s. to S. Africa and New Caledonia.

Dahab, KF

Pixie hawkfish 9 cm
Cirrhitichthys oxycephalus
Pale with close-set brown to red blotches. **Biology:** inhabits lagoon and seaward reefs, 1 to 40 m. Common in areas of rich coral growth. Typically in or near small heads of branching hard corals, occasionally among soft corals or in seagrass beds. **Range:** Red Sea and S. Oman to Panama, n. to S. Japan, s. to S. Africa and New Caledonia.

Jordan, AK

Oman hawkfish 12 cm
Cirrhitichthys calliurus
Orange-brown with brown-spotted white tail. **Biology:** inhabits rocky and coral reefs, 1 to 30 m, usually below 15 m. Sits on hard bottoms, waiting to ambush prey. Solitary or paired. **Range:** S. Red Sea and Gulf of Aden to Gulf of Oman only.

Salalah, KF

CARDINALFISHES

APOGONIDAE

Cardinalfishes are a large family of small carnivorous fishes with two dorsal fins, the first with 6 to 8 spines, two anal fin spines and a large mouth and eyes. They are named for the red coloration of some of the species; although most are rather drab, many are striped and a few are transparent. A few species associated primarily with deeper mud habitats have bioluminescent organs in which the light is produced either by a chemical system of the fish itself (*Apogon* subgenus *Jaydia*) or by bacteria (*Siphamia*). Cardinalfishes are abundant in a variety of habitats in all warm seas. Some species occur in brackish and fresh waters. Most species tend to remain hidden under ledges or in holes during the day, but a few form dense aggregations above or among branching corals or other shelter. They typically disperse over the reef at night to feed, with some species remaining near the bottom to hunt for small benthic crustaceans and others well above the reef to feed on zooplankton. A few of the larger ones may opportunistically feed during the day. Male cardinalfishes typically brood the eggs in their mouths until hatching, but in some species females may do this as well. Most of the species seen by divers are members of the largest subfamily Apogoninae. The subfamily Pseudaminae includes a few small elongate species that remain deep within the reef during the day. At least 54 species occur in the Red Sea.

Mangrove Bay, EL

Orange-lined cardinalfish   8 cm
Archamia fucata
Sides with about 23 narrow orange bars, base of tail with large black spot. **Biology:** inhabits sheltered areas of lagoon and seaward reefs, 2 to 60 m. Usually in small groups among rocks or branching corals. Feeds during night on zooplankton. **Range:** Red Sea to Samoa, n. to Ryukyus, s. to S. Africa. **Similar:** *A. lineolata* with about 13 orange bars (Red Sea only). ▼

Two-lined cardinalfish   5 cm
Archamia bilineata
Translucent with 2 narrow dark stripes, base of tail with yellow-edged black spot. **Biology:** inhabits sheltered areas of lagoon and seaward reefs, 12 to 15 m. Usually in small groups near the base of large coral knolls. **Range:** N. Red Sea only, to at least as far south as Marsa Shagra. **Similar:** Anklet cardinalfish *Apogon pselion* has larger peduncular spot and broad yellowish stripes, also in small groups among shelter, 2 to 35 m (N. Red Sea only; 5 cm). ▼

Marsa Shagra, RM

Flower cardinalfish   13 cm
Apogon fleurieu
Base of tail with large black spot
(subadults) or bar. **Biology:** inhabits
rocky and coral reefs, under 7 to at
least 73 m. In small aggregations near
shelter. A continental species, absent
from oceanic islands. **Range:** Red
Sea and Arabian Gulf to Indonesia,
n. to Hong Kong, s. to S. Africa.
Similar: *A. aureus* (Gulf of Aden
to Tonga; 15 cm) with black bar
narrower in middle than at ends. ▼

Oman, JPH

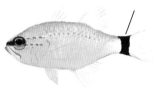

Ringtail cardinalfish   8 cm
Apogon annularis
Pearly-grey with light-margined black
band at base of tail and dark wedge
below eye. **Biology:** inhabits silty
lagoon and protected seaward reefs,
1 to 20 m. Secretive by day, up to
2 m above the reef feeding on large
zooplankton by night. **Range:** Red
Sea only. **Similar:** *A. zebrinus* with
many pale stripes (see below); *A.
guamensis* with narrower eye wedge
and dark saddle at base of tail (Red
Sea to Samoa; 11 cm).

Marsa Shagra, RM

Zebra cardinalfish   10 cm
Apogon zebrinus
Pearly-grey with narrow light bars
and dark wedge below eye.
Biology: inhabits silty lagoon and
protected seaward reefs, 1 to 15 m.
Secretive by day, up to 2 m above
the reef by night, feeding on large
zooplankton. **Range:** Red Sea
and Gulf of Aden only.
Similar: *A. annularis* and
A. guamensis lack conspicuous
narrow pale bars.

Marsa Shagra, RM

Marsa Bareka, EL

Eyeshadow cardinalfish   11cm
Apogon exostigma
Pale with light-edged black stripe
ending in black spot above centre of
stripe. **Biology:** inhabits lagoon and
protected seaward reefs, 2 to 60 m.
Under ledges and in crevices at the
base of reef slopes and isolated coral
heads by day. Iridescent, often lacks
black spot at night when it feeds
on zooplankton close above sand.
Range: Red Sea to W. Samoa, n. to
Ryukyus, s. to S. Africa and GBR.
Very similar: Bridled cardinalfish
A. fraenatus has black spot centred
at end of stripe (Red Sea to Fr.
Polynesia; 10 cm).

Guam, RM

Iridescent cardinalfish   15 cm
Apogon kallopterus
Broad, light-edged dark lateral stripe.
Biology: inhabits clear lagoon and
seaward reefs, 1 to 45 m. Near shelter
of caves during the day. Becomes
iridescent blue-green at night when
it actively forages close above sandy
bottoms. Solitary. **Range:** Red Sea
to Fr. Polynesia, n. to Ryukyus, s. to
S. Africa and New Caledonia.

Abu Galum, EL

Yellow-striped cardinalfish 8 cm
Apogon cyanosoma
Six yellow stripes with narrow pale
interspaces. **Biology:** inhabits lagoon
and seaward reefs, 1 to 50 m. In small
groups near the shelter of branching
corals, *Diadema* sea urchins, and ledges
by day. Common. Disperses at night
to feed on zooplankton. **Range:** Red
Sea to Marshall Is., n. to Arabian Gulf
and Ryukyus, s. to Mozambique.
Similar: Copperstriped cardinalfish
A. holotaenia has dark mid-lateral
stripe (Red Sea to Indonesia; 8 cm). ▼

Twobelt cardinalfish   12.5 cm
Apogon pharaonis
Biology: inhabits shallow silty inner reefs, seagrass beds and mangroves. **Range:** Red Sea and Arabian Gulf to India, s. to S. Africa. **Similar:** Doublebar cardinalfish *A. pseudotaeniatus* lacks ocellus on sides, also inhabits shallow silty reefs to 150 m (Red Sea and Arabian Gulf to PNG, n. to Japan; 14 cm). ▼

Oman, JPH

Dhofar cardinalfish   14 cm
Apogon dhofar
Juv. with 2 narrow dark bars and spot at base of tail. **Biology:** inhabits rocky reefs, 1 to 10 m. Common. **Range:** C. Oman to S. Yemen only. **Similar:** other large robust cardinalfishes have dark pinstripes: Manyline cardinalfish *A. natalensis* (S. Oman to S. Africa; 19 cm), Oman cardinalfish *A. omanensis* (Oman only, 15.5 cm) and Multi-striped cardinalfish *A. multitaeniatus* (Red Sea to Djibouti; 18 cm). ▼

Salalah, KF

Bluestreak cardinalfish   6 cm
Apogon leptacanthus
Deep body, filamentous D fin; translucent with irridescent vertical blue lines on head and front of body. **Biology:** in aggregations among branching corals in sheltered bays and lagoons, 1 to 12 m. Juvs and subadults lack blue lines.
Range: Red Sea to Samoa, n. to Ryukyus, s. to Mozambique, Tonga, and New Caledonia.

Palau, RM

Cook's cardinalfish   10 cm
Apogon cookii
Broad dark stripes, a partial one above mid-lateral stripe. **Biology:** inhabits sheltered, often turbid inshore reefs to 3 m. **Range:** Red Sea to New Caledonia, n. to S. Japan, s. to S. Africa. **Similar:** Blackstriped cardinalfish *A. nigrofasciatus* lacks the partial stripe (Red Sea to Fr. Polynesia; 10 cm). ▼

Kenya, RM

Ruby cardinalfish   5 cm
Apogon erythrosoma
Translucent red. **Biology:** inhabits shallow rocky and coral reefs, 3 to 22 m. Secretive, seen only at night close to the bottom. **Range:** Red Sea to Arabian Gulf only. **Similar:** other very similar species elsewhere in Indo-Pacific. Larger elongate red species found in far reaches of caves: *Apogon evermanni* (Oman; Indo-Pacific and W. Atlantic; 12 cm) and *A. isus* (Red Sea only). ▼

Marsa Bareka, EL

Mottled cardinalfish   5 cm
Fowleria vaiulae
3 to 4 dark bands radiating behind eye. **Biology:** in inner lagoons to seaward reef slopes, 3 to 52 m. Secretive, seen only at night always close to the bottom. **Range:** Red Sea to Society Is., n. to Ryukyus, s. to GBR. **Similar:** other similar species are larger with dark ocellus on operculum: *F. aurita* (Red Sea to Samoa) and *F. marmorata* (Red Sea to Fr. Polynesia) are reddish-brown and *F. variegata* is mottled (Red Sea and Arabian Gulf to Samoa). ▼

Sinai, MH

Tiger cardinalfish   25 cm
Cheilodipterus macrodon
Dark stripes thick, closely spaced;
caudal peduncle pale with diffuse
black spot. **Biology:** inhabits coral-rich
areas of lagoon and seaward reefs,
0.5 to 40 m. Solitary or in small
groups hovering under ledges or
in small caves. Easily approached.
Feeds primarily on small fishes.
Range: Red Sea to Fr. Polynesia,
n. to Arabian Gulf and Ryukyus,
s. to S. Africa and GBR.

Mangrove Bay, RM

Aqaba cardinalfish   15 cm
Cheilodipterus lachneri
Narrow dark stripes alternating in
width; base of tail with dark spot in
yellow patch. **Biology:** inhabits silty
lagoon and protected seaward reefs,
1 to 20 m. Usually hovers near shelter
of caves or overhangs. **Range:** Gulf of
Aqaba to N. Sudan only. Progressively
replaced s. of Gulf of Aqaba by Arabian
cardinalfish *C. lineatus* with more
stripes (s. to S. Africa; 15 cm); Persian
cardinalfish *C. persicus* also has smaller
caudal spot (S. Oman to Arabian
Gulf; 15 cm). ▼

Mövenpick Bay, RM

Fiveline cardinalfish   12 cm
Cheilodipterus quinquelineatus
Five narrow black stripes; yellow
spot on peduncle. **Biology:** inhabits
lagoon and seaward reefs, 1 to 40 m.
Common. Hovers in aggregations in
or near shelter of long-spined
Diadema sea urchins, corals, or rocks.
Diurnal. Feeds on small fishes,
crustaceans
and gastropods. **Range:** Red Sea to
Fr. Polynesia, n. to S. Japan, s. to
Mozambique and SE. Australia.
Very similar: *C. novemstriatus*
(Red Sea to Arabian Gulf; 8 cm)
and *C. pygmaios* (Red Sea only; 6 cm)
have an additional dark spot at top of
base of tail.

Mangrove Bay, KF

TILEFISHES

MALACANTHIDAE

Tilefishes are elongate with long continuous dorsal and anal fins, small conical and villiform teeth in the jaws, and pharyngeal teeth in the throat. They live in burrows in mud, sand or rubble. There are two subfamilies, the true tilefishes (Latilinae), which are large blunt-headed somewhat deep-bodied inhabitants of continental slopes usually below 100 m, and the sand-tilefishes (Malacanthinae), which are smaller, more elongate inhabitants of coral reefs and fore-reef slopes. Sand tilefishes are territorial and haremic or paired. Some species construct a large mound of rubble while others inhabit a hole or burrow in a small patch of sand. Species of *Malacanthus* feed on benthic invertebrates while species of *Hoplolatilus* feed primarily on zooplankton. The two Red Sea species of *Hoplolatilus* are unlikely to be encountered by divers. *H. oreni* is known from Eritrea and *H. geo* from the Sinai at depths of 80 to 116 m.

▲ **Blue blanquillo**   50 cm
Malacanthus latovittatus
Snout elongate; light grey above, white below with broad black band on sides. **Biology:** inhabits sand and rubble bottoms of deep lagoon and seaward reefs, 5 to 70 m. Typically hovers motionless well above the bottom. Shy and hard to approach. Juveniles may mimic juveniles of the wrasse *Hologymnosus annulatus* (p. 160). Adults form monogamous pairs. **Range:** Red Sea to Cook Is., n. to S. Japan, s. to S. Mozambique and New Caledonia. *(Photo: Kenya, KF.)*

juv.

Flagtail blanquillo   30 cm
Malacanthus brevirostris
Snout short and blunt; pale, tail with broad white band with dark margins. **Biology:** uncommon on patch reefs with open barren sandy areas, 5 to 61 m. Usually close to bottom. Solitary or in pairs. Wary, retreats into a hole or under a rock when frightened. Feeds on small benthic invertebrates. **Range:** Red Sea to Panama, n. to S. Japan and Hawaii, s. to S. Africa and SE. Australia; absent from Arabian Sea.

Kenya, KF.

REMORAS
ECHENEIDAE

Sharksucker 1.0 m
Echeneis naucrates
Elongate fishes with first part of
D fin modified into a sucking disc
for attachment to larger animals which
they use to hitch a ride. **Biology:**
usually associated with sharks, rays, or
other large fishes, occasionally with
turtles or free-swimming over coral
reefs, 1 to at least 60 m. May try to
attach to divers or snorkellers. **Range:**
all warm seas. **Similar:** Remora,
Remora remora (50 cm) is uniformly
black, usually associated with sharks
(circumtropical).

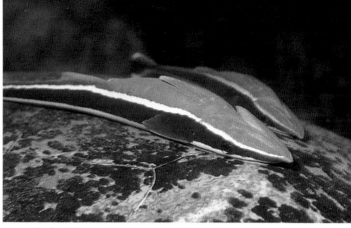

Sipadan, KF

COBIA
RACHYCENTRIDAE

Cobia 1.8 m; 68 kg
Rachycentron canadum
Large and remora-like but first dorsal
fin as detached spines. May resemble
a shark at a distance. Juv. with 2 pale
lateral stripes. **Biology:** primarily
pelagic, occasionally visits coral reefs,
0 to 53 m. Feeds primarily on crabs,
also other crustaceans, small fishes and
squids. May follow sharks and rays,
adults often misidentified as sharks.
Range: continental margins of all
warm waters except E. Pacific.

Aqaba, JN

JACKS AND TREVALLIES   CARANGIDAE

Carangids are typically medium to large silvery laterally compressed to fusiform fishes with two dorsal fins, small (sometimes embedded) cycloid scales, a complete lateral line, a slender caudal peduncle reinforced by a series of bony scutes, and a strongly forked tail. Many species have a streamlined adipose eyelid of clear fatty tissue. The queenfishes (*Scomberoides* spp.) have venomous dorsal and anal fin spines. Some carangids resemble tunas and mackerels (Scombridae, p. 208) but are easily distingushed by the presence of scutes in most species, two stout spines in front of the anal fin, and in most species, a more compressed body and absence of two or more sets of dorsal and ventral finlets. Carangids are strong-swimming open-water carnivores of fishes and crustaceans. Some species form large inactive schools during the day and disperse at night to feed, while others are opportunistic predators that constantly roam the reefs in search of prey. Many species inhabit turbid coastal and deeper continental shelf waters. Many carangids are economically important food fishes, but large individuals of a few may be ciguatoxic in certain areas.

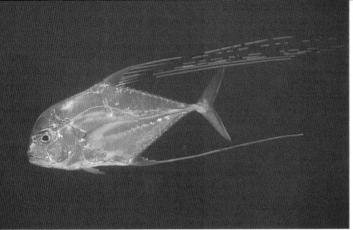

Palau, RM

African pompano   1.1 m; 18 kg
Alectis ciliaris
Highly compressed with steep
head profile; juvs. ovoid with long
filamentous D and A fin rays and
pale bars, adults become somewhat
elongate and lose the first D fin.
Biology: juvs. inhabit coastal and
pelagic surface waters, adults occur
near the bottom to a depth of 100 m.
The long-streamered juvs. mimic
jellyfishes. **Range:** circumtropical.
Similar: *A. indicus* with more angular
head (p. 99).

Lahami Reef, KF

Herring scad   56 cm
Alepes vari
Elongate; pale silvery with dusky spot
at rear of operculum. **Biology:** in
large roving schools in shallow coastal
waters, often near surface, 2 to 50 m.
Feeds on small fishes and larger
planktonic crustaceans. **Range:** Red
Sea and Arabian Gulf to New Guinea,
n. to Ryukyus. **Similar:** several smaller
species: *A. djedaba* with darker oblong
opercular spot (Red Sea to Taiwan;
35 cm); *Alepes kleinii* with deeper
body (S. Oman to New Guinea;
19 cm); and *Atule mate* with more
elongate body and yellowish tail (Red
Sea to Hawaii and Samoa; 30 cm).

Black trevally   80 cm
Caranx lugubris
Head profile slightly concave; ground-
colour variable from silvery to dark
flat grey. **Biology:** occurs in small
groups along steep outer reef slopes or
offshore banks, 12 to 300 m, usually
below 30 m. Often abundant off
isolated offshore and mid-oceanic
banks. Feeds primarily on fishes, also
on crustaceans. Large ones may be
ciguatoxic. **Range:** circumtropical.

Sulawesi, RM

Bluefin trevally   1.0 m
Caranx melampygus
Olivaceous-silver with tiny black
specks, small blue flecks above, and
blue fins. **Biology:** inhabits lagoon
and seaward reefs, 1 to 190 m.
Common, solitary or in small
groups patrolling reef edges and
slopes. Feeds primarily on reef fishes,
occasionally on squids. May shadow
goatfishes and moray eels in search of
prey. Juveniles occur in small groups
along sandy shores of lagoons and
estuaries. Large individuals may
be ciguatoxic. **Range:** Red Sea to
Panama, n. to S. Japan and Hawaii,
. to S. Africa and New Caledonia.

Yap, RM

Bigeye trevally   85 cm
Caranx sexfasciatus
Eye large; D and A fins with white
tips. **Biology:** inhabits lagoon and
seaward reefs, 1 to 90 m. Adults form
large semi-stationary schools by day at
preferred sites along reef edges, in
channel entrances and off
promontories. These schools are often
in a ball or circular shape with the
fishes swimming in a tight circle.
They disperse at night to feed mainly
on fishes. Juvs. inhabit estuaries.
Range: Red Sea to C. America, n. to
S. Japan, s. to S. Africa and New
Caledonia. **Similar:** Blacktip trevally
C. heberi has yellow fins and tail, the
upper tail lobe with a black tip. It is
abundant in S. Oman (Gulf of Aden
to Indonesia; 85 cm).

Maldives, RM

Giant trevally   1.7 m
Caranx ignobilis
Silvery to blackish with steep head
profile. **Biology:** inhabits lagoon
and seaward reefs, 5 to 80 m. Solitary
or in small groups along steep slopes.
Juveniles in schools over sandy inshore
or estuarine areas. Feeds primarily
on fishes and crustaceans. Large
individuals may be ciguatoxic.
Uncommon. **Range:** Red Sea to
Fr. Polynesia, n. to Arabian Gulf
and S. Japan, s. to S. Africa and
New Caledonia.

Egypt, AK

Mangrove Bay, EL

Orangespotted trevally   53 cm
Carangoides bajad
Yellow or silvery to grey with orange spots. **Biology:** inhabits lagoon and seaward reefs, 2 to 50 m. Solitary or in small groups, common along upper edges of slopes. May shadow the yellow phase of the goatfish *Parupeneus cyclostomus* (p. 121) to get closer to prey. **Range:** Red Sea to PNG, n. to Ryukyus and Arabian Gulf, s. to E. Africa.

adult yellow variant

Palau, RM

Barred trevally   70 cm
Carangoides ferdau
Silvery with 5 to 7 faint dusky chevrons, occasionally with a few small yellow spots. **Biology:** inhabits open water of lagoons and outer reef slopes, 1 to 60 m. Usually in small fast-swimming schools along slopes of fringing reefs or over sand flats or seagrass beds. Feeds on small fishes and benthic crustaceans by rooting in sand. Tiny juveniles often associate with jellyfishes. Not shy. Common. **Range:** Red Sea to Fr. Polynesia, n. to Arabian Gulf and S. Japan, s. to S. Africa and GBR. **Similar:** *Uraspis* spp. sometimes have dusky chevrons, but deeper bodies and shorter fins (circumtropical).

Yellow-dotted trevally   1.03 m
Carangoides fulvoguttatus
Silvery with yellow spots on back arranged in bars; up to 3 blackish blotches posteriorly along lateral line, occasionally with dusky bars. **Biology:** occurs in small schools along outer reef slopes, rocky coasts and offshore banks to 100 m. Often away from reefs over open sand and seagrass beds or patrolling along upper slopes. Feeds on fishes and crustaceans. **Range:** Red Sea to New Caledonia, n. to Ryukyus and Arabian Gulf, s. to S. Africa and GBR.

Marsa Shagra, PM

Golden trevally 1.2 m
Gnathanodon speciosus
Yellowish-silver with a few black
spots; subadults with faint dusky bars;
juvs. yellow with black bars. **Biology:**
inhabits deep lagoon and seaward
reefs, 1 to 50 m. Usually in small
schools. Juveniles associate with
groupers, sharks, turtles, rays and even
sea snakes for protection in the same
way as pilotfishes. Roots in sand for
invertebrates and fishes. **Range:** Red
Sea to Panama, n. to Ryukyus and
Arabian Gulf, s. to S. Africa and
New Caledonia.

juv.

Guam, TA

Rainbow runner 1.2 m
Elagatis bipinnulata
Elongate, a detached finlet at rear
of D and A fins; olive with two
blue lateral stripes. **Biology:** pelagic,
occasionally visits deep lagoon and
seaward reefs, 1 to 150 m. In small
to large schools. Feeds on small
fishes and large planktonic
crustaceans. A popular gamefish.
Range: circumtropical.

The Bells, Dahab, EL

Doublespotted queenfish 70 cm
Scomberoides lysan
Silvery with 6 to 8 dark spots in a
double row. **Biology:** inhabits shallow
lagoons, bays and seaward reefs, 1 to
100 m. Juveniles in protected inshore
waters and estuaries, adults solitary,
usually just below the surface in clear
water along beaches. Juveniles feed on
scales torn from small schooling fishes,
adults feed on small fishes and
crustaceans. Dorsal and anal fin spines
venomous, capable of causing painful
wounds. **Range:** Red Sea to Fr.
Polynesia, n. to S. Japan and Arabian
Gulf, s. to S. Africa and SE. Australia.
Also known as Leatherback. **Similar:** *S.
commersonianus* and *S. tol* with single
row of larger round spots (p. 99).

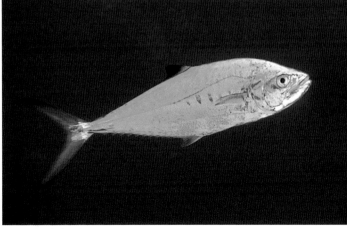

Hurghada, AK

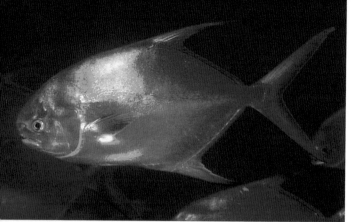

Snubnose pompano 1.1 m
Trachinotus blochii
Head profile steep; fins yellowish; juvs. with elongate D and A fins. **Biology:** inhabits clear lagoons, coastal bays and seaward reefs, 1 to at least 50 m. Solitary or in schools. Juveniles over sandy or muddy shorelines in shallow water, adults often deep. Feeds on gastropods and hermit crabs. **Range:** Red Sea to Samoa, n. to S. Japan and Arabian Gulf, s. to S. Africa and SE. Australia.

Bali, RM

Small-spotted pompano 54 cm
Trachinotus baillonii
Ovoid with falcate D and A fins and small dark spots along lateral line. **Biology:** inhabits near-surface waters of clear lagoons, channels and seaward reefs, 0.5 to 5 m. Most often along protected sandy beaches. Feeds on small fishes. **Range:** Red Sea, Oman to Fr. Polynesia, n. to S. Japan, s. to S. Africa, SE. Australia, and Rapa.

Thailand, RM

Blackbanded jack 70 cm; 5 kg
Seriolina nigrofasciata
Silvery grey with irregular dark blotches in 5 bands. **Biology:** occurs over sandy slopes at depths of 20 to 150 m. Occasionally rests on bottom. Feeds on bottom-dwelling fishes, shrimps and cephalopods. Usually solitary. **Range:** Red Sea to E. Australia, n. to Arabian Gulf and S. Japan, s. to S. Africa.

Mangrove Bay, KF

1. Indian threadfish   1.65 m
Alectis indicus
Biology: juvs. near surface where they mimic jellyfishes, adults usually near bottom below 60 m. **Range:** Red Sea to E. Australia.

2. Bigeye scad   39 cm
Selar crumenopthalmus
Biology: coastal pelagic, in large migratory schools in bays and along outer reef slopes to 170 m. Feeds on zooplankton and small benthic invertebrates. **Range:** circumtropical.

3. Mackerel scad   30 cm
Decapterus macarellus
Biology: in large midwater schools near offshore reefs, to 200 m. Feeds on zooplankton. **Range:** circumtropical. Several similar species, usually far offshore.

4. Pilotfish   70 cm
Naucrates ductor
Biology: accompanies pelagic sharks, rays, turtles, or other large fishes (see p. 24). Juveniles with jellyfishes or floating weed. **Range:** circumtropical.

5. Needlescale queenfish   51 cm
Scomberoides tol
Spines venomous. **Biology:** in small schools in coastal waters, usually near surface, but to 50 m. **Range:** Red Sea and Arabian Gulf to Fiji.

6. Talang queenfish   1.2 m
Scomberoides commersonianus
Spines venomous. **Biology:** midwater above reefs, to 30 m. **Range:** Red Sea and Arabian Gulf to E. Australia.

7. Longrakered jack   1.0 m
Ulua mentalis
Biology: coastal waters. Feeds on zooplankton. **Range:** Red Sea and Arabian Gulf to PNG.

8. Greater amberjack   1.88 m; 81 kg
Seriola dumerili
Lacks scutes. **Biology:** deeper seaward reefs, 20 to 360 m, occasionally enters shallow bays. Feeds on schooling fishes. Occasionally ciguatoxic. A prized gamefish. **Range:** most tropical and warm-temperate seas.

DOLPHINFISHES   CORYPHAENIDAE

9. Dolphinfish, Mahimahi, Dorado 2.1 m; 38 kg
Coryphaena hippurus
Biology: pelagic near surface, occasionally near steep offshore reefs. Feeds on squids and fishes, especially flyingfishes which it can follow from beneath the surface. Juveniles around floating *Sargassum* weed. An important migratory oceanic gamefish. Grows rapidly, maturing and spawning within its first year. The largest individuals in a year's catch are typically only 3 years old, the largest ever caught was 5 years old. **Range:** circumtropical in all seas warmer than 20°C.

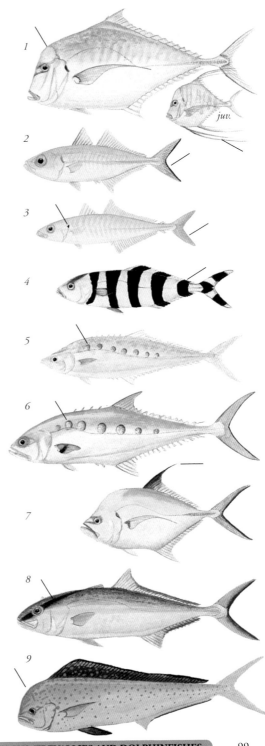

juv.

SNAPPERS

Snappers are robust perch-like fishes with a continuous dorsal fin, large coarse scales, large canine teeth, a maxillary mostly covered by the cheek, and an emarginate to forked tail. Most species feed heavily on crustaceans but some feed primarily on fishes and others on zooplankton. Snappers are among the most important commercial fishes in tropical and subtropical regions. Members of the subfamily Lutjaninae occur in large numbers on coral reefs as well as in estuaries and deeper coastal and continental-shelf waters. Many species of *Lutjanus* occur in large semi-stationary schools during the day, and disperse to feed on benthic invertebrates at night. Some of the larger piscivores may be ciguatoxic. The Twinspot snapper *L. bohar* is among the most frequently toxic species and is banned from sale in certain areas of the Indian and Pacific oceans. It is also among the largest and most aggressive of snappers and eagerly participates in bait-induced shark feeding frenzies. Most members of the subfamily Etelinae are deep-dwelling planktivores of pelagic tunicates and crustaceans normally found at depths of 90 to 360 m, but a few (*Aphareus furca, Aprion virescens, Paracaesio sordidus*) occur in shallow as well as deep water.

◀**One-spot snapper**   60 cm
Lutjanus monostigma
Pastel grey to silvery with yellow fins, a small dark spot on lateral line may disappear in large adults. **Biology:** inhabits lagoon and seaward reefs, 5 to 60 m. Solitary or in small groups, common in wrecks and along reef margins with caves and overhangs. Feeds on fishes primarily during the night. May be ciguatoxic in some areas. **Range:** Red Sea to Fr. Polynesia, n. to Ryukyus, s. to Mozambique. *(Photo: Ras Mohammed, MH.)*

Marsa Shagra, RM

Ehrenberg's snapper   35 cm▶
Lutjanus ehrenbergi
Body with five narrow yellow stripes, the uppermost ending in a large black spot, none through eye. **Biology:** in small aggregations in lagoons, bays and along fringing reefs, 5 to 20 m. Common on rocky reefs in Oman and Yemen. Juveniles in mangroves and estuaries. Easy to approach. **Range:** Red Sea to Moluccas, n. to Ryukyus, s. to S. Africa and GBR.

Longspot snapper   35 cm
Lutjanus fulviflamma
Body with six narrow yellow stripes, one becoming dark and extending forward through eye; mid-lateral spot oblong. **Biology:** inhabits rocky and coral reefs, 3 to 35m. Often in large schools with Bengal and Blue-striped snappers (p. 102). Juveniles occur in brackish and turbid inshore areas. **Range:** Red Sea to Samoa, n. to S. Japan, s. to S. Africa and SE. Australia.

New Caledonia, RM

Mangrove Bay, KF

Blue-striped snapper 34 cm
Lutjanus kasmira
Yellow with 4 blue stripes; belly below
P fin pale; snout usually dusky; 10 D
spines. **Biology:** inhabits lagoon and
seaward reefs, 10 to 265 m. Common,
often in large aggregations around
prominent coral heads or rocky
formations during the day. Disperses
at night to feed on small fishes and
benthic crustaceans. Juveniles
common in seagrass beds or around
patch reefs in protected turbid areas.
An important foodfish. **Range:** Red
Sea to Fr. Polynesia, n. to S. Japan,
s. to S. Africa and SE. Australia.

Oman, JPH

Bengal snapper 30 cm
Lutjanus bengalensis
Yellow with 4 blue stripes, white
below (may be partially yellow at
night); snout yellow; 11 or 12 D
spines. **Biology:** inhabits coastal rocky
and coral reefs, 5 to 30 m. Solitary or
in small groups, usually in turbid
lagoons and bays. **Range:** Red Sea
to Moluccas, n. to Oman, s. to
Mauritius. **Similar:** other blue-striped
snappers in the area have 10 D fin
spines. Blue-striped snapper *L. kasmira*
usually has narrow grey stripes on the
belly and more yellow below the
lowest blue line.

Oman, JR

Bluelined snapper 40 cm
Lutjanus coeruleolineatus
Yellow with 7 or 8 narrow blue lines
and dark spot posteriorly on lateral line.
Biology: inhabits clear-water rocky
and coral reefs, 10 to 20 m. Solitary
or in aggregations. An important
foodfish. **Range:** S. Red Sea e. to
Strait of Hormuz only. **Similar:** other
blue-striped yellow snappers have 4
or 5 blue lines.

Bigeye snapper   30 cm
Lutjanus lutjanus
Body somewhat elongate, eye large;
yellow mid-lateral stripe.
Biology: inhabits lagoons, bays and
semi-exposed slopes of coastal reefs,
to 90 m. Common in continental
waters. Usually in dense aggregations
with other snappers and occasionally
goatfishes by day, disperses to feed at
night. **Range:** Red Sea to Solomon Is.,
n. to Ryukyus, s. to E. Africa and
GBR.

Kenya, RM

Russell's snapper   45 cm
Lutjanus russelli
Grey with 5 oblique brown stripes
(sometimes very pale) and large black
spot posteriorly. **Biology:** inhabits
coastal rocky and coral reefs, 2 to 80
m. Uncommon, solitary or in groups.
Juveniles occur in shallow brackish
estuaries. **Range:** Red Sea to Fiji,
n. to Ryukyus, s. to S. Africa and
E. Australia.

Similan Is., RM

Blacktail snapper   40 cm
Lutjanus fulvus
Yellowish; D and C fins reddish-black,
A and V fins yellow.
Biology: solitary or in small groups
on rocky and coral reefs, 1 to 75 m.
Juveniles in sheltered brackish areas
and seagrass beds. Adults in lagoons
and on semi-protected seaward
reefs. Feeds on benthic invertebrates
and fishes. **Range:** Red Sea to Fr.
Polynesia, n. to S. Japan, s. to
S. Africa and New Caledonia.

Guam, RM

Marsa Shagra, PM

Twinspot snapper   90 cm
Lutjanus bohar
Yellowish to reddish-grey with darker fins. **Biology:** inhabits lagoons, channels and seaward reef slopes, 1 to 70 m. Solitary or in small groups. A voracious predator of smaller reef fishes, participates in feeding frenzies with sharks. Notoriously ciguatoxic in some areas. Juveniles mimic the damselfishes *Chromis flavaxilla* and *C. ternatensis* to approach prey more closely. **Range:** Red Sea to Fr. Polynesia, n. to S. Oman and Ryukyus, s. to S. Africa and GBR.

juv.

Mangrove Bay, RM

Mangrove snapper   1.2 m
Lutjanus argentimaculatus
Reddish-brown to grey with darker fins and dark-centred scales. **Biology:** inhabits coastal rocky and coral reefs from turbid lagoons to seaward reefs, 10 to 120 m. Juveniles in estuaries, adults in rivers as well as ocean. Feeds on fishes and benthic invertebrates. **Range:** Red Sea to Samoa, n. to Arabian Gulf and Ryukyus, s. to S. Africa; migrant to E. Mediterranean.

juv.

Bali, RM

Scribbled snapper   80 cm
Lutjanus rivulatus
Olivaceous-silver with narrow blue lines on head and yellow fins; juv. with white mid-lateral spot. **Biology:** solitary or in small groups on rocky and coral reefs, 2 to 100 m. Ventures into shallow inshore flats. Uncommon and wary, difficult to approach. **Range:** Red Sea and Oman to Fr. Polynesia, n. to Ryukyus, s. to S. Africa and GBR.

juv.

Humpback snapper   50 cm
Lutjanus gibbus
Pale reddish-grey to orange with dark
red fins; adults with concave head
profile and enlarged upper tail lobe.
Biology: inhabits deep lagoons,
channels and sheltered seaward reefs,
1 to 150 m. In dense semi-stationary
aggregations by day. Disperses at night
to feed mainly on crabs. Juveniles in
seagrass beds and shallow sheltered
sandy areas. **Range:** Red Sea to
Fr. Polynesia, n. to Ryukyus, s. to
S. Africa and SE. Australia.

juv.

Bali, RM

Emperor snapper   1.0 m; 27 kg
Lutjanus sebae
Body deep; juvs. and subadults white
with 3 diverging dark red bands; large
adults uniformly red. **Biology:** juvs.
inhabit shallow sheltered reefs and
estuaries, often shelter among spines
of *Diadema* sea urchins. Subadults
occur along the bases of slopes or
outcrops bordering sand. Adults are
rarely seen and occur below 15 m to
at least 100 m. **Range:** continental
shelves from Red Sea and Oman to
PNG, n. to S. Japan, s. to S. Africa
and SE. Australia.

Borneo, KF

Black snapper   66 cm
Macolor niger
Juvenile black and white; adult
uniformly dark grey to black. **Biology:**
inhabits deep lagoons, channels and
steep outer reef slopes, 3 to 90 m.
Juveniles solitary in coral-rich areas,
adults in aggregations along upper
edges of steep slopes by day, disperse at
night to feed on larger zooplankton.
Range: Red Sea to Ryukyus, s. to S.
Africa and New Caledonia.

juv.

Mövenpick Bay, KF

Red Sea, JR

False fusilier snapper   35 cm
Paracaesio sordidus
Dark grey; tail deeply forked.
Biology: solitary or in small schools
along current-swept reefs and dropoffs,
5 to 200 m, usually below 20 m.
Feeds in midwater on planktonic
crustaceans. Often overlooked.
Range: Red Sea and S. Oman to
Samoa, n. to Ryukyus, s. to E. Africa.
Similar: the larger *Pinjalo pinjalo* is
reddish with deeper body and occurs
in schools (Gulf of Oman to PNG; 61
cm).

Fiji, RM

Green jobfish   1.0 m
Aprion virescens
Elongate with blunt snout,
conspicuous canine teeth and
forked tail; grey, olivaceous above.
Biology: solitary or in small groups
in midwater along slopes of deep
lagoons, channels and seaward reefs, 3
to 180 m. A voracious piscivore
that also feeds on octopuses and
crustaceans. May be ciguatoxic.
Flesh highly esteemed. Wary and
difficult to approach. **Range:** Red Sea
to Fr. Polynesia, n. to Arabian Gulf
and Ryukyus, s. to S. Africa and SE.
Australia.

Guam, RM

Smalltooth jobfish   40 cm
Aphareus furca
Elongate body with large mouth and
deeply forked tail. **Biology:** solitary
or in small groups in midwater
close above bottom of clear lagoons,
channels and seaward reefs, 1 to
122 m. Feeds on fishes and
crustaceans. Often curious and
approachable. **Range:** Red Sea
to Panama, n. to Ryukyus, s. to
S. Africa and SE. Australia.

▲ Red Sea fusilier   25 cm
Caesio suevica
Tail with white-edged black tips.
Biology: in schools in midwater of
deep lagoons, channels and seaward
reefs, 2 to 25 m. Often curious and
approachable, frequently visits cleaning
stations. Feeds on zooplankton.
Common. **Range:** Red Sea only. *(Photo:
Marsa Shagra, RM.)*

FUSILIERS   CAESIONIDAE

Fusiliers are closely related to snappers, but are adapted to a pelagic,
planktivorous life. They have a more elongate fusiform body, smaller
scales, a small terminal mouth with a very protrusible upper jaw, small
conical teeth and a deeply forked tail. During the day they occur in large,
often mixed-species schools that roam the waters above the reef, feeding
on zooplankton. They are most abundant along steep outer reef slopes.
At night they shelter within the reef and often assume a red colour. In
some areas they are important foodfishes.

Lunar fusilier   35 cm
Caesio lunaris
Blue, tail with black tips; juv. with
yellow band posteriorly. **Biology:** in
schools in midwater along steep slopes
of deep lagoon and seaward reefs, 1 to
40 m. Feeds on zooplankton. Easily
approached. Uncommon in the N.
Red Sea. **Range:** Red Sea to Solomon
Is., n. to Arabian Gulf and Ryukyus, s.
to Seychelles and GBR.

Thailand, RM

Lahami, EL

Striated fusilier 18 cm
Caesio striata
4 black stripes on back; juv. with yellow spot on upper peduncle.
Biology: in schools in midwater of deep lagoons, sheltered bays and seaward reefs, 1 to 40 m. Common on semi-protected coral and rocky reefs, often around cleaning stations. Juveniles in large aggregations in sheltered bays. Easily approached.
Range: Red Sea only.

Arabian Gulf, AK

Yellow-lined fusilier 25 cm
Caesio varilineata
4 or 5 yellow stripes, sometimes the lowest ones missing; tips of tail lobes black. **Biology:** in schools in midwater of lagoons, sheltered bays and seaward reefs, 1 to 25 m. Feeds on zooplankton. Easily approached.
Range: Red Sea to Sumatra, n. to Arabian Gulf, s. to Seychelles.
Similar: *C. caerulaurea* lacks yellow stripes below the lateral line (Red Sea to Samoa). Uncommon. ▼

Goldband fusilier 18 cm
Pterocaesio chrysozona
Golden band below lateral line; tail lobe tips red-brown. **Biology:** in schools along reef slopes of lagoons, bays and seaward reefs, 3 to 25 m. Feeds in midwater on zooplankton. Common and easily approached. Prefers continental waters.
Range: Red Sea and Gulf of Oman to PNG, n. to China, s. to S. Africa and GBR.

Bali, RM

MONOS
MONODACTYLIDAE

Silver mono   22 cm
Monodactylus argenteus
Diamond-shaped, highly
compressed fishes with tall D and A
fins, small mouth and small scales;
silvery with black tips on D and A
fins. **Biology:** in schools in brackish
rivers, estuaries, lagoons and turbid
protected coastal reefs, 0 to 15 m.
Feeds on plankton. **Range:** Red Sea
and Oman to Samoa, n. to Ryukyus,
s. to S. Africa and New Caledonia.

Eritrea, AK

MOJARRAS
GERREIDAE

Small silvery fishes characterized by a
highly protrusible mouth, single dorsal
fin and deeply forked tail. Feeds by
sifting benthic invertebrates from sand.

Smallscale mojarra   37 cm
Gerres longirostris
Silvery, front of D fin elongate, tail
margins dusky. **Biology:** common
over shallow, sheltered sandy areas
of clear bays and lagoons, 0 to 20 m.
Often close to coral reefs. Solitary.
Feeds on benthic invertebrates.
Range: Red Sea and Oman to
Marshall Is., n. to S. Japan, s. to
S. Africa and New Caledonia.

Marsa Shagra, RM

Blacktip mojarra   20 cm
Gerres oyena
Tip of D fin black. **Biology:** inhabits
protected sandy flats and edges of
reefs, 2 to at least 20 m. Also in
estuaries and among mangroves.
Juveniles in small groups along
beaches. Solitary adults prefer deeper
water. Roots in sand for benthic
invertebrates. **Range:** Red Sea to
Samoa, n. to Arabian Gulf and
S. Japan, s. to S. Africa and New
Caledonia.

Sudan, JR

◄ Blackspotted sweetlips 45 cm
Plectorhinchus gaterinus
Biology: inhabits deep lagoons, bays and semi-sheltered seaward reefs, 3 to 55 m. Common. Usually in stationary aggregations at outcrops, under coral ledges, or in caves by day. Spawns in the late afternoon. **Range:** Red Sea to Mauritius, n. to Arabian Gulf, s. to S. Africa. *(Photo: Fury Shoals, TE.)*

juv.

SWEETLIPS AND GRUNTS
HAEMULIDAE

Sweetlips and grunts are closely related to snappers. They differ by having smaller mouths located somewhat lower on the head, small conical teeth instead of canines, thickened lips, and pharyngeal teeth (in the throat). Grunts (subfamily Haemulinae) get their common name from their ability to make grunting sounds by grinding their pharyngeal teeth and amplifying the sound with their gas bladder. Many sweetlips (subfamily Plectorhinchinae) are colourful and undergo dramatic changes in coloration with growth. Juveniles swim with an undulating motion and may be distasteful to predators. Most grunts (*Pomadasys* spp.) are generally a drab silver, often with darker marks dorsally. Haemulids are primarily nocturnal and are not as wary as most other large fishes. During the day they hover under or near overhangs or tabular corals, and at night they disperse to feed on benthic invertebrates. Many species are important foodfishes.

Goldspotted sweetlips 60 cm
Plectorhinchus flavomaculatus
Body with orange spots, head with orange stripes. **Biology:** inhabits sheltered waters of deep lagoons, bays and silty fringing reefs, 3 to 35 m. Usually solitary and uncommon. Juveniles in shallow protected weedy areas. **Range:** Red Sea and Oman to GBR, n. to S. Japan, s. to S. Africa and SE. Australia.

Kenya, RM

Giant sweetlips 1.0 m
Plectorhinchus albovittatus
Dark brown to grey with black-edged fins; juv. with 3 cream stripes and yellow belly. **Biology:** inhabits deep lagoons, bays and steep seaward reef slopes, 2 to 50 m. Often in or near large caves by day. Juveniles in estuaries and brackish areas. The largest species of sweetlips. **Range:** Red Sea to Fiji, n. to S. Japan, s. to Mauritius and New Caledonia. **Similar:** *P. pictus* is silvery with dark spots on back (Oman to Sri Lanka; 60 cm).

juv.

Palau, RM

Hanish Is., EL

Gibbus sweetlips   70 cm
Plectorhinchus gibbosus
Head profile steep, body deep; grey
with a dark eye-band. **Biology:**
inhabits sheltered rocky or coral reefs,
3 to 30 m. Prefers turbid coastal reefs.
Solitary or in small groups. Easily
approached. **Range:** Red Sea to Samoa,
n. to S. Japan, s. to S. Africa and GBR.

juv.

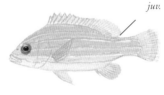

Seychelles, EL

Minstrel sweetlips   80 cm
Plectorhinchus schotaf
Dark grey, edge of operculum red.
Biology: inhabits rocky and coral
reefs, 1 to 25 m. Usually in groups
around rocky outcrops and prominent
coral heads. Emits a croaking sound.
Common in large aggregations in
S. Oman and Yemen. Juveniles in
tidepools. **Range:** S. Red Sea and
Arabian Gulf to Sri Lanka, s. to
S. Africa.

Oman, JPH

Black sweetlips   30 cm
Plectorhinchus sordidus
Dark silvery-grey, edge of operculum
black. **Biology:** inhabits rocky and
coral reefs close to weedy areas, 2 to
25 m. Solitary or in small groups.
Abundant in the S. Red Sea and
Yemen. Shy, difficult to approach.
Range: Red Sea to Seychelles and
Mauritius, s. to Madagascar.

juv.

White-barred sweetlips 50 cm
Plectorhinchus playfairi
Dark grey with 4 narrow white bars
above, white below, fins black.
Biology: inhabits rocky and coral
reefs, tidepools to 80 m. Solitary.
Feeds on worms, benthic crustaceans
and small fishes. Active by day.
Range: Red Sea, Gulf of Aden,
and S. Oman to S. Africa, e. to
Seychelles and Mauritius.

Oman, KF

Painted sweetlips 90 cm
Diagramma pictum
Grey with dusky spots, juv. striped.
Biology: inhabits lagoons and coastal
reefs, 1 to 30 m. Usually solitary or in
small stationary groups around bases
of patch reefs and coral slopes by day.
Disperses at night to feed on benthic
invertebrates. **Range:** subspecies
punctatum: Red Sea (and Gulf of
Aden?) only; subspecies *cinerascens*:
Arabian Gulf and Gulf of Oman to
Sri Lanka; others from E. Africa to
New Caledonia.

juv.

Shams Alam Reefs, EL

Bronze-striped grunt 25 cm
Pomadasys taeniatus
Silvery with 7 brown stripes.
Biology: inhabits rocky reefs or mixed
rock and sand bottoms with scattered
living corals, 3 to 20 m. Usually
in large aggregations close to
bottom. **Range:** Oman to Yemen.
Similar: *P. punctulatus* with dusky
pinstripes and black-margined fins
(Red Sea to Oman); *P. argenteus* with
small dark spots on back (Red Sea to
W. Pacific); *P. kaakan* with dusky
blotches on back; *P. stridens* with
stripes on back (Red Sea to S. Africa);
P. aheneus (Gulf Aden and Oman
only; 27 cm). ▼

Oman, KF

Sulawesi, RM

Nabq, RM

Maldives, RM

EMPERORS
LETHRINIDAE

Bigeye emperor   60 cm
Monotaxis grandoculis
Snapper-like fishes; this species with
large eye. **Biology:** common over
sandy patches of lagoons, bays and
seaward reefs, 1 to 100 m. Solitary or
in small groups along reef edges. Feeds
at night on hard-shelled invertebrates.
Range: Red Sea to Fr. Polynesia, n. to
Hawaii and Ryukyus, s. to S. Africa.

juv.

Mahsena emperor   65 cm
Lethrinus mahsena
Dusky bars on body. **Biology:** inhabits
lagoons, bays and seaward reefs, 2 to
100 m. Solitary along coral-rich
slopes, or over sand and seagrass.
Feeds on echinoderms including
Diadema sea urchins, as well as
crustaceans and fishes. Common,
but wary. **Range:** Red Sea to Sri
Lanka, n. to Arabian Gulf, s. to
Mauritius.

Yellowlip emperor   60 cm
Lethrinus xanthochilus
Lips yellow, base of P fin red.
Biology: inhabits outer reef flats and
lagoon and seaward reefs, 2 to 50 m.
Solitary or in small groups in areas
of mixed sand, rubble, coral or
seagrass. Juveniles in seagrass beds.
Feeds on hard-shelled invertebrates
and small fishes. Curious, but wary.
Range: Red Sea to Fr. Polynesia,
n. to Ryukyus, s. to S. Africa and
New Caledonia.

Snubnose emperor 40 cm
Lethrinus borbonicus
Silvery greenish with yellow patch
above eye. **Biology:** common in
lagoons, sheltered bays and seaward
reefs, 1 to 40 m. Usually hovers above
and along the bases of coral slopes.
Enters reef flats during the night to
feed on hard-shelled invertebrates.
Curious and approachable.
Range: Red Sea to Mauritius, n. to
Arabian Gulf, s. to Mozambique.

Mangrove Bay, EL

Spangled emperor 85 cm
Lethrinus nebulosus
Head profile slightly concave, head
with blue bands. **Biology:** inhabits
lagoons, sheltered bays and seaward
reefs, 1 to 75 m. Common, often
over seagrass beds or sand near reefs.
Also among mangroves. Solitary
or in small groups. Feeds on hard-
shelled invertebrates. Curious, often
approachable. **Range:** Red Sea to
Samoa, n. to S. Japan, s. to S. Africa
and New Caledonia.

Mövenpick Bay, EL

Orange-stripe emperor 45 cm
Lethrinus obsoletus
Grey with a golden lateral stripe
which may disappear at will.
Biology: inhabits reef flats, lagoons,
bays and seaward reefs near areas
of sand and seagrass, 1 to 30 m.
Solitary or in small aggregations.
Feeds on sand-dwelling invertebrates,
even in very shallow areas. Juveniles
in seagrass beds. Occasionally wary,
often approachable. **Range:** Red Sea
to Samoa, n. to Ryukyus, s. to
Mozambique and New Caledonia.

Mangrove Bay, KF

Eritrea, AK

Smalltooth emperor   70 cm
Lethrinus microdon
Olivaceous to grey with elongate snout
often assumes a mottled pattern
rendering it nearly invisible. **Biology:**
inhabits lagoon and seaward reefs, 1 to
80 m. Common, usually solitary,
occasionally in small groups,
sometimes with *L. olivaceus*. Feeds on
fishes, cephalopods and crustaceans.
Large individuals may be ciguatoxic.
Range: Red Sea to PNG, n. to
Ryukyus, s. to NW. Australia. **Similar:**
Longface emperor *L. olivaceus* (Red Sea
to Samoa, 60 cm) nearly identical
except has more scale rows above lateral
line (5½ vs 4½ in *L. olivaceus*) and
longer snout.

Bali, RM

Blackspot emperor   50 cm
Lethrinus harak
Mid-lateral black blotch usually
present. **Biology:** over sand, rubble
and seagrass of protected lagoons,
reef flats and channels, occasionally
among mangroves, 1 to 25 m. A
common inshore species that feeds
primarily on benthic invertebrates.
Usually solitary, aggregates to spawn.
Range: Red Sea to Samoa, n. to S.
Japan, s. to Mauritius and New
Caledonia.

Mangrove Bay, RM

Blue-lined large-eye bream
Gymnocranius grandoculis   80 cm
Silvery-grey with blue lines on snout,
sometimes with network of darker
wavy lines on sides. **Biology:** inhabits
deep lagoons, bays and seaward slopes,
10 to 100 m. On sand and rubble
flats away from coral slopes. Feeds on
benthic invertebrates and small fishes.
Range: Red Sea and Gulf of Aden
to Fr. Polynesia, n. to S. Japan, s. to
S. Africa and SE. Australia.

SPINECHEEKS
NEMIPTERIDAE

Bream-like fishes with a small
rear-projecting spine below eye.

Arabian spinecheek   18 cm
Scolopsis ghanam
Biology: inhabits sandy areas of
lagoons, bays and sheltered seaward
reefs, 1 to 20 m. Solitary or in small
groups. Common, hangs motionless
above sand, then swims in short
spurts. Feeds on small invertebrates
picked from sand or rubble. Shadows
goatfishes to capture prey. Easily
approached. **Range:** Red Sea to
Andaman Is., n. to Arabian Gulf,
s. to Madagascar.

Mövenpick, RM

juv.

Blackstreak spinecheek   36 cm
Scolopsis taeniatus
Silvery-yellow with broad brown
stripe. **Biology:** shallow coastal reefs,
3 to 18 m. Prefers silty and sandy
areas close to coral or rocky reefs.
Common on the Arabian Sea coast.
Solitary or in groups. **Range:** Red Sea
and Arabian Gulf to Sri Lanka, s. to
Gulf of Aden. **Similar:** Thumbprint
spinecheek *S. bimaculatus* with a
shorter dark stripe (Red Sea to Sri
Lanka, s. to E. Africa; 31 cm). ▼

Salalah, KF

Whitecheek spinecheek   25 cm
Scolopsis vosmeri
Reddish-brown with white bar
through rear of head. **Biology:**
inhabits sand or mud bottoms of
estuaries and coastal rocky reefs, 1 to
45 m. Solitary or in pairs. Feeds on
benthic invertebrates. **Range:** Red Sea
to PNG, n. to Arabian Gulf and
Ryukyus, s. to S. Africa and GBR.

Thailand, KF

Shams Alam Reefs, EL

Mangrove Bay, RM

Salalah, KF

SEA BREAMS
SPARIDAE

Stocky fishes with steep forehead, single dorsal fin, coarse scales, and many with molariform teeth for crushing hard-shelled invertebrates.

Doublebar bream   50 cm
Acanthopagrus bifasciatus
Silvery with two black bars on head.
Biology: inhabits outer reef flats, deep lagoons, bays and seaward reefs, 2 to 20 m. Prefers reef flats with deep holes and reef margins, often in surgy areas. Solitary or in small semi-stationary groups near shelter. **Range:** Red Sea to Arabian Gulf, s. to S. Africa and Mauritius.

Picnic sea bream   75 cm
Acanthopagrus berda
Uniformly silvery. **Biology:** inhabits lagoons, channels and seaward reef slopes, 2 to 50 m. Common in turbid soft-bottom areas near reefs. Juveniles in estuaries. Feeds on wide variety of benthic invertebrates. An excellent foodfish. **Range:** Red Sea to New Caledonia, n. to Arabian Gulf and S. Japan, s. to S. Africa. **Similar:** Arabian pinfish *Diplodus noct* with round black spot at base of tail; sometimes absent (Red Sea, 25 cm). ▼

Zebra bream   30 cm
Diplodus cervinus omanensis
Silvery with four curved black bars.
Biology: inhabits coastal rocky reefs, 2 to 40 m. Solitary or in groups in open water near reefs. Spawns in mangrove areas. Shy. **Range:** S. Oman only; the subspecies *D. c. hottentotus* from S. Mozambique to S. Africa.
Similar: Santer seabream *Cheimerius nufar* from depths of 60 to 100 m has five dusky-pink bars (Red Sea to Madagascar and Mauritius; 60 cm); Yellowfin bream *Rhabdosargus sarba* has yellowish P, A, V and tail fins (Red Sea to Mauritius; 60 cm).

Striped boga   25 cm
Boops lineatus
Blue-green with six narrow brown
stripes. **Biology:** in schools on
shallow coastal rocky reefs, 1 to 20 m.
Range: Yemen to Gulf of Oman only.

MULLETS
MUGILIDAE

Silvery elongate cylindrical fishes with
short snout, small mouth, two dorsal
fins, and large scales. Mullets travel in
schools and feed on fine algae,
diatoms, and detritus from bottom
sediments. Several species inhabit
estuarine and fresh waters. Many are
important foodfishes.

Oman, JPH

Fringelip mullet   55 cm
Crenimugil crenilabis
Dark spot at base of P fin, tail forked.
Biology: common in sandy areas of
estuaries, silty lagoons, harbours and
reef flats, 0.2 to 15 m. In small
schools, often close to shore. Feeds on
fine algae from surface of bottom
sediments. **Range:** Red Sea to Fr.
Polynesia, n. to S. Japan, s. to S.
Africa and SE. Australia. **Similar:**
Bluespot mullet *Moolgarda seheli* has
blue spot on upper P fin base (Red
Sea to Samoa; 50 cm).

Bali, RM

Foldlip mullet   25 cm
Oedalechilus labiosus
Silvery; upper lip with folds and lobes
Biology: inhabits reef flats, lagoons
and protected bays with sandy
stretches close to coral reefs, 1 to at
least 20 m. Usually in small schools.
Expels sediments through gills.
Common. **Range:** Red Sea and Gulf
of Oman to Marshall Is., n. to S.
Japan, s. to S. Africa. **Similar:**
Squaretail mullet *Ellochelon vaigiensis*
has truncate tail, yellowish tail and
fins, and black P fins in juveniles. It
occurs along shallow sandy shorelines
(Red Sea to Fr. Polynesia; 63 cm). ▼

Oman, JPH

GOATFISHES

Goatfishes are elongate soft-bodied fishes with a prominent pair of long barbels under the chin, two widely separated dorsal fins, a small mouth with a slightly protruding upper jaw, a forked tail, and relatively large finely ctenoid scales. The barbels contain chemosensory organs and are used to probe sand or holes in the reef for prey, primarily benthic invertebrates or small fishes. When not in use, they are tucked between the gill covers. Foraging goatfishes are often accompanied by an entourage of other predatory fishes that take advantage of the disturbance to ambush escaping prey. Goatfishes foraging in sand may stir up small clouds of silt. Some species of goatfishes (particularly *Mulloidichthys vanicolensis*) are active primarily at night and tend to hover in stationary aggregations or rest on coral ledges by day. Others are most active by day, and some are active by day or night. Goatfishes are relatively soft bodied and highly esteemed as a food by certain predatory fishes as well as humans. They are also favoured clients of the Bluestreak cleaner wrasse *Labroides dimidiatus* (p. 162), which will provide services to them before attending other fishes. Goatfishes being serviced or soliciting service, spread their fins and orient themselves in an upright tail-standing posture. During courtship, male goatfishes rapidly wriggle their barbels.

Hamata, EL

Red Sea goatfish 28 cm
Parupeneus forsskali
Pale with broad dark stripe from snout through eye to dark spot at upper base of tail; tail yellow.
Biology: inhabits open stretches of sand and coral rubble of lagoon and seaward reefs, 1 to 30 m. Feeds on benthic invertebrates. Often accompanied by other fishes including wrasses, *Scolopsis ghanam* and young *Variola louti*. The latter adopts a similar colour pattern and may mimic the goatfish to get closer to the unsuspecting juvenile fishes too slow for the goatfish to catch. Common and easily approached. **Range:** Red Sea and Gulf of Aden only.

Bali, RM

Dash-and-dot goatfish 50 cm
Parupeneus barberinus
Pale with broad dark stripe from upper jaw through eye to dark spot at upper base of tail. **Biology:** inhabits sandy areas of reef flats and lagoon and seaward reefs to 100 m. Forages in small groups by day, assumes mottled pattern and rests on reef by night. **Range:** Gulf of Aden and S. Oman to Fr. Polynesia, n. to S. Japan, s. to S. Africa and SE. Australia.

Longbarbel goatfish   30 cm
Parupeneus macronema
Pale to rosy with broad dark stripe
from eye to dark spot at upper base of
tail; base of soft D fin with dark band.
Biology: inhabits stretches of sand or
rubble of lagoons, bays and sheltered
seaward reefs, 2 to 30 m. Common,
prefers areas near coral heads. Solitary
or in small groups, often at cleaning
stations. Feeds on benthic invertebrates.
Easily approached. **Range:** Red Sea
to PNG, n. to Gulf of Oman and
Philippines, s. to Sodwana Bay,
S. Africa and Indonesia.

Mövenpick Bay, EL

Two-barred goatfish   35 cm
Parupeneus trifasciatus
Pale with two broad dark bars, one
below each D fin; front ⅔ of nuptial
♂ may be entirely dark. Identified
as *P. bifasciatus*, an invalid name, in
most recent works. **Biology:** inhabits
lagoon and seaward reefs, 1 to over
30 m. Often rests on rocks or corals
by day. Feeds on benthic invertebrates
and fishes. **Range:** S. Oman to
S. Indonesia, s. to Sodwana Bay,
S. Africa; recently shown to be
distinct from the very similar
P. crassilabris in the W. Pacific
(Indonesia to Fiji) and *P. insularis*
in Oceania (Marianas to Hawaii
and SE. Polynesia).

Maldives, RM

Yellowsaddle goatfish   50 cm
Parupeneus cyclostomus
Blue with yellow saddle on caudal
peduncle or uniformly yellow.
Biology: inhabits reef flats and lagoon
and seaward reefs, 1 to 95 m. On
corals, rubble and sand. Solitary or
in groups, sometimes accompanied by
jacks, groupers, or the Sling-jaw
wrasse (p. 157). Feeds primarily on
fishes flushed out by thrusting the
barbels into holes. Common. **Range:**
Red Sea to Hawaii and Fr. Polynesia,
n. to Ryukyus, s. to S. Africa and New
Caledonia. *Yellow phase.* ▼

Marsa Bareka, RM

Indian goatfish 35 cm
Parupeneus indicus
Pale with large yellow spot on mid upper sides and smaller dark spot on upper base of tail. **Biology:** inhabits shallow sandy or silty areas of lagoon and coastal reefs, 0.5 to over 20 m. **Range:** Gulf of Aden to Samoa, n. to India and Philippines, s. to S. Africa and Tonga.

Bali, RM

Redspot goatfish 36 cm
Parupeneus heptacanthus
White or yellowish to rosy with a red spot on mid-lateral line. **Biology:** inhabits seagrass beds and sandy areas of lagoons, bays and seaward reefs, 5 to 350 m. Juveniles in large aggregations over sandy flats. Adults more common in deeper waters. Solitary or in small groups. Feeds on sand-dwelling invertebrates. **Range:** Red Sea to Samoa, n. to Arabian Gulf and S. Japan, s. to S. Africa and SE. Australia.

Mövenpick Bay, RM

Rosy goatfish 42 cm
Parupeneus rubescens
Pale to rosy with pale-edged darker band from snout through eye to below end of 2nd D fin; a black saddle preceded by white patch on caudal peduncle. **Biology:** inhabits silty areas of coral or rocky reefs with rubble, seagrass and sand, 5 to 35 m. Solitary or in groups. Often in mixed aggregations with other goatfishes. Shy. **Range:** Red Sea to Gulf of Oman, s. to S. Africa and Réunion. **Similar:** replaced by *P. margaritatus* with indistinct dark saddle at base of tail from Arabian Gulf to C. Oman.

Kenya, RM

Yellowstripe goatfish   43 cm
Mulloidichthys flavolineatus
Pale with yellow lateral stripe,
often with dark blotch in middle.
Biology: inhabits shallow sandy
areas of lagoon and seaward reefs,
1 to 35 m. Common, often in large
inactive semi-stationary aggregations
by day. When feeding, the yellow
stripe is replaced by a black blotch.
Often with other goatfishes. Easily
approached. **Range:** Red Sea to
Hawaii and Fr. Polynesia, n.
to Ryukyus, s. to S. Africa,
SE. Australia and Rapa.

Mangrove Bay, EL

Yellowfin goatfish   38 cm
Mulloidichthys vanicolensis
Pale with yellow to orange
mid-lateral stripe and yellow fins.
Biology: inhabits reef flats and lagoon
and seaward reefs, 1 to 50 m. Hovers
in large semi-stationary aggregations
along reef edges or near shelter
of large coral mounds by day.
Disperses at night to feed on benthic
invertebrates. Common and easily
approached. **Range:** Red Sea to
Hawaii and Fr. Polynesia, n. to
S. Japan, s. to S. Africa and New
Caledonia.

Mangrove Bay, EL

Blackstriped goatfish   30 cm
Upeneus tragula
Pale, mottled with brown to
red specks and blotches; tips of
D fins dark, tail with dark bars.
Biology: inhabits shallow sandy to
silty areas of lagoon and protected
seaward coastal reefs. **Range:** Red Sea
and Arabian Gulf to New Caledonia,
n. to S. Japan, s. to S. Africa and SE.
Australia. **Similar:** several similar
Upeneus spp. with bars on tail: *U.
sundaicus* with single broad bright
yellow stripe (Red Sea to PNG,
22 cm); *U. vittatus* with 4 narrow
yellow stripes (Red Sea to Fr.
Polynesia; 28 cm). ▼

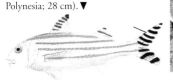

Oman, KF

Mangrove Bay, KF

SWEEPERS
PEMPHERIDAE

Compressed fishes with deep body tapering posteriorly and large eyes.

Golden sweeper   10 cm
Parapriacanthus ransonneti
Translucent coppery with yellow head; digestive tract with bioluminescent organ. **Biology:** in dense swarms in caves, crevices or under overhangs, in lagoons, bays and sheltered seaward reefs, 0.3 to 40 m. Disperses at night to feed on plankton above the reef. Common. **Range:** Red Sea and Gulf of Oman to Fiji, n. to S. Japan, s. to S. Africa.

Vanikoro sweeper   20 cm
Pempheris vanicolensis
Adults bronze with base of A fin and tip of D fin dusky. **Biology:** inhabits lagoon and sheltered seaward reefs, 2 to 40 m. In schools in or near shelter of caves, crevices and overhangs by day, disperses at night to feed on planktonic invertebrates and small fishes. Easily approached. **Range:** Red Sea and Gulf of Oman to Samoa, n. to Philippines, s. to Mozambique.

Kenya, RM

Yellowtail sweeper   15 cm
Pempheris schwenkii
Margin of A fin dusky, base of P fin unmarked; tail yellowish. **Biology:** inhabits sheltered rocky and coral reefs, 1 to 40 m. In groups in or near shelter of caves, crevices and overhangs by day. **Range:** Red Sea to Fiji, s. to S. Africa and GBR. **Similar:** several other spp. reported from Red Sea, but occurrence uncertain; family in need of revision.

Kenya, RM

RUDDERFISHES
KYPHOSIDAE

Medium-sized fishes with small head, small terminal mouth with incisiform teeth, single D fin, small scales and slightly forked tail. Omnivores of exposed seaward reefs.

Highfin rudderfish   45 cm
Kyphosus cinerascens
Soft dorsal and anal fin high.
Biology: inhabits outer lagoon and seaward reefs, 1 to 24 m. In aggregations along surf-swept reef margins and around shipwrecks. Feeds on detached as well as attached algae. **Common. Range:** Red Sea to Hawaii and Fr. Polynesia, n. to S. Japan, s. to S. Africa and SE. Australia.

Bali, RM

Brassy rudderfish   60 cm
Kyphosus vaigiensis
Soft D and A fin low; brassy moustache; may exhibit mottled pattern. **Biology:** inhabits lagoons, bays and seaward reefs, 1 to 30 m. In aggregations along surf-swept reef margins, slopes with moderate currents and around shipwrecks. Feeds on detached as well as attached algae. **Common. Range:** Red Sea to Hawaii and Fr. Polynesia, n. to S. Japan, s. to S. Africa and Rapa.

Bali, RM

Grey rudderfish   70 cm
Kyphosus bigibbus
Occasionally uniformly yellow; otherwise distinguishable from Brassy rudderfish by dark vertical fins and lower number of D fin rays (12 vs. 14). **Biology:** occurs in aggregations primarily around offshore islands, 1 to at least 30 m. **Range:** anti-equatorial: Red Sea to Oman and Ryukyus to S. Japan in N. hemisphere, S. Africa to Rapa in S. hemisphere. Very closely related to *K. sandvicensis* which is anti-equatorial in the Pacific.

Mangrove Bay, KF

Butterflyfishes are among the most colourful and conspicuous of coral reef fishes. They are characterized by highly compressed disc-like body covered with small scales, a small protractile mouth with small brush-like teeth, a single continuous dorsal fin and a rounded to truncate tail. Butterflyfishes are typically diurnal and some ranging. At night they rest among corals while assuming a sombre coluor. Many species are generalists that feed on a combination of cnidarian polyps or tentacles, small invertebrates, fish eggs and filamentous algae while others are specialists. Corallivores are territorial and restricted to areas near their food corals. Plank-ivores typically occur in midwater aggregations. Butterflyfishes appear to be gonochorists (sex predetermined and permanent) with most species in stable pairs for years, if not life. Spawning generally occurs at dusk at the apex of an ascent during which the male nudges the female's abdomen. The eggs are planktonic and the larval stage probably lasts for several weeks. Settlement occurs at night and juveniles typically occur in shallow sheltered areas. Coloration typically changes little with growth and nearly all species are easily recognized by colour pattern alone.

◀ Orangeface butterflyfish

Chaetodon larvatus   12 cm
Grey with thin white chevrons and rust-red head. **Biology:** inhabits sheltered areas of well-developed coral reefs, 2 to 20 m. Only in areas with *Acropora* staghorn or tabletop corals. In pairs, feeds exclusively on *Acropora* coral polyps. Maintains a home range. Easily approached. Rare in N. Red Sea. **Range:** Red Sea to Gulf of Aden only. (*Photo: Shams Alam Reefs, EL.*)

Masked butterflyfish   23 cm ▶

Chaetodon semilarvatus
Yellow with dark blue-grey eye-patch. **Biology:** inhabits coral-rich areas of lagoons, bays and seaward reefs, 1 to 20 m. Usually in pairs or small groups, often hovers under coral ledges. Sometimes in spectacular spawning aggregations. Common. **Range:** Red Sea to Gulf of Aden only.

Red Sea raccoon butterflyfish

Chaetodon fasciatus   22 cm
Orange with diagonal black lines; short white band above eye-bar. **Biology:** inhabits coral-rich areas of reef flats, lagoons, bays and seaward reefs, 1 to 25 m. Usually in pairs or loose aggregations. Feeds on coral polyps, algae and small invertebrates. Common. **Range:** Red Sea to Gulf of Aden only. **Very similar:** Raccoon butterflyfish *C. lunula* has large black mark on tail base (Gulf of Aden to Polynesia; 20 cm).

Fury Shoals, EL

Marsa Shagra, RM

Exquisite butterflyfish   14 cm
Chaetodon austriacus
Orange with thin curved black stripes; A and tail fins black.
Biology: inhabits coral-rich areas of lagoons, bays and seaward reefs, 0.5 to 20 m. Common, adults usually in pairs that patrol a home range. Feeds primarily on coral polyps and anemone tentacles. Juveniles among coral branches. **Range:** Red Sea to S. Oman only. **Similar:** Redfin butterflyfish *C. trifasciatus* (Gulf of Aden to Indonesia; 15 cm) has orange bands on tail and A fins.

juv.

Arabian butterflyfish   13 cm
Chaetodon melapterus
Orange with black D, A and tail fins.
Biology: inhabits coral-rich areas of shallow reefs, 1 to 20 m. Usually in pairs or in loose aggregations. Feeds exclusively on coral polyps within a home range. Easily approached. Juveniles among branches of corals on outer reef flats and reef margins. **Range:** S. Red Sea to Gulf of Oman and Arabian Gulf only.

Blackbacked butterflyfish
Chaetodon melannotus   15 cm
Biology: inhabits coral-rich areas of reef flats, lagoons, bays and seaward reefs, 0.3 to 20 m. Common around staghorn coral thickets, but rare on exposed coral reefs. Home-ranging and usually in pairs, feeds primarily on polyps of soft and hard corals. **Range:** Red Sea to Samoa, n. to S. Japan, s. to S. Africa and SE. Australia.

Crown butterflyfish   14 cm
Chaetodon paucifasciatus
White with black chevrons, abruptly
ed posteriorly. **Biology:** inhabits reef
lats, lagoons, bays and sheltered
eaward reefs, 1 to 30 m. Common
n coral-rich areas, but also on patch
eefs in seagrass or rubble areas.
Usually in pairs or small groups.
Feeds on hard and soft corals, algae,
crustaceans and polychaetes within
a large home-range. **Range:** Red Sea
o Gulf of Aden only. Replaced by
C. madagaskariensis in Indian Ocean
and *C. mertensii* in Pacific.

Mövenpick Bay, RM

Threadfin butterflyfish   23 cm
Chaetodon auriga
White with diagonal black lines in
wo directions; orange posteriorly;
adults with D fin filament.
Biology: inhabits areas of mixed
and, rubble and coral of shallow
eef flats and lagoon and sheltered
eaward reefs, 1 to 35 m. Feeds by
earing pieces from polychaetes,
anemones, coral polyps and algae.
Solitary or in pairs within a home
ange. Common. Not Shy.
Range: subspecies *auriga*: Red Sea
only; subspecies *setifer* (with black
pot on D fin): Gulf of Aden to Fr.
Polynesia, n. to S. Japan, s. to S.
Africa.

Mangrove Bay, EL

Painted butterflyfish   18 cm
Chaetodon pictus
White with diagonal black lines in
wo directions; D, A and tail mostly
yellow with black basal band joining
eye-bar. **Biology:** coral and rocky
eefs, including turbid areas with dead
corals, 1 to 30 m. Feeds primarily on
coral polyps and algae. Solitary or in
pairs, timid. Abundant in Gulf of
Aden and S. Oman. **Range:** S. Red
Sea to Gulf of Oman only. Replaced
by similar Vagabond butterflyfish
C. vagabundus from E. Africa to
Fr. Polynesia. ▼

Oman, HF

Mövenpick Bay, EL

Lined butterflyfish   30 cm
Chaetodon lineolatus
White with thin black bars joining
black band at base of yellow soft D
fin. **Biology:** inhabits coral-rich areas
of lagoons, bays and seaward reefs,
2 to 50 m. Uncommon. Solitary or
in pairs, occasionally in spawning
aggregations. Feeds on coral polyps,
small anemones, small invertebrates
and algae. Difficult to approach.
The largest species of butterflyfish.
Range: Red Sea and Gulf of Aden to
Hawaii and Fr. Polynesia, n. to S.
Japan, s. to S. Africa and SE.
Australia; absent from Oman to
Arabian Gulf.

Salalah, KF

Somali butterflyfish   18 cm
Chaetodon leucopleura
White, becoming dusky above and
along margins; D, A and tail fins
yellow. **Biology:** inhabits rocky and
coral reefs, 10 to 75 m, usually over
open bottoms at base of coral-rich
slopes below 20 m. Often solitary or in
pairs. Rare in S. Red Sea, common in
S. Oman and Yemen. **Range:** S. Red
Sea and S. Oman to Seychelles and E.
Africa. **Similar:** Zanzibar butterflyfish
C. zanzibariensis inhabits coral-rich
areas of lagoon and seaward reefs, 3 to
40 m (Gulf of Aden to S. Africa, e. to
Chagos Is.; 12 cm). ▼

Gardiner's butterflyfish   17 cm
Chaetodon gardineri
White, becoming black posteriorly;
D fin black with yellow margin.
Biology: inhabits deep rocky reefs,
5 to 50 m, usually below 15 m.
Solitary or in pairs. Feeds on benthic
invertebrates and algae. **Range:** Gulf
of Aden to Gulf of Oman, e. to Sri
Lanka.

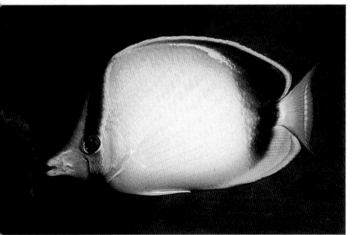

Oman, JPH

Whiteface butterflyfish 13 cm
Chaetodon mesoleucos

Grey with 15 thin black bars; head off-white. **Biology:** inhabits coral-rich sheltered seaward and offshore reefs, 3 to 20 m. Also on algae-covered rocky reefs, coral rubble and wrecks. Usually in pairs within a home range. Uncommon in central Red Sea, common in Yemen and Eritrea. Easily approached. **Range:** central Red Sea to Gulf of Aden only. **Similar:** Jayakar's butterflyfish *C. jayakari* occurs at 168 m in the Red Sea, but as shallow as 33 m off Oman (Red Sea to India; 17 cm). ▼

Eritrea, AK

Collared butterflyfish 16 cm
Chaetodon collare

Dark grey with pale-centred scales and white bars on head, tail mostly red. **Biology:** inhabits shallow rocky and coral reefs from rocky shores to coral-rich outer reef slopes, 1 to 20 m. Feeds primarily on coral polyps and polychaetes, occasionally on algae. Usually in pairs, occasionally in groups. Common in Gulfs of Aden and Oman. Easy to approach. **Range:** Gulf of Aden and Gulf of Oman to Philippines, s. to Maldives and Indonesia.

Oman, KF

Oman butterflyfish 18 cm
Chaetodon dialeucos

Grey with broad white bar behind head. **Biology:** inhabits algae-covered rocky reefs, sometimes on coral patch reefs, 1 to 10 m. Solitary or in pairs. Feeds on small benthic invertebrates and filamentous algae. Common in S. Oman, rare in Gulf of Aden (Sidka Is.). **Range:** Gulf of Aden to S. Oman only.

juv.

Salalah, KF

Mangrove Bay, EL

Chevron butterflyfish 18 cm
Chaetodon trifascialis
White with thin black chevrons; D and A fins angular and orange. **Biology:** inhabits coral-rich areas of reef flats, and lagoon and semi-protected seaward reefs, 1 to 30 m. Common and highly territorial around tabletop and staghorn *Acropora* corals. Feeds exclusively on their polyps and mucus. **Range:** Red Sea to Fr. Polynesia, n. to S. Japan, s. to S. Africa and New Caledonia.

juv.

Red Sea bannerfish 20 cm
Heniochus intermedius
Pale yellow becomong white above with 2 broad diagonal black bands; anterior D fin a long filament. **Biology:** inhabits reef flats, lagoons and sheltered seaward reefs, 3 to 50 m. Common, often hovers in aggregations under coral heads or around prominent coral formations. Feeds in open water on zooplankton. Territorial. Not shy. Juveniles at base of slopes in large aggregations. **Range:** Red Sea and W. Gulf of Aden only.

Mövenpick Bay, RM

Schooling bannerfish 18 cm
Heniochus diphreutes
White with 2 broad diagonal black bands; anterior D fin a long filament; breast rounded. **Biology:** inhabits seaward reefs, 3 to 210 m, usually below 15 m. Generally shallower in areas of cool upwelling (at Bab El Mandeb). Usually in aggregations, the adults in midwater, juveniles around isolated coral heads or patch reefs. Feeds primarily on zooplankton. **Range:** Red Sea to Hawaii, n. to S. Japan, s. to S. Africa. **Similar:** Longfin bannerfish *H. acuminatus* (Gulf of Aden to Fr. Polynesia; 25 cm) has smaller eye, longer snout, more rounded A fin and usually 11 D fin spines (vs. 12 in *diphreutes*).

Salalah, HF

▲ Zebra angelfish 20 cm
Genicanthus caudovittatus
♂ with narrow black bars; ♀, juv.
pale cream with black bar above eye.
Biology: inhabits coral-rich seaward
reef slopes and dropoffs, 15 to 70 m.
Shallower in upwelling areas.
Common, usually in small groups.
♂ maintains a harem of 4 to 9 ♀s in
a territory of about 25 m². Feeds in
aggregations on zooplankton with
males often high in water and females
closer to the bottom. Juveniles secretive.
Usually indifferent towards divers.
Range: Red Sea to Maldives, s. to
Mozambique. *(Photo: Mövenpick Bay, EL.)*

♀

ANGELFISHES

Angelfishes resemble butterflyfishes and share their splendid colours,
brush-like teeth and small scales. They differ by having a prominent
spine on the preopercle and swim primarily with their pectoral fins. The
juvenile and adult colour patterns of species of *Pomacanthus* are strik-
ingly different. Angelfishes are active by day and prefer well-developed
coral reefs or rocky areas with abundant shelter in the form of caves,
holes and crevices. Some large *Pomacanthus* emit 'knocking' sounds. All
species are protogynous hermaphrodites with haremic social systems.
Males defend a territory containing 2 to 8 females. Territory size ranges
from a few square metres (*Centropyge*) to more than 1000 square metres
(*Pomacanthus*). Reproduction is typically seasonal, from April to
November, perhaps longer in the south. Spawning is paired and occurs
around sunset at the apex of an upward rush following courtship displays and nuzzling by the male. Hatching occurs within 24 hours and
the pelagic larval stage lasts about 3 to 4 weeks. Species of *Centropyge*
feed primarily on small benthic invertebrates and filamentous algae.
Species of *Genicanthus* feed primarily above the reef on zooplankton.
Most other species feed mainly on sponges and to a lesser extent on soft-
bodied invertebrates, algae and fish eggs. Most *Centropyge*, *Genicanthus*
and few species of *Pomacanthus* do well in captivity, but others are
difficult to maintain.

Marsa Shagra, RM

Regal angelfish   25 cm
Pygoplites diacanthus
Biology: inhabits coral-rich areas of
lagoon and seaward reefs, 1 to 80 m.
Common, usually solitary or in pairs,
often near caves or crevices. Feeds
on sponges and tunicates. Juveniles
secretive. **Range:** Red Sea to Fr.
Polynesia, n. to Ryukyus, s. to
S. Africa.

juv.

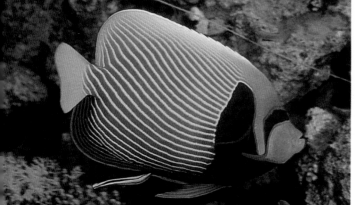

Marsa Bareka, EL

Emperor angelfish   40 cm
Pomacanthus imperator
Biology: inhabits coral-rich areas of
lagoon and seaward reefs, 3 to 70 m.
Common, usually near shelter. Has
a large home range. Haremic, but
usually solitary or in pairs. Feeds
on sponges and tunicates. Makes a
'knocking' sound. Juveniles usually
in holes in clear shallow areas.
Range: Red Sea to Fr. Polynesia,
n. to S. Japan, s. to S. Africa.

juv.

Salalah, KF

Semicircle angelfish   38 cm
Pomacanthus semicirculatus
Biology: inhabits coastal reefs with
heavy coral growth or algae-rich
rocky reefs, 1 to 40 m. Solitary or
in pairs, often around wrecks and
in well-encrusted caves. Juveniles in
shallows among rocks or corals. Shy.
Range: S. Red Sea and Oman to
Samoa, n. to S. Japan, s. to S. Africa
and SE. Australia.

juv.

Yellowbar angelfish   50 cm
Pomacanthus maculosus
Biology: inhabits shallow lagoons, silty bays, and coral-rich seaward reefs, 2 to 60m. Usually singly or in pairs. Feeds on sponges and algae. Juvenile secretive among corals. Not shy.
Range: Red Sea and Arabian Gulf to Seychelles.

juv.

Mangrove Bay, RM

Arabian angelfish   40 cm
Pomacanthus asfur
Biology: inhabits turbid inshore reefs or silty offshore reefs with moderate coral growth, 3 to 40 m. Uncommon in central Red Sea, but very common around Hanisch Is., S. Yemen with up to 35 individuals per 5,000 m².
Range: C. and S. Red Sea to Oman only.

juv.

Eritrea, AK

Red Sea angelfish   15 cm
Apolemichthys xanthotis
Biology: inhabits semi-protected coral or rocky reefs, 10 to 65 m. Occurs on rubble slopes with whip and soft corals or algae-covered rocks. On deep coral-rich slopes on the Sinai coast. Solitary or in small groups feeding on sponges, algae and invertebrates.
Range: Red Sea to Oman only.

juv.

Nuweiba, EL

Kenya, RM

African cherubfish   8 cm
Centropyge acanthops
Deep blue with orange head and back.
Biology: inhabits areas of coral or
rubble of seaward reefs, 6 to 40 m.
Often in small groups among the
shelter of branching corals and rubble.
In the open for short intervals when
feeding on filamentous algae. Shy,
but approachable. Juvenile resembles
adult. **Range:** Gulf of Aden and
S. Oman (rare) to Maldives, s. to
S. Africa and Mauritius.

Maldives, RM

Manyspine cherubfish   14 cm
Centropyge multispinis
Dark brown with black 'earspot',
usually with brilliant blue fin margins.
Biology: inhabits isolated coral heads
and coral-rich or rubble areas of reef
flats, and lagoon and protected
seaward reefs, 1 to 30 m. Territorial
and haremic, often in small groups.
Feeds on filamentous algae. Common,
but very shy, darts into shelter when
alarmed. Juvenile resembles adult.
Range: Red Sea to W. Thailand, s. to
S. Africa and Mauritius.

DAMSELFISHES   POMACENTRIDAE

Damselfishes are small, often colourful fishes with moderately deep
compressed bodies, small terminal mouths with conical or incisiform
teeth, moderately large scales, continuous dorsal fins and an interrupted
lateral line. They are conspicuous and abundant on all coral and rocky
reefs. The family includes herbivores that are typically highly territorial
and pugnacious (*Plectroglyphidodon*, *Stegastes* and some *Abudefduf*),
omnivores that occur in small groups near shelter (many *Chrysiptera* and
Pomacentrus) and planktivores that may aggregate high in the water
(*Abudefduf*, *Amblyglyphidodon*, *Chromis*, *Dascyllus* and *Neopomacentrus*).
The anemonefishes (p. 148) live in close association with large sea
anemones. Damselfishes lay demersal eggs that are guarded by the male.
Spawning generally occurs in the morning. In many species reproductive activity peaks on a lunar cycle, and in many tropical areas occurs
more frequently in the early summer. At least 48 species occur around
the Arabian Peninsula, with 37 species in the Red Sea and 29 in Oman.

Indo-Pacific sergeant   20 cm ▶
Abudefduf vaigiensis
Pale olivaceous becoming yellowish
above with five black bars. **Biology:**
inhabits rocky or coral reefs, 0.3 to
12 m. Common. In aggregations
along upper seaward reef margins,
along jetties and piers. Juveniles often
in tidepools or under floating
Sargassum weed. Feeds high in the
water on zooplankton, or on benthic
algae. Ground colour becomes bluish
during courtship. The eggs are laid on
a cleared hard surface and guarded by
both parents. **Range:** Red Sea to Fr.
Polynesia, n. to S. Japan, s. to S.
Africa and SE. Australia. *(Photo: Tiran
Is., EL.)*

Scissortail sergeant   19 cm
Abudefduf sexfasciatus
Pale olivaceous with five black
bars and black caudal lobes.
Biology: inhabits coral-rich areas
of upper lagoon and seaward reef
slopes, 1 to 15 m. Common,
usually in aggregations. Feeds on
zooplankton, and occasionally on
benthic algae. **Range:** Red Sea to
Fr. Polynesia, n. to S. Japan, s. to S.
Mozambique and SE. Australia.

Brothers, AK

Blackspot sergeant   23 cm
Abudefduf sordidus
Olivaceous with broad dusky bars,
black spot on upper caudal peduncle.
Biology: inhabits rocky lagoons,
jetties, and inshore reefs with mild
surge, 0.3 to 5 m. Juveniles often in
exposed tidepools. Highly territorial,
chases away competitors. Feeds on
benthic algae. Common but difficult
to approach. Shy. **Range:** Red Sea to
Fr. Polynesia, n. to S. Japan, s. to
S. Africa and SE. Australia.

Bali, RM

Yellowtail sergeant   17 cm
Abudefduf notatus
Dark brown with narrow pale bars;
tail pale yellow. **Biology:** inhabits
rocky shores, 1 to 12 m, usually in
areas of moderate surge. In roving
aggregations that are somewhat
difficult to approach. **Range:** Gulf
of Aden and S. Oman (rare) to
Indonesia, n. to S. Japan, s. to
S. Africa.

Palau, RM

   DAMSELFISHES

Half-and-half chromis 9 cm
Chromis dimidiata
Black in front, abruptly white behind.
Biology: inhabits shallow reef crests
of patch reefs and lagoon and seaward
reefs, 0.5 to 36 m. In large aggregations
along reef tops and upper edges of
slopes, often with *Pseudanthias
squamipinnis*. Feeds on zooplankton.
Very common and easily approached.
Range: Red Sea to Java, n. to
Thailand, s. to S. Africa.

Mangrove Bay, RM

Dusky chromis 14 cm
Chromis pelloura
Pale greenish-white with black tail
base. **Biology:** inhabits coral-rich
dropoffs and seaward slopes with
abundant holes and crevices,
30 to 70 m. Solitary or in small
midwater aggregations. Feeds on
zooplankton. Uncommon and shy.
Observed in several places on the
Sinai coast. **Range:** Gulf of Aqaba
only. Closely related to *C. axillaris*
(Indian Ocean).

Marsa Bareka, EL

Yellow-edge chromis 13 cm
Chromis pembae
Brown with white tail and black
D and A fins; juvs. light grey with
yellow tail. **Biology:** inhabits steep
rocky or coral slopes of deep outer
lagoon and seaward reefs, 3 to 50 m,
usually below 25 m. Juveniles at
base of reef slopes. Usually in small
aggregations. Feeds on zooplankton.
Uncommon in most areas, but easily
approached. **Range:** Red Sea and
Oman to Chagos Is., s. to E. Africa
and Seychelles.

Marsa Shagra, RM

Marsa Abu Dabab, RM

Bluegreen chromis   9 cm
Chromis viridis
Metallic blue-green with a forked tail.
Biology: in large aggregations along
coral-rich fringing reefs, in sheltered
bays and on patch reefs, 0.3 to 15 m.
Abundant, usually near branching
Acropora corals into which they retreat
when frightened. Juveniles closely tied
to individual coral heads. Courting ♂
becomes golden with black dorsal fins.
Range: Red Sea and S. Gulf of Aden
to Fr. Polynesia, n. to Ryukyus, s. to
Mauritius and New Caledonia.

Abu Galum, EL

Weber's chromis   12 cm
Chromis weberi
Olivaceous to bluish-grey; preopercle
with dark bar, tips of tail lobes black.
Biology: inhabits patch reefs and the
upper edges of lagoon and seaward
reefs, 1 to 25 m. Often around
coral heads with *Chromis viridis* and
C. dimidiata. Solitary or in small
groups. Feeds on zooplankton, often
high in the water column. Common
and easily approached. **Range:** Red
Sea and S. Oman to Pitcairn Is.,
n. to S. Japan, s. to S. Africa and
New Caledonia.

Marsa Shagra, RM

Arabian chromis   7 cm
Chromis flavaxilla
Olive-brown with black tail lobes;
P axil yellow. **Biology:** occurs in small
aggregations near branching corals
and along the upper edges of lagoon
and seaward reefs, 1 to 15 m. Feeds
on zooplankton. **Range:** Red Sea
to Arabian Gulf only. **Similar:** *C.
ternatensis* (E. Africa to Samoa,
10 cm).

Three-spot dascyllus 14 cm
Dascyllus trimaculatus
Body grey to black. **Biology:** inhabits most reef zones, 0.3 to 20 m. Usually in small groups around prominent coral heads or large rocks. Juveniles associated with large sea anemones, sharing them with anemonefishes and immune to the stinging cells. They may also shelter between the spines of *Diadema* urchins and branching corals. **Range:** Red Sea to Pitcairn Is., n. to S. Japan, s. to S. Africa and SE. Australia.

Ras Mohammed, EL

juv.

Red Sea dascyllus 6 cm
Dascyllus marginatus
Olivaceous; D and A fin margins black. **Biology:** inhabits patch reefs and sheltered lagoon and seaward reefs, 1 to 15 m. In small aggregations associated with branching *Acropora*, *Porites* and *Stylophora* corals into which they retreat when frightened. Common. Feeds above the corals on zooplankton. Courting male develops bright blue D fin spines. Spawns around full moon. **Range:** Red Sea to Gulf of Oman only.

Ras Mohammed, RM

Humbug dascyllus 8 cm
Dascyllus aruanus
White with three broad black bars converging at black D fin. **Biology:** inhabits shallow protected lagoons, reef flats, bays and patch reefs, 0.3 to 20 m. Common, in colonies closely tied to branching *Acropora* and *Pocillopora* corals. Aggregate above corals to feed on zooplankton, and wedge themselves among the branches when frightened. **Range:** Red Sea to Fr. Polynesia, n. to Ryukyus, s. to S. Africa and SE. Australia.

Nabq, EL

Mangrove Bay, RM

Onespot demoiselle   8 cm
Chrysiptera unimaculata
Pale brown with black spot at base of D fin and yellow P fin. **Biology:** inhabits outer reef flats and shallow lagoons, bays and harbours, 0.3 to 2 m. Common in exposed areas just inside reef margins with slight wave action and coral rubble. Territorial. **Range:** Red Sea to Fiji, n. to Ryukyus, s. to Mauritius and GBR.

juv.

Safaga, HHH

Blackbarred demoiselle   9 cm
Chrysiptera annulata
White with 5 black bands. **Biology:** occurs around isolated coral heads or rocks of shallow sheltered areas including seagrass beds and mangroves, to about 2 m. Uncommon. **Range:** Red Sea and C. Oman to Seychelles, s. to S. Africa and Mauritius.

Mövenpick Bay, EL

Black damselfish   16 cm
Neoglyphidodon melas
Uniformly black. **Biology:** inhabits lagoon and protected seaward reefs, 1 to 12 m. Uncommon, usually near *Tridacna* clams and soft corals. Feeds on soft corals and perhaps *Tridacna* clam faeces. Juveniles in shallow lagoons, usually around staghorn *Acropora* or soft corals. Solitary or in pairs. Territorial.
Range: Red Sea to Vanuatu, n. to S. Japan, s. to S. Africa and GBR.

juv.

Pale damselfish   9 cm
Amblyglyphidodon indicus
Deep-bodied; pale blue-green with
dusky bar on cheek, margins of soft
D fin and tail black. **Biology:** inhabits
coral-rich areas of lagoon and
sheltered seaward reefs and isolated
patch reefs in clear water, 2 to 35 m.
solitary or in small groups near
prominent coral formations. Feeds
on zooplankton and floating organic
material. Common and easily
approached. **Range:** Red Sea to
Andaman Sea, s. to Madagascar and
Chagos Archipelago. Closely related
to *A. leucogaster* (W. Pacific).

Mangrove Bay, RM

Yellow-side damselfish   10 cm
Amblyglyphidodon flavilatus
Pale grey becoming pale yellow
posteriorly. **Biology:** inhabits
coral-rich areas of lagoon and
sheltered seaward reefs and isolated
patch reefs in clear water, 3 to 20 m.
Common, solitary or in small loose
groups. Feeds on zooplankton well
above the bottom, often together
with Pale damselfish. **Range:** Red Sea
and Gulf of Aden only.

Marsa Bareka, PM

Jewel damselfish   11 cm
Plectroglyphidodon lacrymatus
Brown with blue dots, numerous in
juveniles, nearly lost in large adults;
eye yellow. **Biology:** inhabits areas of
mixed coral and rubble on clear
lagoon and seaward reefs, 0.3 to 40 m.
Somewhat pugnacious, defends an
algae-covered territory and feeds
primarily on benthic algae and
associated small invertebrates and fish
eggs. **Range:** Red Sea to Fr. Polynesia,
n. to Ryukyus, s. to S. Africa and SE.
Australia.

juv.

Nabq, EL

Mövenpick Bay, RM

Whiteband damsel 12 cm
Plectroglyphidodon leucozonus
Brown with broad pale bar lost
in large adults. **Biology:** inhabits
surge-swept shorelines, reef margins,
and outer reef flats with rubble and
algae-coated spaces, 0.3 to 3 m.
Juveniles in intertidal pockets, adults
on ridges between surge channels.
Territorial, feeds primarily on benthic
algae. **Range:** subsp. *cingulum* in Red
Sea only; subsp. *leucozonus* from E.
Africa to Fr. Polynesia, n. to S. Japan,
s. to S. Africa and SE. Australia.

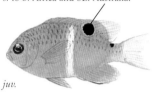

juv.

Regal demoiselle 10 cm
Neopomacentrus cyanomos
Dark bluish-grey with white spot
at rear base of D fin and yellow
rear margins of D, A and tail fins.
Biology: inhabits sheltered fringing
reefs, patch reefs and bays, 5 to
30 m. Usually around coral heads and
near caves, feeds on zooplankton.
Range: Red Sea to Oman, e. to
Solomon Is., n. to S. Japan, s. to
S. Africa and GBR.

GBR, RM

Miry's demoiselle 11 cm
Neopomacentrus miryae
Pale tan to grey becoming olivaceous
above with white spot on upper
peduncle. **Biology:** inhabits sheltered
patch reefs and sheltered lagoon and
seaward reefs, 1 to 25 m. Usually in
loose aggregations near corals. Feeds
on zooplankton close above the
bottom, often with other damselfish
species. Easily approached. **Range:**
Red Sea to Gulf of Oman only.

Nabq, RM

Sulphur damsel   11 cm
Pomacentrus sulfureus
Yellow with black spot at base of
P fin. Juv. with black ocellus in D fin.
Biology: inhabits patch reefs and
sheltered fringing reefs with rich coral
growth, 1 to 14 m. Usually in small
loose groups near coral heads and
Millepora fire corals near reef edges.
Common. **Range:** Red Sea to
Seychelles, s. to E. Africa and
Mauritius.

Mangrove Bay, RM

Threeline damsel   10 cm
Pomacentrus trilineatus
Tan to dark grey with blue lines
on head and black spot on caudal
peduncle; tail pale. **Biology:** inhabits
shallow protected rocky and coral
reefs, 0.3 to 4 m. Often in turbid
inner areas. Usually solitary or in
pairs. **Range:** Red Sea and S. Oman
to Seychelles, s. to Mozambique and
Madagascar.

Nabq, RM

juv.

Caerulean damsel   8 cm
Pomacentrus caeruleus
Brilliant blue with a yellowish tail.
Biology: inhabits lagoon and seaward
reef slopes, usually over rubble near
base of reefs, 1 to 20 m. In loose
aggregations close to bottom. Shy.
Range: Red Sea, Gulf of Aden
(common) and S. Oman to Maldives,
s. to S. Africa.

Kenya, KF

Nabq, EL

Reticulated damsel   11 cm
Pomacentrus trichrourus
Greyish-black with white tail.
Biology: a common inhabitant of
lagoon and sheltered seaward reefs
and patch reefs, 1 to 43 m. Sometimes
on reef flats with seagrasses and rocks.
Solitary or in loose groups. Feeds
on algae, ostracods, fish eggs and
sponges. **Range:** Red Sea to Oman,
s. to S. Africa.

juv.

Slender damsel   7 cm
Pomacentrus leptus
Dark brown with white tail; rear of
D fin and tail yellowish. Gulf of
Oman population is dark blue.
Biology: inhabits shallow inshore
reefs, 1 to 10 m. In large aggregations
near bottom, occasionally solitary.
Range: S. Red Sea and Gulf of
Aden to Gulf of Oman, s. to
E. Africa. **Similar:** Whitefin damsel *P.
albicaudatus* (p. 147) which has white
tail.

Oman, JPH

Dusky gregory   13 cm
Stegastes nigricans
Greyish-brown with black spot below
rear base of D fin. **Biology:** inhabits
reef flats, bays and lagoons, 0.3 to
10 m. Prefers thickets of branching
Acropora and *Porites* corals. Also
known as a 'farmer fish' because
it cultivates patches of algae on
dead coral branches by removing
undesirable species and encouraging
the growth of desirable ones.
Extremely pugnacious and territorial,
may even bite divers. **Range:** Red Sea
to Fr. Polynesia, n. to Ryukyus, s. to
Mauritius and New Caledonia.

Marsa Shagra, RM

1. Triplespot chromis 6 cm
Chromis trialpha
Dark grey with white spot at rear bases of soft D and A fins and black tail lobes. **Biology:** inhabits seaward reefs, 3 to 50 m. Occurs in small groups around caves and overhangs. Remains close to shelter. Uncommon. **Range:** Red Sea only.

2. Red Sea demoiselle 6 cm
Neopomacentrus xanthurus
Bluish-grey; tail and rear margin of soft D fin yellow. **Biology:** inhabits coral reefs, 1 to 15 m, in groups under overhangs. **Range:** S. Red Sea and Gulf of Aden only.

3. Whitefin damsel 7 cm
Pomacentrus albicaudatus
Greyish-brown with yellowish hue on lower sides and white tail. **Biology:** inhabits inshore and offshore coral, 1 to 12 m. **Range:** Red Sea only. **Similar:** Slender damsel *P. leptus* (p. 146) which has yellowish tail.

4. Dark damsel 12 cm
Pomacentrus aquilus
Juvs. yellow with 2 blue stripes from upper head to D fin ocellus. **Biology:** inhabits coral and rocky reefs, 1 to 15 m. **Range:** Red Sea and Arabian Gulf, s. to Kenya.

5. Bluedotted damsel 11 cm
Pristotis cyanostigma
Pale blue with dark-centred scales; upper head and back olivaceous. **Biology:** inhabits inshore and offshore coral reefs, 5 to 10 m. **Range:** Red Sea and Gulf of Aden only. **Similar:** Gulf damsel *P. obtusirostris* has dark blue spot on upper P fin axil.

6. Gulf damsel 14 cm
Pristotis obtusirostris
Pale blue with dark-centred scales; dark spot at upper base of P fin. **Biology:** inhabits patch reefs or open sand or rubble bottoms of lagoon and coastal reefs, 2 to 80 m. **Range:** Red Sea and Arabian Gulf, e. to New Caledonia.

7. Bluntsnout gregory 13 cm
Stegastes lividus
Brown to nearly black with light-margined spot at rear base of D fin. **Biology:** inhabits shallow reef flats and lagoon reefs, 0.2 to 5 m, usually in groups among partially dead staghorn *Acropora* corals. Extremely aggressive and pugnacious, will chase all intruders, even will rush a diver. Also known as 'farmer fish' for its habit of weeding its patch of filamentous algae by removing undesirable species in order to promote the growth of preferred ones. Makes loud staccato sounds. **Range:** Red Sea to Fr. Polynesia, n. to Ryukyus, s. to New Caledonia and Tonga.

8. Jordan's damsel 14 cm
Teixeirichthys jordani
Pale blue, darker above with narrow pale stripes below; more elongate than *Pristotis* spp. **Biology:** inhabits seagrass beds and sandy bottoms, 10 to 20 m. Primarily on continental coasts. **Range:** Red Sea to S. Japan, s. to E. Africa.

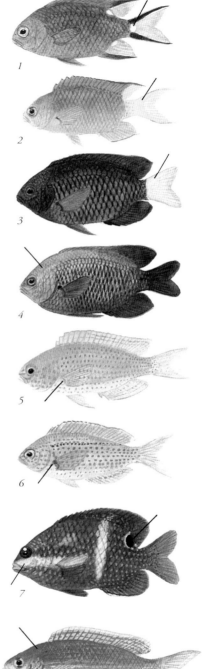

1

2

3

4

5

6

7

8

ANEMONEFISHES   POMACENTRIDAE Subfamily Amphiprioninae

Anemonefishes live in close association with 10 species of large sea anemones. Normally each anemonefish species has a preferred anemone host. They never live without an anemone and are protected from the anemone's stinging cells by their mucus which carries the anemone's chemical signature. This is acquired by the larvae during settlement and inhibits the anemone's nematocysts from firing. Large anemones often host a semi-permanent monogamous pair of adult anemonefishes and several small juveniles. Anemonefishes are protandrous hemaphrodites, that is all individuals mature first as males and all females are sex-reversed males. Sex and growth are controlled by the dominant female. In the absence of a female, the largest male will turn into a female and the largest juvenile will mature into a male within a week. The growth of the remaining juveniles is stunted by the adult pair. Spawning usually takes place around the full moon, all year round in the tropics, but only during the warmer months in the subtropics. Typically 500 to 1,500 adhesive eggs are laid on a patch of cleared rock near the base of the anemone and cared for by the male. Any intruder that is perceived as a danger to the anemone is aggressively driven away, even divers may be attacked. Hatching occurs at night after about a week, and the larvae drift in the plankton for 16 days or more before settling and seeking an anemone host. Anemonefishes feed primarily on zooplankton, especially copepods, and filamentous algae. All species are popular aquarium fishes and several are easy to spawn and raise in captivity.

◄ & ►Red Sea anemonefish
Amphiprion bicinctus　　14 cm
Colour variable from yellow to dark
brown with 2 broad white bars; body
bar ends at base of D fin. **Biology:**
inhabits patch reefs, sheltered bays,
protected fringing reefs, and seaward
reefs below the surge zone, 0.5 to
30 m. Associated with the anemones
*Stichodactyla gigantea, Entacmaea
quadricolor, Heteractis aurora, H.
crispa* and *H. magnifica*. Territorial
and rarely more than 2 m from host
anemone. **Range:** Red Sea, Gulf of
Aden and Chagos Archipelago only.
(Photo: Tiran Is., EL.)

Mangrove Bay, RM

Oman anemonefish　　15 cm
Amphiprion omanensis
Brownish-orange with two narrow
white bars; tail forked. **Biology:**
inhabits sheltered rocky coastal reefs,
2 to 23 m. Uncommon, always close
to the host anemones *Heteractis crispa*
and *Entacmaea quadricolor*.
Range: S. Oman only. **Similar:**
Yemen anemonefish *Amphiprion* sp.
has broader bars (S. Red Sea? and
Yemen coast of Gulf of Aden; 12 cm).
Further study is required to determine
if it is a distinct species. Lives in the
host anemone *Heteractis crispa*. ▼

Salalah, KF

Sebae anemonefish　　14 cm
Amphiprion sebae
Dark brown above, orange below
with two broad white bars, the second
extending along top of soft D fin;
tail rounded. **Biology:** inhabits clear
lagoons, bays and coastal reefs, 2 to
35 m. Usually on rubble and sand
flats. Uncommon, inhabits only one
host anemone species, *Stichodactyla
haddoni*. **Range:** Gulf of Aden and
Gulf of Oman to Java, s. to Maldives.
Similar: Clark's anemonefish *A.
clarkii* has forked tail (Oman to W.
Pacific).

Oman, PN

ANEMONEFISHES　　149

WRASSES

LABRIDAE

Wrasses are among the most diverse of reef fishes in both size and form. They typically have an elongate body, terminal mouth usually with thickened lips and one or more pairs of protruding canine teeth, nodular pharyngeal teeth (in the throat), a continuous or interrupted lateral line and a single unnotched dorsal fin. Most species have complex and often brilliant colour patterns that change with growth or sex. Sex-reversal appears to be universal and is protogynous with females having the capacity to turn into males. Most species have an **initial phase** (IP) consisting either entirely of females or of both females and non-sex-reversing males, and a brightly coloured **terminal phase** (TP) consisting of males that were once females. Wrasses are pelagic spawners. Initial phase males spawn in large groups, terminal males are usually territorial and successively pair spawn with females of their choice. The family is divisible into various tribes based primarily on dentition and diet. Most wrasses are carnivores of benthic invertebrates or fishes. Many use their protruding canines to pluck small hard-shelled invertebrates from the bottom, then use their pharyngeal teeth to crush them. Some are planktivores (*Cirrhilabrus, Paracheilinus*), corallivores (adult *Larabicus*), or cleaners that feed on the ectoparasites of other fishes (*Labroides*; juv. *Larabicus*). All wrasses are inactive at night and most of the smaller species sleep beneath sand. Some are highly specialized for diving into sand and moving beneath the surface (*Iniistius*). At least 96 species are known from the Arabian Peninsula with 82 in the Red Sea and 43 in Oman.

Napoleon wrasse 2.3 m; 191 kg
Cheilinus undulatus
Pair of short black lines behind eye; adults with hump on forehead.
Biology: inhabits lagoon and seaward reefs, 1 to 60 m. Juveniles in protected areas among branching corals. Adults usually have a home cave to hide and sleep in. Feeds on molluscs and armoured invertebrates, including toxic sea urchins and Crown-of-thorns starfishes. May be ciguatoxic. Intelligent and wary, except where protected. Populations in many areas decimated by unscrupulous cyanide fishing driven by high demand for live fishes in SE. Asian restaurants. **Range:** Red Sea to Fr. Polynesia, n. to Ryukyus, s. to S. Africa and New Caledonia; absent from Gulf of Aden to Arabian Gulf.
(Photo: Palau, RM.)

juv.

Bluespotted wrasse   42 cm

Anampses caeruleopunctatus
TP ♂ blue-green with light green bar.
Biology: inhabits exposed seaward
reefs, 1 to 30 m. Prefers coral-rich
surge zone. Uncommon and wary.
Sleeps beneath sand during the
night. IP in groups, TP ♂'s
solitary and territorial. Feeds on
crustaceans, molluscs and polychaetes.
Range: Red Sea to Easter Is., n. to S.
Japan, s. to S. Africa and SE.
Australia.

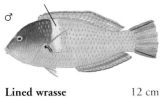

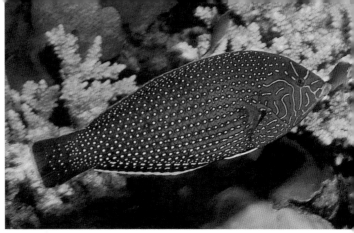

IP, Lahami, KF

Lined wrasse   12 cm

Anampses lineatus
Juv. black with white spots and dashes
and white tail base; TP ♂
with yellowish wash on sides.
Biology: inhabits steep lagoon
and seaward reef slopes, 15 to 45 m.
Solitary or in groups over coral and
rubble. Uncommon. Sleeps under
sand during the night. Feeds on
benthic invertebrates. **Range:** Red Sea
and S. Oman to Bali, s. to S. Africa.

juv.

IP, Mövenpick Bay, EL

Spotted wrasse   22 cm

Anampses meleagrides
Juv., IP black with white spots and
yellow tail. **Biology:** inhabits lagoons,
bays and seaward reefs, 1 to 60 m.
Common in areas of mixed coral,
rubble, rock and sand. Solitary or in
small groups. Feeds on benthic
invertebrates and small fishes.
Range: Red Sea and S. Oman to
Fr. Polynesia, n. to S. Japan, s. to
S. Africa and SE. Australia.

juv, ♀

♂, Similan Islands, RM

Yellowbreasted wrasse 18 cm
Anampses twistii
Olivaceous with yellow breast and blue spots; TP ♂ loses ocelli. **Biology:** inhabits lagoons, bays, and seaward reefs in areas of mixed coral, rubble and sand, 0.5 to 30 m. Very common on protected fringing reefs. Stays close to bottom. Feeds on benthic invertebrates. **Range:** Red Sea to Fr. Polynesia, n. to Ryukyus, s. to Mauritius and GBR.

juv.

Ras Mohammed, EL

Lyretail hogfish 21 cm
Bodianus anthioides
Orange-brown in front, white with black spots behind; tail lunate. **Biology:** inhabits deep lagoons, bays and seaward reefs, 6 to 60 m. Common, juveniles in coral-rich areas around gorgonians and black corals, adults usually in areas of mixed sand, coral and rubble. Solitary, feeds on benthic invertebrates. Often shadows goatfishes. **Range:** Red Sea to Fr. Polynesia, n. to S. Japan, s. to S. Africa and New Caledonia.

juv.

Dahab, EL

Axilspot hogfish 20 cm
Bodianus axillaris
Brown in front, yellowish-white below and behind; P axil black. **Biology:** inhabits bays, lagoons and seaward reefs, 3 to 40 m. Common, often around coral heads and patch reefs with caves and overhangs. Adults and juveniles sometimes act as cleaners. Solitary, swims close to bottom. **Range:** Red Sea and S. Oman to Marquesas and Pitcairn, n. to S. Japan, s. to S. Africa and SE. Australia.

juv.

Eritrea, AK

Diana's hogfish 25 cm
Bodianus diana
Brown with yellowish sides and 4 white spots along back. **Biology:** inhabits rocky and coral areas of deep lagoon and seaward reefs, 3 to 50 m. Common in coral-rich areas of steep slopes. Juveniles act as cleaners around gorgonians and black corals. **Range:** Red Sea and Oman to W. Indonesia, s. to S. Africa; Pacific population is a distinct undescribed species.

juv.

Abu Galum, RM

Giant hogfish 80 cm
Bodianus macrognathos
Juv., IP dark brown with broad white dorsal stripe and belly; TP ♂ olivaceous with broad pale bars. **Biology:** inhabits exposed rocky reefs with scattered algae and corals, 2 to 65 m. Solitary and wary, feeds on molluscs and corals. **Range:** Gulf of Oman to Kenya; not in Red Sea.

♂

♀*Salalah, KF*

Red-striped hogfish 18 cm
Bodianus opercularis
Elongate with alternating broad red and white stripes. **Biology:** inhabits steep seaward slopes and dropoffs with caves and black corals, 17 to 70 m. Solitary or in pairs. Seems to be territorial. Feeds on benthic invertebrates. Uncommon, but approachable. **Range:** Red Sea to Marshall Is., s. to Mauritius.

Abu Galum, RM

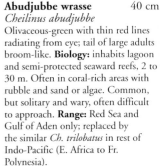

Abudjubbe wrasse   40 cm
Cheilinus abudjubbe

Olivaceous-green with thin red lines radiating from eye; tail of large adults broom-like. **Biology:** inhabits lagoon and semi-protected seaward reefs, 2 to 30 m. Often in coral-rich areas with rubble and sand or algae. Common, but solitary and wary, often difficult to approach. **Range:** Red Sea and Gulf of Aden only; replaced by the similar *Ch. trilobatus* in rest of Indo-Pacific (E. Africa to Fr. Polynesia).

♂, Mövenpick Bay, RM

Broomtail wrasse   50 cm
Cheilinus lunulatus

Green with small red spots anteriorly; TP dark blue posteriorly with ragged tail and yellow P fins. **Biology:** inhabits outer reef flats, and lagoon and seaward reefs, 0.5 to 30 m. Typically solitary along coral-rich reef margins with deep holes and sand or rubble patches. Feeds on bottom-dwelling invertebrates, primarily molluscs and crustaceans. Common, but wary, particularly large males. During afternoon high tides, males territorial along reef margins with harems of several females. Spawns during afternoon high tides. **Range:** Red Sea to Arabian Gulf only.

♂, Mangrove Bay, RM

Redbreasted wrasse   36 cm
Cheilinus quinquecinctus

Face green becoming red behind followed by broad pale black and white bars; rear margin of tail of large adults ragged. **Biology:** inhabits lagoon and seaward reefs, 4 to 40 m. Common in areas of mixed coral, rubble and sand. Feeds on benthic invertebrates. Often swims near diver's fins to prey on exposed invertebrates. Territorial. Easily approached. **Range:** Red Sea only; replaced in rest of Indo-Pacific by the very similar *C. fasciatus* which does not have ragged tail margin (E. Africa to Samoa, n. to Ryukyus, s. to Madagascar and New Caledonia).

Mangrove Bay, KF

Cigar wrasse   50 cm
Chelio inermis
Very elongate; colour variable from
green to brown to yellow. **Biology:**
inhabits seagrass beds and rocky and
coral reefs with heavy algal growth, 1
to 35 m. Feeds on molluscs, crabs,
urchins and shrimps. Often
accompanies goatfishes and other
wrasses to exploit exposed prey.
Common, but wary and difficult
to approach. **Range:** Red Sea to
Polynesia, n. to S. Japan and Hawaii,
s. to SE. Australia.

♂

♀, *Mangrove Bay, KF*

Purple-boned wrasse   14 cm
Cirrhilabrus blatteus
TP ♂ with red head and yellow
body with red mid-lateral stripe
and lanceolate tail; IP red.
Biology: inhabits rock and coral
bottoms of deep reef slopes, 40 to
70 m. Males territorial around a
harem of females. Feeds on
zooplankton in the water column.
Rarely seen. **Range:** Red Sea only.

♀

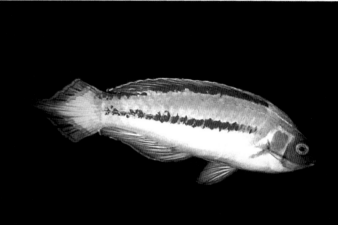
♂, *Saudi Arabia, JR*

Social wrasse   6.5 cm
Cirrhilabrus rubriventralis
TP ♂ red, white ventrally with
elongate red pelvic fins; IP red
with narrow bluish-white stripes.
Biology: in aggregations over seagrass
beds and exposed rubble bottoms,
3 to 43 m. Male maintains a harem.
Male displays by elevating the dorsal
fin, lowering the pelvic fins and
fluttering the tail. Feeds on
zooplankton. **Range:** Red Sea and
S. Oman to Sri Lanka and Sumatra.

♀

♂, *Jordan, JN*

♂, Hurghada, AK

Clown coris 1.0 m
Coris aygula
IP off-white to pale green with small dark spots on head; TP ♂ dark green with broom-like tail and bulbous forehead. **Biology:** inhabits exposed outer reef flats, lagoons and seaward reefs, 1 to 50 m. Common over mixed coral, rubble and sand. Feeds on hard-shelled invertebrates which are crushed by the pharyngeal teeth. Males patrol reef edges. **Range:** Red Sea to Fr. Polynesia, n. to S. Japan, s. to S. Africa and New Caledonia.

juv.

♂, Mövenpick Bay, RM

Spottail coris 20 cm
Coris caudimacula
Olivaceous, IP with brown stripes above; TP ♂ bluer with dark patch on elevated front of D fin. **Biology:** inhabits areas of sand and rubble near corals of lagoons, bays and protected seaward reefs, 2 to 25 m. Always close to bottom searching for hard-shelled invertebrates and accompanied by other species. Common. **Range:** Red Sea to Gulf of Oman, e. to Bali and NW. Australia, s. to S. Africa.

♂, Mangrove Bay, EL

African coris 38 cm
Coris cuvieri
Juv. red with black-edged white saddles; adult greenish-brown with green bands on head. **Biology:** inhabits exposed lagoon and seaward reefs, 0.5 to 50 m. Uncommon, usually over sand and rubble near corals, juvs. in shallow sandy pockets. Turns over rocks to feed on small hard-shelled invertebrates. Solitary and fast-swimming. **Range:** Red Sea and S. Oman to Bali, s. to S. Africa.

juv.

ueen coris   60 cm
oris formosa
iology: inhabits areas of mixed sand,
bble and coral of exposed reefs, 2 to
m. Uncommon. **Range:** S. Red Sea
d Gulf of Oman to Sri Lanka, s. to
Africa.

♀ ♀

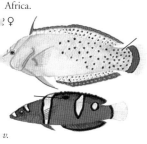

v.

♂, Eritrea, EL

apple coris   20 cm
oris variegata
le green with thin pale bars above,
ack specks on sides; juv. with
rge ocellus on mid-D fin.
iology: inhabits lagoons, bays
d sheltered seaward reefs, 1 to
) m. Solitary or in small loose
oups, common over sand or
bble bottoms near reefs. Feeds
small crustaceans and gastropods.
ange: Red Sea only. Replaced by
e very similar *C. batuensis* from
Africa to Tonga.

v.

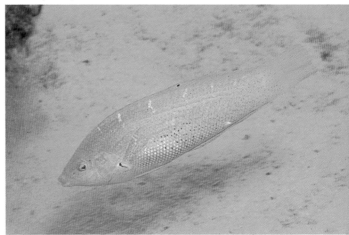

Shams Alam Reefs, EL

ingjaw wrasse   35 cm
pibulus insidiator
ws highly extensible; juv. with
in white bars; ♀ yellow or brown;
with white head, angular D and
fins. **Biology:** inhabits coral-rich
eas of lagoon and seaward reefs, 2
45 m. Uses tubular mouth to feed
small coral-dwelling shrimps, crabs
d fishes. Common. **Range:** Red Sea
Fr. Polynesia, n. to S. Japan and
W. Hawaiian Is., s. to S. Africa and
ew Caledonia.

yellow phase ♀

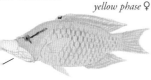

♂, Marsa Shagra, RM

Indian bird wrasse   28 cm
Gomphosus caeruleus
Tiny juv. pale with dusky stripe
through eye, shorter snout. **Biology:**
inhabits coral-rich areas of lagoons, bay:
and seaward reefs, 1 to 30 m. Females
in small groups, males solitary. Uses
snout to capture small invertebrates
from interstices of rocks or corals.
Common, but shy. **Range:** subsp.
klunzingeri in Red Sea only; subsp.
caeruleus from Oman to Andaman Sea,
s. to Mauritius and S. Africa.

♀

♂, Marsa Shagra, RM

Chequerboard wrasse   27 cm
Halichoeres hortulanus
Chequerboard pattern on sides;
♂ greener with yellow dorsal spot.
Biology: inhabits clear lagoon and
seaward reefs, 1 to 30 m. Common
over large expanses of sand and rubble
near reefs and seagrass beds. Juveniles
in bottoms of surge channels or over
sand or rubble under ledges. Feeds
on sand-dwelling invertebrates.
Male territorial over a large area.
Range: Red Sea and S. Oman to
Fr. Polynesia, n. to S. Japan, s. to
S. Africa and SE. Australia.

juv.

Dahab, EL

Rainbow wrasse   11 cm
Halichoeres iridis
Reddish-brown with yellow head and
white back; juv., ♀ with ocelli at mid
and rear D fin base. **Biology:** inhabits
sand and rubble near rocky and coral
reefs, 6 to 43 m, usually below 20 m.
Solitary, remains close to bottom. Shy.
Range: S. Red Sea to Gulf of Oman,
e. to Maldives, s. to Madagascar and
Mauritius.

Kenya, RM

Dusky wrasse 17 cm
Halichoeres marginatus
P brown with dark pinstripes, ocellus
n mid-D fin base. **Biology:** inhabits
oral-rich areas of clear lagoons, bays
nd seaward reefs, 1 to 30 m. Common
long upper edges of coral formations.
Haremic, male territorial in an area
vith several females. Juveniles
ecretive on reef margins close to
orals. **Range:** Red Sea to Polynesia,
. to S. Japan, s. to S. Africa and
GBR.

uv.

♂, Bali, RM

Nebulous wrasse 12 cm
Halichoeres nebulosus
Green to brown with complex
attern; ♀ with pink blotch on breast;
P ♂ with more pronounced red and
reen bands on head. **Biology:** inhabits
hallow exposed rocky and coral reefs,
.3 to 40 m. Common on weedy reef
lats and rocky shorelines near small
oral heads or dead coral rubble.
eeds on benthic invertebrates.
Range: Red Sea and S. Oman to
NG, n. to S. Japan, s. to S. Africa
nd SE. Australia. **Similar:** Flag
vrasse, *H. signifer* has broad pale
ateral stripe (S. Red Sea to Oman;
0 cm).

Bali, RM

Zigzag wrasse 20 cm
Halichoeres scapularis
ale with dark zigzag dorsal stripe;
P ♂ greener with pink spots on
ead. **Biology:** inhabits sandy areas
f lagoon and sheltered seaward reefs,
.5 to 15 m. Common. Often
ccompanies goatfishes or emperors to
aid the invertebrates exposed by their
igging. Male maintains a loose
arem of females. **Range:** Red Sea to
Vanuatu, n. to S. Japan, s. to S. Africa
nd GBR.

uv., IP

Nabaa Bay, EL

Mangrove Bay, RM

Red Sea thicklip wrasse   50 cm
Hemigymnus sexfasciatus
Wedge-shaped black bars on sides.
Biology: inhabits areas of mixed
sand, rubble and corals of lagoon
and sheltered seaward reefs, 1 to
20 m. Uncommon, usually
solitary, occasionally in small groups.
Extracts small invertebrates such
as foraminifera, crustaceans and
molluscs from sand. Juveniles among
corals or urchin spines. **Range:** Red
Sea only; replaced by the similar
H. fasciatus from E. Africa to SE.
Polynesia.

juv.

Half-and-half thicklip   50 cm
Hemigymnus melapterus
Front pale, abruptly dark behind.
Biology: inhabits areas of mixed sand,
rubble and corals of lagoon and
sheltered seaward reefs, 1 to 30 m.
Juveniles among coral or rubble.
Feeds primarily on hard-shelled
benthic invertebrates. Uncommon
and somewhat wary. **Range:** Red Sea
to Society Is., n. to S. Japan, s. to
S. Africa and SE. Australia.

Maldives, RM

juv.

Ring wrasse   40 cm
Hologymnosus annulatus
Elongate; juv. black with broad cream
stripe above and below; IP brown
with dark bars; TP ♂ dark green with
bluish bars. **Biology:** inhabits areas of
mixed sand, rubble and corals of clear
lagoons, bays and seaward reefs, 1 to
25 m. Uncommon. Swims well above
bottom. Male patrols large territory.
Juveniles near bottom, may mimic
juv. *Malacanthus latovittatus* (p. 92).
Range: Red Sea to Pitcairn Is., n. to
S. Japan, s. to Mozambique and SE.
Australia.

♂, Shams Alam, EL

juv.

Longface wrasse 38 cm
Hologymnosus doliatus
Elongate; IP pale with thin green bars.
Biology: inhabits clear lagoons and
seaward reefs, 3 to 30 m. Occurs in
areas of sand and rubble, particularly
sandy terraces. Juveniles in closely knit
small groups. Male patrols large
territory containing harem of females.
Feeds primarily on fishes and
crustaceans. **Range:** S. Red Sea and S.
Oman to Line Is. and Samoa, n. to S.
Japan, s. to S. Africa and SE. Australia.

juv.

Indonesia, RM

Blue razorfish 41 cm
Iniistius pavo
1st 2 D spines forming elongate
detached finlet that shortens with
age; rarely all black. **Biology:** inhabits
large expanses of open sand of clear
lagoon and seaward reefs, juveniles
from 2 to 100 m, adults below 20 m.
Feeds on benthic invertebrates. Dives
and moves beneath the sand when
frightened. Bites when handled.
Range: Red Sea and Gulf of Aden
to Panama, n. to S. Japan and Hawaii,
s. to S. Africa and SE. Australia.

juv.

Jordan, EL

Bicolor cleaner wrasse 14 cm
Labroides bicolor
Biology: inhabits clear lagoon and
seaward reefs, 2 to 40 m. Feeds by
picking parasites off other fishes.
Juveniles establish stations at fixed
sites at the base of overhanging corals.
Adults wander over a large area of reef
in search of clients. **Range:** Gulf of
Aden and S. Oman to Fr. Polynesia,
n. to S. Japan, s. to S. Africa and
SE. Australia.

juv.

Indonesia, RM

Mangrove Bay, EL

♂, Marsa Shagra, RM

♂, Ras Mohammed, RM

Bluestreak cleaner wrasse 11 cm
Labroides dimidiatus
Biology: in all coral reef zones from reef flats to lagoon and seaward reefs, 1 to 40 m. Common. Lives exclusively on parasites, mucus and injured tissue cleaned from other fishes; generally immune from predators. Establishes cleaning stations around prominent coral formations. Advertises its trade with a distinctive up-and-down dance. Haremic, adults often in pairs. Juveniles in crevices or under ledges. Mimicked by the aggressive fin-nipping blenny *Aspidontus taeniatus* (p. 182). **Range:** Red Sea and Arabian Gulf to SE. Polynesia, n. to S. Japan, s. to S. Africa and SE. Australia.

juv.

Fourline wrasse 11 cm
Larabicus quadrilineatus
Biology: inhabits coral-rich areas of lagoons, bays and seaward reefs, 0.5 to 22 m. Common. Juveniles and subadults are cleaners that occur in small groups around small coral heads, but without a fixed station. Adults are territorial and feed on coral polyps. **Range:** Red Sea and Gulf of Aden only.

juv. IP

Vermiculate wrasse 13 cm
Macropharyngodon bipartitus
Juv., IP brown with white spots, blackish on belly. **Biology:** inhabits areas of mixed coral, rubble and sand of lagoon and seaward reefs below effects of surge, 2 to 25 m. Stays close to bottom. Feeds on small benthic invertebrates. Male territorial in large area containing harem of females. **Range:** subsp. *marisrubri*: Red Sea only; subsp. *bipartitus*: Gulf of Aden (?) and Oman to Maldives, s. to S. Africa.

IP

Minute wrasse   6 cm
Minilabrus striatus
IP pale with dusky stripes; TP ♂ blue
and yellow with red stripes and mid-
lateral black spot. **Biology:** inhabits
upper reef slopes, 1 to 14 m.
Aggregates above mixed sand and
rubble zones of reef crests to feed on
zooplankton. Uncommon and shy.
Range: C. and S. Red Sea to Gulf of
Aden only.

♂, S. Egypt, DE

Rockmover wrasse   32 cm
Novaculichthys taeniourus
IP with dark lines radiating from
eye. **Biology:** inhabits semi-exposed
lagoon and seaward reefs, 2 to 45 m.
Prefers areas of mixed rubble and
sand patches, often with mild surge.
Turns over rubble to prey on exposed
invertebrates. Juveniles mimic clump
of detached drifting algae. Usually
solitary, adults sometimes paired and
wary. Bites if handled. **Range:** Red Sea
and S. Oman to Panama, n. to Ryukyu
and Hawaiian Is., s. to S. Africa, SE.
Australia and SE. Polynesia.

Mangrove Bay, RM

juv.

Bandcheek wrasse   35 cm
Oxycheilinus digrammus
Variable: greenish with red belly or
broad pale stripes; red streaks on
head and throat. **Biology:** inhabits
coral-rich areas of lagoons, bays and
seaward reefs, 2 to 60 m. Common,
usually solitary and often curious
towards divers. A voracious predator
of small fishes. Often swims well
above the bottom. **Range:** Red Sea to
Samoa, n. to Ryukyus, s. to S. Africa
and New Caledonia; absent from
Gulf of Aden to Arabian Gulf.

Mangrove Bay, EL

Jordan, EL

Mental wrasse 24 cm
Oxycheilinus mentalis
Greenish with fine white spots
and lines, brown lateral stripe,
white blotch on caudal peduncle.
Biology: inhabits coral-rich areas of
lagoons, bays and seaward reefs, 1 to
25 m. Adults common around small
coral heads, subadults near gorgonians
and soft corals. Solitary. Curious and
slow, easily approached. **Range:** Red
Sea to Maldives, s. to Madagascar;
absent from Oman to Arabian Gulf.
Similar: *O. arenatus* inhabits caves of
steep dropoffs (Red Sea to Samoa;
19 cm). ▼

Red Sea flasher wrasse 9 cm
Paracheilinus octotaenia
Orange-red; IP with 4–5 dark stripes;
TP ♂ with red D, A and tail fins and
8 dark stripes becoming blue on head.
Biology: inhabits outer lagoon and
seaward reef slopes and patch reefs,
5 to 30 m. Common, usually in
loose aggregations of a few males
and their harems of 6 to 12 females.
Feeds well above coral formations on
zooplankton. Male erects bright red
fins in conspicuous display.
Range: Red Sea only.

♂, Mangrove Bay, RM

Striated wrasse 8 cm
Pseudocheilinus evanidus
Pale dirty red with fine faint
pinstripes, silver streak on cheek.
Biology: inhabits deep lagoon and
seaward reefs, 5 to 45 m. Common
in patches of rubble and small
branching corals. Secretive, always
near the bottom and often overlooked.
Solitary and shy. Feeds on small
benthic invertebrates. **Range:** Red
Sea and Gulf of Aden to Polynesia,
n. to S. Japan and Hawaii, s. to
S. Africa and GBR.

Marsa Shagra, RM

Sixline wrasse 8 cm
Pseudocheilinus hexataenia
Purplish with 6 orange stripes above belly. **Biology:** inhabits coral-rich areas of lagoon and seaward reefs, 2 to 35 m. Common, but secretive among short-branched corals or near large anemones. Leaves shelter only briefly to move from coral to coral. Feeds on small crustaceans in rubble or corals. **Range:** Red Sea and Oman to Fr. Polynesia, n. to Ryukyus, s. to S. Africa and SE. Australia.

Sulawesi, RM

Chiseltooth wrasse 25 cm
Pseudodax moluccanus
Strong chisel-like teeth; juv. black with 2 thin blue lines forming wedge. **Biology:** in coral-rich areas of clear lagoon and seaward reefs with caves and crevices, 2 to 40 m. A fast swimmer. Feeds on encrusting invertebrates. Juveniles secretive, usually on steep dropoffs close to shelter. Sometimes observed as cleaners. **Range:** Red Sea and Gulf of Aden to Fr. Polynesia, n. to S. Japan, s. to S. Africa and GBR.

juv.

♂, Marsa Shagra, KF

Cryptic wrasse 9.5 cm
Pteragogus cryptus
Mottled olive to reddish with ocellus on opercle. **Biology:** inhabits seagrass beds and algal flats of shallow sheltered fringing reefs and lagoons, 0.5 to 20 m. Common, but secretive among seagrasses and algae. Appears in open only for brief intervals. Usually solitary or in pairs. **Range:** Red Sea to Samoa, s. to New Caledonia. **Similar:** *P. pelycus* (Red Sea to Mauritius, 15 cm) has rounder, more horizontally oriented ocellus on opercle; *P. flagellifer* is green to brown and larger, to 20 cm (Red Sea and Arabian Gulf to W. Pacific).

Hurghada, DE

♂, Sinai, AK

Bluelined wrasse   12 cm
Stethojulis albovittata
P axil orange. **Biology:** inhabits reef
flats and lagoon and seaward reefs to
21 m. Common in shallow areas of
mixed sand, rubble and corals or
seagrasses. Males territorial with harem
of females. Swims rapidly, stopping to
pick invertebrates from bottom,
expelling sand from the gill openings.
Range: Red Sea and S. Oman to Java,
Thailand, s. to S. Africa.

           ♀, *II*

♂, Bali, RM

Cutribbon wrasse   13 cm
Stethojulis interrupta
TP ♂ green to brown with 2
blue stripes on head, and separate
dorsal stripe extending to tail.
Biology: inhabits areas of mixed
rubble, sand, and corals of lagoon
and coastal reefs, 1 to 31 m, usually
shallow, sometimes in tidepools. Male
territorial in large area containing
several females. A rapid swimmer.
Range: Red Sea and Arabian Gulf
to Solomon Is., n. to Philippines,
s. to S. Africa and SE. Australia.

           ♀, *IF*

♂, Bali, RM

Crescent wrasse   27 cm
Thalassoma lunare
Tail lunate, the central rays yellow; IP
green with red bands on head; TP ♂
blue-green with purple head and fins.
Biology: inhabits lagoon and seaward
rocky and coral reefs, 0.5 to 20 m.
Abundant on upper slopes. Constantly
on the move, but curious and
approachable. Feeds on benthic invertebrates, sometimes on small fishes.
Often around diver's fins to hunt for
exposed invertebrates. Male territorial
with harem. **Range:** Red Sea and
Arabian Gulf to Line Is., n. to S. Japan,
s. to S. Africa and N. New Zealand.

juv.

Sunset wrasse 25 cm
Thalassoma lutescens
Juvenile brown above, white below; IP
yellow. **Biology:** inhabits clear outer
lagoon and seaward reefs to 30 m.
Over open sand and rubble as well
as coral, common in exposed areas.
Feeds on small benthic invertebrates.
Range: Gulf of Aden to Hawaiian
Islands and Fr. Polynesia, n. to
Ryukyus, s. to E. Africa and SE.
Australia. **Similar:** Sixbar wrasse
T. hardwicke has 6 dark bars along
back and broad pink lines radiating
from eyes (Gulf of Aden to Fr.
Polynesia; 21 cm). ▼

♂, Guam, RM

Surge wrasse 43 cm
Thalassoma purpureum
Biology: occurs almost exclusively in
surge zone of reefs and exposed rocky
shores, very rarely to 20 m. Always on
the move, seeks shelter in extremely
shallow turbulent water. Very shy and
difficult to photograph. Feeds on
benthic invertebrates. Male territorial
with a harem, IP in small groups.
Range: Red Sea to Marquesas and
Easter Is., n. to S. Japan and Hawaii,
s. to S. Africa and SE. Australia.

♂, Hawaii, RM

♀, IP

Klunzinger's wrasse 20 cm
Thalassoma rueppellii
Biology: inhabits outer reef flats,
patch reefs and coral-rich lagoon
and seaward reefs, 0.5 to 20 m.
Very common along upper reef slopes.
An active swimmer, often inquisitive
towards divers. Large male territorial
and haremic. Sleeps in corals rather
than under sand. Feeds mainly on
benthic invertebrates and small fishes.
Range: Red Sea only.

♂, Mövenpick Bay, RM

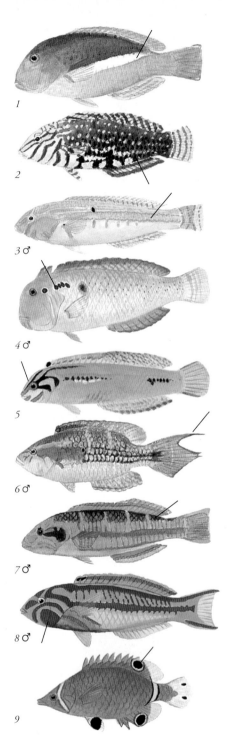

1. Robust tuskfish   35 cm
Choerodon robustus
Biology: inhabits areas of sand and rubble of coastal rocky and coral reefs to 250 m. **Range:** Red Sea and Arabian Gulf to Indonesia, s. to Mozambique and Mauritius.

2. Jewelled wrasse   14 cm
Halichoeres lapillus
Resembles *Macropharyngodon* spp. **Biology:** inhabits rocky weedy areas, 5 to 15 m. **Range:** S. Red Sea and Oman to S. Africa and Mauritius.

3. Goldstripe wrasse   20 cm
Halichoeres zeylonicus
Biology: inhabits sandy areas near isolated reefs, usually below 20 m. **Range:** Red Sea and S. Oman to Bali, s. to S. Africa.

4. Fivefinger razorfish   25 cm
Iniistius pentadactylus
Juv., IP have pale belly with pink scale margins and lack red spots behind eye. **Biology:** inhabits coastal sand slopes, 2 to 18 m. Feeds on sand-dwelling benthic invertebrates. Dives beneath the sand when frightened. **Range:** Red Sea to Line Is., n. to S. Japan, s. to S. Africa and GBR. **Similar:** Baldwin's razorfish *I. baldwini* is pale with a large black blotch beneath the front third of the dorsal fin (Red Sea to Hawaii).

5. Seagrass razorfish   15 cm
Novaculoides macrolepidotus
Biology: inhabits shallow seagrass beds and protected sand flats with abundant algae, to 4 m. **Range:** Red Sea to New Caledonia, n. to Ryukyus, s. to S. Africa and SE. Australia.

6. Twospot wrasse   15 cm
Oxycheilinus bimaculatus
Biology: inhabits clear lagoon and seaward reefs, 2 to 110 m. Usually in weedy areas of sand and rubble. Male highly territorial with harem of females. **Range:** E Africa and Gulf of Oman, S Oman to Hawaii and Fr. Polynesia, n. to S. Japan, s. to S. Africa and SE. Australia.

7. Slantband wrasse   18 cm
Thalassoma loxum
Biology: inhabits coastal rocky shores and reefs. The most abundant wrasse in S. Oman. **Range:** Oman only, abundant in south, rare in Gulf of Oman.

8. Fivestripe wrasse   16 cm
Thalassoma quinquevittatum
Red bands turn blue in nuptial ♂. **Biology:** inhabits exposed outer lagoon and seaward reefs, shoreline to 40 m, usually above 15 m. Feeds on small benthic invertebrates and fishes. **Range:** Gulf of Aden and Oman to Hawaiian Islands (rare), n. to Ryukyus, s. to S. Africa and SE. Australia.

9. Blackspot pygmy wrasse   8 cm
Wetmorella nigropinnata
Biology: inhabits lagoon and seaward reefs, 1 to 30 m. Very secretive, in caves and crevices. **Range:** Red Sea to Marquesas and Pitcairn, n. to Ryukyus, s. to New Caledonia.

▲♂ **Bicolor parrotfish**   80 cm
Cetoscarus bicolor
Head elongate; colour distinct at all
sizes, juv. white with orange band
on head, ocellus on front of D fin.
Biology: inhabits deep lagoon and
seaward reefs, 1 to 30 m. Common
along upper reef slopes. Male
territorial and haremic. Juveniles
solitary, in sandy holes among corals.
Range: Red Sea to Fr. Polynesia,
n. to S. Japan, s. to Mauritius and
New Caledonia. *(Photo: Dahab, AK.)*

♀

juv.

PARROTFISHES   SCARIDAE

Parrotfishes are among the most recently evolved reef fishes. They closely resemble wrasses but differ by having teeth fused into beak-like plates,
and pavement-like plates of pharyngeal teeth in the throat. All have large
cycloid scales and a continuous dorsal fin. They are herbivores and most
scrape filamentous algae from dead coral rock. A few feed from the surface of coarse sand, or eat leafy algae, seagrasses, or living corals. Bits of
coral rock mixed with algae are crushed by the mill-like pharyngeal
plates. This aids digestion by breaking down cell walls, and makes parrotfishes the biggest producers of sand on coral reefs. Many species of
parrotfishes school together, often with surgeonfishes, and may travel
long distances between feeding and sleeping grounds. They sleep at
night, wedged into holes and crevices. Some secrete a mucous cocoon
which may inhibit the sense of smell of predators. Most parrotfishes
change colour with growth and sex. Most mature into a drab grey to
brown **initial phase** (IP) which may be either female or male in some
species (diandric), or only female in others (monandric). Some females
change sex into a brilliant green to blue **terminal phase** (TP) male. Juveniles and initial phase fish of some species are difficult to distinguish
from one another, but the terminal phase is generally more distinct. Parrotfishes are important food fishes but have soft flesh and don't keep well.

Sipadan Island, KF

Bumphead parrotfish   1.3 m
Bolbometapon muricatum
Steep head profile with large hump on forehead; young with more normal head and white spots. **Biology:** in closely knit groups on coral-rich outer reef flats and deep lagoon and seaward reef slopes, 1 to 50 m. Feeds on living corals, producing an audible crunching sound. Uses forehead to break corals. Sleeps in groups in crevices. Wary and rare in most areas. The largest species of parrotfish, reaches at least 70 kg, possibly 115 kg. **Range:** Red Sea and E. Africa to Fr. Polynesia, n. to Taiwan, s. to Madagascar and New Caledonia.

♂, Dahab, EL

Dotted parrotfish   27 cm
Calotomus viridescens
Teeth distinct but fused black dots on sides; IP brown; TP ♂ green with red spots on face. **Biology:** inhabits reef flats and semi-protected lagoon and seaward rocky and coral reefs, 1 to 35 m. Common in areas of seagrasses or algae with coral heads. Feeds on seagrasses and leafy and epiphytic algae. Easy to approach when feeding. Male solitary, female in groups. **Range:** Red Sea only; replaced elsewhere by similar *C. carolinus* (Gulf of Aden to W. Pacific, 50 cm) which lacks black spots.

♂, Jordan, EL

Seagrass parrotfish   35 cm
Leptoscarus vaigiensis
Slender; grey to green, TP ♂ with blue spots and mid-lateral white stripe. **Biology:** common in seagrass and algae beds, 0.3 to 15 m. Rare on coral reefs. Shy and secretive, adept at swimming beneath the canopy of seagrasses. Feeds on seagrasses. The only parrotfish species known to not undergo sex reversal. Spawns in pairs or in groups during falling tides. Shy. **Range:** Red Sea and S. Oman to Easter Is., n. to Ryukyus, s. to S. Africa and N. New Zealand.

Red Sea steephead parrotfish
Chlorurus gibbus   70 cm
Biology: inhabits coral-rich areas of deep lagoon and seaward reefs, 1 to 35 m. Common along upper reef slopes. TP ♂ with large territory, often swims well above the bottom, IP in groups. Wary. Sleeps in mucous cocoon at night. **Range:** Red Sea only; replaced by very similar *C. strongylocephalus* from Oman to Java and *C. microrhinos* in Pacific.

IP

♂, Dahab, EL

Bullethead parrotfish   40 cm
Chlorurus sordidus
Head smoothly rounded; small juvs. striped; IP variable with white spots or pale tail base with dark spot. **Biology:** inhabits reef flats and lagoon and seaward reefs, 1 to 25 m. The most common parrotfish of outer reef flats and upper reef slopes. Juveniles on coral rubble and in seagrasses. May migrate long distances from feeding to sleeping grounds. Not shy. **Range:** Red Sea to Fr. Polynesia, n. to Ryukyu and Hawaiian Is., s. to Mozambique and SE. Australia.

IP

♂, Marsa Bareka, RM

Purplestreak parrotfish   30 cm
Chlorurus genozonatus
Head smoothly rounded; IP reddish-brown, TP ♂ with large purple patch on lower cheek. **Biology:** inhabits deep lagoons, bays and seaward reef slopes with rich coral growth, 5 to 25 m. Common, but often overlooked. TP with large territory on reef slopes. IP in groups, often over mixed sand, rubble and corals. Fast and wary. **Range:** Red Sea and Gulf of Aden only. **Similar:** very similar to *C. sordidus*.

IP

♂, Mövenpick Bay, EL

Indian longnose parrotfish
Hipposcarus harid   75 cm
Head pointed, pale; TP ♂ pale
blue-green with elongate tail lobes.
Biology: inhabits deep lagoons, bays
and semi-protected seaward reefs,
1 to at least 30 m. Common over sand
and rubble near coral reefs, where they
feed on filamentous algae. In mobile
groups of one TP and several IP fish.
Occasionally in large daily migrations.
Juvenile secretive in rubble and
branching *Acropora* coral thickets.
Range: Red Sea to Java, s. to
Mozambique.

♂, *Mövenpick Bay, RM*

Rusty parrotfish   40 cm
Scarus ferrugineus
IP brown with yellow tail; TP ♂
yellowish-green with distinct green
marks on snout. **Biology:** inhabits
coral-rich areas of lagoon and
seaward reefs, 1 to 60 m. Very
common on large patch reefs and
well-developed coral slopes with
small sandy stretches. TP ♂ territorial
and haremic. Easily approached.
Range: Red Sea to Arabian Gulf only.

IP

♂, *Mangrove Bay, RM*

Bridled parrotfish   47 cm
Scarus frenatus
IP tan with dusky stripes and reddish
fins; TP ♂ dark green with red
vermiculations becoming abruptly
bright green posteriorly; juv. light
blue posteriorly. **Biology:** inhabits
clear lagoon and seaward reef margins
and crests, 0.3 to 25 m. Common.
Feeds on outer reef flat during high
tide. Usually solitary, TP ♂ sometimes
in groups. **Range:** Red Sea and Oman
to Fr. Polynesia, n. to S. Japan, s. to
Mauritius and SE. Australia.

IP

♂, *Maldives, RM*

Purple-brown parrotfish   38 cm
Scarus fuscopurpureus
IP with broad dark chevrons;
TP ♂ usually with a pale bar. **Biology:**
inhabits bays, channels and seaward
reefs, 2 to 20 m. Usually on sandy areas
near patch reefs or coral slopes. Solitary
or in small groups, often with *H. harid*.
TP ♂ territorial and haremic. Can
change colour rapidly. **Range:** Red Sea
to Arabian Gulf only; replaced by *S.
russelli* in Indian Ocean.

IP

♂, Shams Alam Reefs, EL

Bluebarred parrotfish   75 cm
Scarus ghobban
IP tan with blue scales in broad
bars; TP ♂ blue-green with pink
scale margins and fin bands.
Biology: inhabits rocky and coral
reefs with stretches of sand and
rubble, 1 to 40 m. Uncommon,
often in turbid areas. Solitary or
in small groups. Juveniles in groups
in algae beds. **Range:** Red Sea and
Arabian Gulf to Panama, n. to
S. Japan, s. to S. Africa, Australia.

IP

♂, Bali, RM

Dusky parrotfish   40 cm
Scarus niger
IP reddish-brown with dark stripes
on sides; TP ♂ dark greenish-blue
with green stripe through eye.
Biology: common on coral-rich
slopes of clear lagoons and bays,
and seaward reefs, 0.5 to 20 m. Usually
solitary except during courtship when
TP ♂ swims rapidly with tail bent
upward and the A fin extended down.
IP in small haremic groups. **Range:**
Red Sea to Fr. Polynesia, n. to Ryukyus,
s. to GBR.

IP

♂, Mövenpick Bay, EL

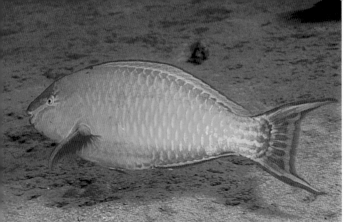

Greenband parrotfish   33 cm
Scarus collana
IP pale brown to greenish; TP ♂
pale green with 3 green stripes on
tail base and pink bands on fins.
Biology: inhabits areas of open sand
or rubble in lagoons, bays and on
sheltered seaward reefs, 1 to 25 m.
Common, in small slow-swimming
groups near coral heads and patch
reefs. Browses on filamentous algae
growing on sand. **Range:** Red Sea
only.

♂, Nabq, EL

Palenose parrotfish   33 cm
Scarus psittacus
TP ♂ blue-green with purplish cap on
forehead, pink bands on blue fins.
Biology: inhabits reef flats, and
lagoon and seaward reefs, 1 to 25 m.
Common on shallow flats and open
hard bottoms with scattered coral
heads. TP ♂ usually solitary, IP in
mixed-species schools. Secretes a
mucous cocoon at night. **Range:** Red
Sea and Arabian Gulf to Fr. Polynesia,
n. to S. Japan and Hawaii, s. to S.
Africa.

IP

♂, Shams Alam, EL

Redlip parrotfish   75 cm
Scarus rubroviolaceus
IP light brown to reddish with
distinct reticulations; TP ♂
yellowish-green with angular snout,
red lines above and below mouth;
sometimes bicoloured, abruptly
darker in front. **Biology:** inhabits
seaward rocky and coral reefs, 1 to
30 m. Solitary or in large groups.
Range: S. Red Sea to Panama, n. to
Ryukyus and Hawaii, s. to S. Africa
and SE. Australia.

IP

♂, Bali, RM

Greenbelly parrotfish 60 cm
Scarus falcipinnis
Juv. mottled black and pale yellow;
IP rosy brown with small white spots
posteriorly; TP ♂ with broad green
band from snout and continuing
along lower sides. **Biology:** inhabits
coastal reefs and steep seaward reef
slopes, 6 to 20 m. **Range:** S. Oman
to S. Mozambique and Mauritius,
e. to Chagos Is.

IP

♂, Kenya, RM

Dusky-capped parrotfish 37 cm
Scarus scaber
Biology: occurs on reef flats and
shallow lagoon reefs with rich coral
growth, 1 to 20 m. **Range:** Gulf of
Aden and S. Oman to Sumatra, s. to
S. Africa and Mauritius.

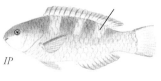

IP

Dhofar parrotfish 50 cm
Scarus zufar
IP resembles IP of *S. fuscopurpureus*.
Biology: common along exposed
rocky coasts. **Range:** Dhofar coast of
S. Oman only. **Similar:** Arabian
parrotfish *Scarus arabicus* inhabits
exposed rocky shores and coral-rich
areas of coastal reefs (Mukallah to Gulf
of Oman only; 45 cm). ▼

♂, Maldives, GG

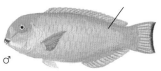

♂

Similar: Persian parrotfish *Scarus
persicus* is uncommon in S. Oman (S.
Oman to Arabian Gulf only; 50 cm). ▼

♂

♂, Oman, JPH

♂, ♀, Nuweiba, PM

SANDPERCHES
PINGUIPEDIDAE

Speckled sandperch　　28 cm
Parapercis hexophthalma
Large black spot on tail; ♀ with black spots on head. **Biology:** inhabits lagoon and semi-sheltered seaward reefs, 2 to 22 m. Common on sand at the base of reefs. Hides and sleeps in depression under piece of rubble. Male territorial with harem of 2 to 5 females. Courtship and spawning occurs before sunset. Feeds on benthic invertebrates. **Range:** Red Sea to Fiji, n. to S. Japan, s. to S. Africa and SE. Australia.

♂, Ras Mohammed, MB

SAND-DIVERS
TRICHONOTIDAE

Red Sea sand-diver　　12 cm
Trichonotus nikei
Extremely elongate with pointed snout; first 2 D spines of ♂ filamentous. **Biology:** uncommon on protected sandy slopes, 2 to 90 m. Usually in groups that hover 1 to 3 m above sand. Feeds on zooplankton. Dives into sand when alarmed and to sleep. A protogynous hermaphrodite. Male territorial and haremic, spends much of his time displaying to females by spreading D and V fins. **Range:** Red Sea only. **Similar:** *T. arabicus* which lacks filamentous D spines (Oman to Arabian Gulf, 14 cm).

Mövenpick Bay, KF

STARGAZERS
URANOSCOPIDAE

Whitemargin stargazer　　38 cm
Uranoscopus cf *sulphureus*
Huge head, body tapers towards tail; black spot on D fin. **Biology:** inhabits sand or mud bottoms of lagoon and seaward reefs, 5 to 150 m. Typically buried in sand with eyes and lips exposed. An ambushing predator of small fishes attracted by worm-like tentacle wriggled enticingly from lower jaw. **Range:** Red Sea to Samoa, n. to Marianas, s. to GBR and Tonga. **Similar:** *U. dahlakensis* (s. Red Sea) and *U. dollfusi* (Red Sea to Arabian Gulf) and other species in deeper water differ in details of colour and in number of scale rows.

TRIPLEFINS
TRIPTERYGIIDAE

Triplefins are tiny elongate fishes with 3 dorsal fins, a large head and small scales. They are highly cryptic and feed on tiny benthic invertebrates. Males use their first dorsal fin to signal to females during courtship.

Red triplefin   5.6 cm
Helcogramma steinitzi
♂ red with black head below eye.
Biology: on corals and rocks of exposed reef margins, 0.3 to 10 m.
Range: Red Sea to Oman and Arabian Gulf only.

Marsa Shagra, BE

Red Sea triplefin   2 cm
Enneapterygius sp. (undescribed)
1st D fin of ♂ a distinctive dark red and yellow. **Biology:** inhabits corals and rocks of exposed reef margins to at least 12 m. **Range:** Red Sea only.
(Photo: Nuweiba, MH.) ▶
Similar: Scaley-head triplefin *Norfolkia brachylepis* gets much larger and inhabits corals and rocks of exposed reef margins to 7 m (Red Sea and Oman to Fiji, s. to New Caledonia; 7.3 cm). ▼

Nuweiba, MH

BLENNIES

BLENNIIDAE

Blennies are a large of group of small elongate blunt-headed, scaleless fishes with a long continuous dorsal fin, anal fin with 2 spines (first may be imbedded in females), and pelvic fins with 1 spine and 2 to 4 rays. Most are bottom-dwelling territorial fishes that lay adhesive eggs often guarded by the male. They are subdivided into five tribes distinguished primarily by dentition and diet. Two are important on coral-reefs: the **Fangblennies** (tribe Nemophini) which are small-mouthed, large-fanged carnivores and the **Combtooth blennies** (tribe Salariini) which are feeble-toothed, wide-mouthed, blunt-headed herbivores. Fangblennies have greatly reduced gill openings and small mouths with stout teeth. Most species are active swimmers with greatly enlarged canine teeth in the lower jaw. Some are specialized predators that attack larger fishes to feed on fins, scales, skin or mucus. Some mimic cleaner wrasses or other fangblennies. The lower jaw fangs of most fangblennies are used for defence and those of the genus *Meiacanthus* are venomous (p. 183). Combtooth blennies have a wide downturned mouth with a single row of close-set incisiform teeth in each jaw used for scraping filamentous algae from the surfaces of rock or dead coral. Many species have tentacles or cirri on the head, usually above the eye; some have a fleshy crest on top of the head. Most species are well-camouflaged, often similarly patterned to one another, and cryptic inhabitants of rocky shorelines, reef flats, or shallow seaward reefs. Some species of *Ecsenius* mimic certain venomous fangblennies.

◄ Fringed blenny 5.5 cm
Mimoblennius cirrosus
Biology: inhabits coral-rich areas of semi-protected lagoon and seaward reefs, 3 to 25 m. Uncommon, usually seen peering out of a small hole in live or dead corals, rarely in the open. Shy. **Range:** Red Sea to Arabian Gulf, s. to Mozambique. *(Photo: Sinai, MH.)*

Red Sea mimic blenny ▶
Ecsenius gravieri 8 cm
Biology: inhabits sheltered lagoon and seaward reefs, 5 to 20 m. Often rests on dead or live coral branches. Mimics the venomous fangblenny *M. nigrolineatus* (p. 183) in colour and behaviour. Common. **Range:** Red Sea to Gulf of Aden only.

Mangrove Bay, RM

Smoothfin blenny 8 cm
Ecsenius frontalis
3 colour morphs. **Biology:** inhabits coral and rocky reefs, 3 to 27 m. Usually rests on mixed corals, rubble, or rocks. Uncommon in N. Red Sea, very common in Yemen. **Range:** Red Sea to Gulf of Aden only.

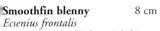

striped nigrovittatus *form*

Ras Mohammed, EL

Aron's blenny 5.5 cm
Ecsenius aroni
Yellowish-brown with a black spot on caudal peduncle. **Biology:** inhabits coral-rich areas of lagoon and seaward reefs below the influence of surge, 2 to 35 m. Uncommon, in areas with small patches of dead coral rock. Shelters in empty tube or hole when disturbed. Sometimes in pairs. **Range:** Red Sea only. **Similar:** *E. pulcher* (S. Oman to NW. India; 11 cm). ▼

Mangrove Bay, RM

Marsa Shagra, RM

Dentex blenny 6 cm
Ecsenius dentex
Pale brown with white spots, black spot behind eye. **Biology:** inhabits protected reefs, 1 to 15 m. Rests fully exposed on dead or live corals. Retreats to shelter of a small hole when disturbed. Common and approachable. **Range:** Gulf of Aqaba only. **Similar:** replaced by the very similar *E. nalolo* from C. Red Sea to East Africa, which often shows peppering of dark spots on cheeks (Indian Ocean population).

Kenya, RM

Midas blenny 13 cm
Ecsenius midas
Has two colour morphs, yellow and bluish-grey, but always with a bluish eye. **Biology:** inhabits coral-rich lagoons and seaward reefs, 2 to 35 m. Yellow phase is a social mimic of Lyretail anthias (p. 75). It gets protection among the large numbers of similar-looking fishes in dense aggregations. Retreats to a small hole when disturbed. Common in some areas but often unnoticed. **Range:** Red Sea to Polynesia, n. to Philippines, s. to S. Africa and GBR.

♀ *Nabq, PM*

Chestnut blenny 12 cm
Cirripectes castaneus
Colour variable, usually dark brown, ♀ with pale polygonal spots, ♂ with red lines on head, pale bars on sides and yellowish tail. **Biology:** inhabits exposed outer reef flats and reef margins and crests, 0.3 to 7 m. Feeds mainly on benthic algae, often in the territories of herbivorous damselfishes. Exposed for brief periods only. Shy, rapidly retreats to shelter when approached. **Range:** Red Sea and Gulf of Aden to Tonga, n. to S. Japan, s. to S. Africa and SE. Australia.

Leopard blenny   14 cm
Exallias brevis
Pale with close-set dark reddish-brown
spots forming distinct clumps; ♀ less
colourful than ♂. **Biology:** inhabits
lagoon and seaward reefs, 0 to 20 m.
Common in areas of good coral
growth. Lives on *Acropora, Pocillopora*
and *Porites* corals and on *Millepora*
fire corals and feeds on their polyps.
Males are territorial and prepare the
nest. Several females lay eggs in this
nest, guarded by the male. Solitary or
in small groups. Shy. **Range:** Red Sea
to Hawaii and Fr. Polynesia, n. to
Ryukyus, s. to S. Africa and
New Caledonia.

Tanzania, KF

Jewelled blenny   14 cm
Salarias fasciatus
Olivaceous to brown with dusky bars
and whitish spots. **Biology:** common
on reef flats with mixed rubble, sand
and corals of protected fringing reefs
and shallow lagoons, 0 to 8 m.
Usually seen perching on tops of
rocks, dead corals or clumps of algae.
Shy. **Range:** Red Sea to Samoa,
n. to Ryukyus, s. to Mozambique
and GBR. **Similar:** many similar
species living primarily in rocky
intertidal areas and exposed reef flats.

Jeddah, JK

Bluebelly blenny   4.5 cm
Alloblennius pictus
Pale with narrow diagonal white
streak below eye and row of reddish
blotches along back; courting ♂
with blue throat. **Biology:** inhabits
sheltered coral slopes and patch reefs,
3 to 25 m. Common in certain areas.
Lives on dead coral with small holes
for shelter. Reteats into hole when
alarmed. Feeds on fish eggs when
given the opportunity. Courting males
develop a brilliant throat. Not shy.
Range: Red Sea only.

♂, Dahab, MH

Ras Mamla, RM

Red Sea rockskipper   12 cm
Istiblennius unicolor
Cryptic with complex pattern of lines
and spots; head of ♂ with small crest.
Biology: inhabits rocky intertidal
areas and jetties. Will jump from pool
to pool. **Range:** Red Sea only from
Aqaba to Eritrea. **Similar:** many
species of *Istiblennius*, *Blenniella* and
Entomacrodus which also skip among
tidepools. *B. periophthalmus* has
numerous small red spots (Red Sea to
Indonesia). ▼

Lance fangblenny   11.5 cm
Aspidontus dussumieri
Pale with dark lateral stripe, tail
lanceolate with filaments. **Biology:**
inhabits sheltered lagoons or bays, 1
to 20 m. Has large territories over
sand, rubble and algal reefs, always
close to shelter of abandoned worm or
vermetid mollusc tubes. Feeds on
algae, zooplankton and detritus. The
large canines are used for defence.
Seems to rarely prey on skin or fins of
other fishes. Solitary and shy. **Range:**
Red Sea to Fr. Polynesia, n. to S.
Japan, s. to S. Africa and SE.
Australia.

Bali, RM

Cleaner mimic   12 cm
Aspidontus taeniatus
Colour identical to Bluestreak cleaner
wrasse (p. 162), blenny differs in
shape of snout which extends past
mouth. **Biology:** inhabits lagoons,
bays and sheltered seaward slopes,
1 to 20 m. A remarkable mimic
of the cleaner wrasse in colour and
behaviour. This charade enables it to
approach naive fishes in order to nip
pieces of fins, skin and scales as well as
to get protection from predators.
Lives in pairs in empty worm tubes.
Uncommon. **Range:** Red Sea to
Fr. Polynesia, n. to S. Japan, s. to
S. Africa and SE. Australia.

Guam, RM

Blackline fangblenny   10 cm
Meiacanthus nigrolineatus
Pale buish-grey becoming yellow
posteriorly; black line behind eye
sometimes broken posteriorly.
Biology: inhabits sheltered areas
of lagoons, bays and seaward reefs,
to 25 m. Commoner in areas with
sparse corals than in coral-rich areas.
Hovers close above bottom, feeds
on zooplankton. Uses its venomous
fangs in defence against inexperienced
predators. Ignored by other fishes.
Mimicked by the blennies
Plagiotremus townsendi and *Ecsenius
gravieri* (p. 179). **Range:** Red Sea to
Gulf of Aden only.

Abu Galum, RM

Townsend's fangblenny   4 cm
Plagiotremus townsendi
Pale bluish-grey becoming yellow
posteriorly; ♂ with red border below
black band on dorsal fin. Oman
population brown in front. **Biology:**
inhabits protected reef slopes and
algae-covered rocky reefs, 5 to 55 m.
Solitary, hovers close to bottom. An
aggressive mimic of the venomous
fangblenny *Meiacanthus nigrolineatus*.
By closely resembling the non-
aggressive model, it is able to sneak up
on larger fishes. It then darts in to
feed on pieces of skin and fins.
Uncommon.
Range: Red Sea to Gulf of Oman
only.

Nabq, RM

Piano fangblenny   14 cm
Plagiotremus tapeinosoma
Broad dark lateral stripe with light
bars along back. **Biology:** inhabits
lagoon and seaward reefs, 1 to 20 m.
Common along upper edges of reef
slopes. Swims well above bottom
with a distinctive wriggling motion.
A 'hit-and-run' nipper of fins and
scales of other fishes, even bites divers.
Quickly retreats into empty worm
tube when alarmed. **Range:** Red Sea
to Fr. Polynesia, n. to S. Japan, s. to S.
Africa and GBR.

Shams Alam Reefs, EL

Abu Galum, RM

Bluestriped fangblenny 12 cm
Plagiotremus rhinorhynchus
Dark brown to yellow with two blue stripes. **Biology:** inhabits clear coral-rich areas of lagoon and seaward reefs, 1 to 40 m. Common along upper reef slopes. Feeds by nipping fins and scales of other fishes. Juvenile mimics juvenile Bluestreak cleaner wrasse (p. 162) in order to get closer to prey. Less aggressive to divers than Piano fangblenny. Seeks shelter and nests in empty worm tubes or other small holes. May engage in a dance-like swim during courtship.
Range: Red Sea to Fr. Polynesia, n. to S. Japan, s. to S. Africa and SE. Australia.

Sinai, MH

Highfin fangblenny 7 cm
Petroscirtes mitratus
Cryptic with high D fin, the first 2 spines forming long extension. **Biology:** lives between clumps of algae and seagrass in protected inner reef flats, bays and lagoons, 1 to 12 m. Swims using only its clear P fins with tail curled to resemble drifting weed. Nests in empty mollusc shells. Common, but often overlooked.
Range: Red Sea to Samoa, n. to Ryukyus, s. to Mozambique and GBR.

Nuweiba, MT

Snake blenny 60 cm
Xiphasia setifer
Extremely elongate and eel-like; olivaceous with broad dark bars. **Biology:** inhabits sand and mud bottoms, sometimes in seagrass beds, 1 to 54 m. Shelters in tube-like burrows. Feeds on crustaceans, polychaetes, foraminifera and small fishes. Seems to be nocturnal, venturing high in water during the night. Occasionally preyed on by pelagic gamefishes. **Range:** Red Sea to Vanuatu, n. to S. Japan, s. to S. Africa and GBR. **Similar:** *X. matsubarai* (Red Sea to Samoa) has fewer D fin spines (11 vs. 13–14).

CLINGFISHES
GOBIESOCIDAE

Feather star clingfish   4.5 cm
Discotrema lineatum
Tiny elongate scaleless fishes with
V fins modified into suction disc.
Biology: lives among the arms of
crinoids (p. 342). **Range:** Red Sea and
Arabian Gulf to W. Pacific. **Similar:**
Urchin clingfish, *Diademichthys
lineatus* has elongate snout, is free-
swimming or among the spines of
Diadema sea urchins (Oman to New
Caledonia, 5 cm). ▼

Sinai, MH

DRAGONETS
CALLIONYMIDAE

Small depressed fishes with a broad
head, large serrated spine on preopercle (cheek), and a small hole for
gill opening. First dorsal fin of males
enlarged and intricately patterned.
Most are highly cryptic and live on
sand, a few live on hard bottoms.
They feed on minute benthic
invertebrates.

Stellate dragonet   7.5 cm
Synchiropus stellatus
Pale with stellate red and white
markings; 1st D fin of ♂ enlarged and
ornate. **Biology:** inhabits sand and
rubble patches of lagoon and seaward
reefs, 1 to 40 m. **Range:** S. Oman to
NW. Australia, s. to S. Africa.

♀, *Seychelles, DE*

Randall's dragonet   12 cm
Diplogrammus randalli
Cryptic with numerous small blue
spots, those on head vertically elongate;
♂ more colourful with elongate 1st D
spine. **Biology:** inhabits sand patches
of tidepools and shallow protected
areas. **Range:** Gulfs of Aqaba and
Suez only; replaced s. of Hurghada
to Mozambique and Mauritius by
the very similar Blue-spotted dragonet
D. infulatus. **Similar:** many similar
species, most below 40 m. Filament
dragonet *Callionymus filamentosus*
larger with 1st D spine of ♂ detached
(Red Sea to PNG; 16.5 cm).

♀, *Sinai, AK*

Sipadan Is., KF

DARTFISHES
PTERELEOTRIDAE

Elongate goby-like fishes with oblique mouth, two dorsal fins, and tiny embedded scales.

Blackfin dartfish 14 cm
Ptereleotris evides
Biology: inhabits clear lagoon and seaward reef slopes, 2 to 30 m. Common, hovers 1 to 2 m above sand or rubble bottom to feed on zooplankton. Moves away, then darts into hole when alarmed. Juvs. in aggregations, adults often paired. **Range:** Red Sea to Fr. Polynesia, n. to Ryukyus, s. to S. Africa and New Caledonia.

Similan Is., RM

Spot-tail dartfish 12 cm
Ptereleotris heteroptera
Pale; tail with large central black blotch. **Biology:** inhabits lagoon and seaward reefs, 7 to 46 m. Occurs over sand or rubble near base of coral slopes. Feeds as high as 3 m above bottom. Adults paired, juvs. in groups. **Range:** Red Sea to Polynesia, n. to Ryukyus, s. to S. Africa and SE. Australia. **Similar:** Pearly dartfish *P. microlepis* lacks dark spot on tail (Red Sea to Fr. Polynesia, 12 cm); Arabian dartfish *P. arabica* has pair of tail filaments (N. Red Sea and Arabian Gulf only, 13 cm); Zebra dartfish *P. zebra* has narrow pink bars and inhabits surge-swept seaward reefs (Red Sea to Fr. Polynesia; 11 cm). ▼

Decorated dartfish 9 cm
Nemateleotris decora
Pale with red and purple fins and purple midline on face.
Biology: inhabits sand and rubble patches and chutes of steep seaward reef slopes below 52 m in Red Sea, in as little as 24 m elsewhere. Hovers a short distance above hole, flicking its first dorsal fin. Feeds on zooplankton. Adults often paired. **Range:** Red Sea to Fiji, n. to Ryukyus, s. to Mauritius and New Caledonia.

Sulawesi, RM

▲ Citron coral goby 6.5 cm
Gobiodon citrinus
Yellow with bluish stripes on head and dark spot on opercle. **Biology:** inhabits sheltered shallow lagoon and seaward reefs, 1 to 25 m. Usually in small groups in large *Acropora* table corals, occasionally in other branching *Acropora* or *Millepora* corals. Wedges between coral branches when frightened. Species of *Gobiodon* produce a toxic bitter-tasting mucus to deter predators. **Range:** Red Sea to Samoa, n. to S. Japan, s. to Mozambique and GBR.
Similar: *G. reticulatus* is orange with pale bars on head and tiny pale spots on body (Red Sea to Arabian Gulf only; 2.1 cm). *(Photo: Jeddah, JK.)*

GOBIES

GOBIIDAE

Gobies are small elongate blunt-headed fishes with large mouths with conical teeth, pelvic fins close together or connected to form a disc, one or two dorsal fins, no lateral line and usually small scales. They are the largest family of marine fishes and the largest family in our area with over 2,000 species worldwide, and at least 126 around the Arabian Peninsula. The world's smallest vertebrate is the goby *Trimmaton nanus* which matures at a length of 10 mm. Most gobies are cryptic bottom-dwelling carnivores of small invertebrates, but many are detritivores or planktivores. A few are quite colourful. Gobies lay demersal eggs guarded by the male. Most goby species inhabit soft bottoms near reefs, but a large number live on or within the reef, or hover in groups in or near caves. A large number of goby species are symbiotic with one or more species of nearly blind alpheid shrimps. The shrimp maintains a burrow while the goby acts as a lookout. This association is extremely close. The shrimp and the goby are always found together, and may even settle from the plankton together as juveniles. While at the burrow entrance, the shrimp maintains contact with the goby with one of its antennae and the goby warns the shrimp of potential danger with a flick of its tail. The gobies feed on minute invertebrates exposed by the shrimps' excavations. Shrimps and gobies are usually paired as adults; occasionally more than one species of goby may occupy the same burrow. The burrows may be quite extensive, up to 50 cm deep with two or more entrances. Many other goby species construct burrows of their own or utilize burrows constructed by invertebrates.

Jeddah, JK

Steinitz' shrimpgoby   8 cm
Amblyeleotris steinitzi
Pale with 5 narrow slightly
diagonal brown bars; D fin with
tiny yellow spots. Face pale or dark,
obscures bands on eyes when dark.
Biology: inhabits expanses of fine
sand of clear lagoon and seaward
reefs, 6 to 30 m. Lives in burrows
with the shrimps *Alpheus djeddensis,
A. djiboutensis* and *A. bellulus*
(p. 326). Common. Wary, but
approachable. **Range:** Red Sea
to Samoa, s. to E. Africa and GBR.

Hurghada, AK

Magnus' shrimpgoby   10 cm
Amblyeleotris sungami
Pale with 5 narrow slightly diagonal
orange bars, the last bending
backwards onto lower tail, blue and
yellow spots on upper body and D fin
larger than in *A. steinitzi*; face pale or
dark. **Biology:** inhabits expanses of
fine sand of clear lagoon and seaward
reefs, 3 to 25 m. Lives in burrows
with alpheid shrimps *Alpheus* spp.
(p. 326). Common. Named after
its discoverer Professor Magnus
spelled backwards. **Range:** Red Sea to
S. Oman, possibly more widespread.

Palau, RM

Red-barred shrimpgoby   10 cm
Amblyeleotris fasciata
Pale yellowish with broad wine-red
bars, interspaces with pale blue
spots. **Biology:** inhabits expanses
of fine to coarse sand of clear lagoon
and exposed seaward reefs, 5 to
30 m. Lives in burrows with the
shrimps *Alpheus ochrostriatus* and
A. djiboutensis (p. 326). Common.
Range: Red Sea to Fiji, n. to
Ryukyus, s. to S. Africa and GBR.
Similar: Blotchy shrimpgoby
A. periophthalma with bars
more fragmented (p. 196).

Black shrimpgoby   9.5 cm
Cryptocentrus fasciatus
Head with fine pale dashes in
diagonal rows; ground colour nearly
black to pale with broad dark bars.
Biology: inhabits areas of silty to
coarse sand, from inner protected
lagoons and margins of seagrass beds
to exposed seaward reefs, 0.5 to 20 m.
Usually in pairs living with the shrimp
Alpheus bellulus (p. 326). Uncommon.
Range: Red Sea to Melanesia, s. to
Zanzibar and GBR.

Similan Is., RM

Red Sea shrimpgoby   13 cm
Cryptocentrus caeruleopunctatus
Brown with blue-edged red spots
anteriorly and on D fins and
thin pale diagonal bars posteriorly.
Biology: inhabits expanses of silty
to coarse rubbly sand from turbid
lagoons to semi-protected seaward
reefs, 2 to 24 m. Usually in areas of
little current. Lives in burrows with
the shrimps *Alpheus ochrostriatus* and
A. rapax (p. 326). Usually quite wary
and difficult to approach.
Range: Red Sea only.

Jeddah, JK

Luther's shrimpgoby   11 cm
Cryptocentrus lutheri
Brown with bright blue spots on
head and broad dark bars posteriorly.
Biology: inhabits expanses of fine
to coarse sand of protected lagoons,
bays and seaward reefs, 1 to 20 m.
Often near seagrass beds. Shy, but
approachable. Lives in burrows with
the shrimps *Alpheus bellulus* and
A. djiboutensis (p. 326). **Range:** Red
Sea to Arabia Gulf, s. to Tanzania.

Mövenpick Bay, RM

Marsa Shagra, RM

Ninebar shrimpgoby   13 cm

Cryptocentrus cryptocentrus

Greyish-brown with tiny blue dots and larger red-edged black spots anteriorly; body with 9 narrow pale bars. **Biology:** inhabits expanses of fine silty sand near edges of sheltered reefs and seagrass beds, 0.5 to 15 m. Lives in burrows with the shrimp *Alpheus djiboutensis* (p. 326). Uncommon. **Range:** Red Sea to Chagos Archipelago, s. to S. Africa. **Similar:** *Psilogobius randalli* lacks blue spots and is smaller and more cryptic (Gulf of Aqaba only).

Marsa Shagra, RM

Red Sea shrimpgoby   7 cm

Ctenogobiops maculosus

Off-white with rows of golden-brown spots, those on head forming diagonal lines; white spot at P fin base. **Biology:** inhabits areas of coarse sand of lagoon and seaward reefs, 1 to 15 m. Common. Uses a burrow with several entrances made by the shrimps *Alpheus bellulus, A. rapax* and *A. rubromaculatus.* **Range:** Red Sea only. **Similar:** Sandy shrimpgoby *C. feroculus* has diffuse spots instead of lines on head (Red Sea to Ryukyus).

Mangrove Bay, PN

Graceful shrimpgoby   4 cm

Lotilia graciliosa

Black with white nape and saddles, ocellus on first D fin, fan-like P fins clear with black and white spots. **Biology:** inhabits patches of fine coral sand along bases of clear lagoon or semi-protected seaward reefs, or in shallow caves along dropoffs, 2 to 20 m. Typically hovers above entrance to burrow, waving its fan-like P fins. Lives in burrows made only by the shrimp *Alpheus rubromaculatus* (p. 326). Very wary and difficult to approach. **Range:** Red Sea to Fiji, n. to Ryukyus, s. to Mozambique and GBR.

Ornate shrimpgoby   8 cm
Vanderhorstia ornatissima
Clusters of black spots and
electric-blue spots, lines and circles.
Biology: inhabits shallow silty
protected areas, 1 to 6 m. Common
in seagrass beds. **Range:** Red Sea to
Fr. Polynesia, n. to Ryukyus, s. to
Mozambique and SE. Australia.
Similar: Merten's shrimpgoby
V. mertensi has small yellow spots
on head, upper body and fins
(p. 196); Ambonoro shrimpgoby
V. ambanoro lacks small blue rings
and hovers above burrow entrance
(Red Sea to Samoa; 13 cm). ▼

Marsa Abu Dabab, RM

Maiden goby   14 cm
Valenciennea puellaris
Pale with rows of elongate orange
spots and narrow reddish-orange bars
dorsally. **Biology:** inhabits expanses of
fine or coarse sand of clear lagoons,
bays and sheltered seaward reefs, 2 to
40 m. Constructs burrow under large
piece of rubble by carrying sand in
mouth. Usually in pairs near burrow.
This and other species in genus feed
on small invertebrates by sifting sand
through their gill-rakers. Moderately
wary. **Range:** Red Sea to Samoa, n. to
S. Japan, s. to Mauritius and GBR.

Egypt, AK

Sixspot goby   16 cm
Valenciennea sexguttata
Pale with blue spots on head,
black spot near tip of 1st D fin.
Biology: inhabits expanses of fine
to silty sand of lagoons, bays and
sheltered seaward reefs, 1 to 10 m.
Constructs a burrow under a rock or
piece of dead coral. Adults often in
pairs. **Range:** Red Sea to Line Is.,
n. to Yaeyamas, s. to Mozambique
and GBR. **Similar:** Gulf goby
V. persica with lateral series of blue
spots along upper sides (p. 196).

Egypt, AK

Nuweiba, PM

Tailspot goby 18 cm
Amblygobius albimaculatus
Olive to grey with five dark bars and
dark spot above P fin and at base of
tail. **Biology:** inhabits areas of silty
sand of protected lagoons and bays,
0.3 to 20 m. Common in somewhat
turbid areas at the bases of reef slopes
and in seagrass beds. Builds a burrow
by shovelling sand with its mouth.
Feeds on tiny invertebrates sifted from
sand through its gill-rakers. Typically
hovers just above the bottom, adults
usually in pairs near their burrow.
Somewhat wary. **Range:** Red Sea,
n. to Arabian Gulf, s. to S. Africa.

Abu Galum, RM

Hector's goby 6.5 cm
Koumansetta hectori
Dark with narrow yellow bars and
3 ocelli: on 1st D fin, at base of
2nd D fin, and on upper tail base.
Biology: inhabits sand patches
along the bases of sheltered lagoon
and seaward reefs, 3 to 20 m. Usually
solitary under overhangs, hovering
just above rubble or sand. Feeds on
plankton. Territorial. **Range:** Red Sea
to Caroline Is., n. to Ryukyus, s. to
Mozambique.

Ras Mamlah, PM

Longspine sandgoby 7 cm
Coryphopterus longispinus
Translucent with orange spots;
♂ with greatly elongate 1st D spine.
Biology: inhabits sandy floors of
caves of seaward reefs, 1 to 18 m.
Occasionally on lagoon patch reefs.
Range: N. Red Sea only. Replaced
by the very similar innerspotted goby
C. inframaculatus from Arabian Gulf
and Oman s. to Mauritius. **Similar:**
Several similar species: Neophyte
sandgoby *C. neophytus* (7 cm) and
Twospot sandgoby *C. duospilus*
(6 cm) translucent with tiny black
specks, the latter with 1 or 2 dark
spots on 1st D fin.

Decorated goby   12.5 cm
Istigobius decoratus
Pale with rows of dark brown spots.
Biology: a common inhabitant of
protected bays, lagoons and mangroves,
0 to 2 m. Feeds on small invertebrates
sifted from sand. Remains motionless
for short periods. Easily approached.
Range: Red Sea to Fiji, n. to Taiwan,
s. to New Caledonia and Madagascar.
Similar: Ornate goby *I. ornatus* has
broad yellowish D fin margin (Red
Sea to Samoa, 11 cm).

Hurghada, AK

Mud reef goby   13 cm
Exyrias bellisimus
Fins large and fan-like, 1st D fin
spines filamentous; brown with small
pale spots and larger dark blotches.
Biology: inhabits protected silty
lagoons and bays, 0.5 to 20 m.
Feeds on small invertebrates sifted
from mud. Remains motionless for
short periods. Easily approached.
Range: Red Sea to Samoa, n. to
Yaeyamas, s. to GBR.

Mangrove Bay, RM

Spinecheek goby   8 cm
Oplopomus oplopomus
Pale with tiny black and pale blue
spots; ♂ more colourful than ♀.
Biology: inhabits silty protected
bays, lagoons and mangroves, 0.5
to 20 m. Shelters in holes made by
an invertebrate. **Range:** Red Sea to
Fr. Polynesia, n. to Yaeyamas, s. to
GBR. **Similar:** *O. caninoides* is more
elongate with tiny white and red
specks and inhabits current-swept
sand usually below 20 m (Red Sea
to W. Pacific, 6 cm); *Silhouettea* spp.
(6 cm) are more elongate with eyes
oriented more upward and can bury
in mud.

Marsa Alam, BE

Jeddah, JK

Harlequin coral goby   3 cm
Gobiodon rivulatus
Identification provisional, typically
green with red marks not forming
circles. **Biology:** true *G. rivulatus* lives
among the branches of living corals,
primarily *Acropora* spp. Species of
Gobiodon produce a toxic bitter-
tasting mucus to deter predators.
Range: Red Sea to Indonesia.
Similar: Redhead goby *Paragobiodon
echinocephalus* has short papillae
on head. It occurs in groups among
branches of *Stylophora* corals (Red
Sea to Fr. Polynesia; 3 cm). ▼

Guam, RM

Eyebar goby   8 cm
Gnatholepis anjerensis
Pale with fine dark speckles, dark bar
through eye. **Biology:** inhabits sand
near shelter of coral or rubble of
sheltered inshore and seaward reefs,
1 to 46 m. Common. Territorial and
probably haremic. **Range:** Red Sea
to Polynesia, n. to Ryukyus, s. to
S. Africa and Rapa. **Similar:** many
other similar small cryptic species
occur in shallow protected sandy
areas: *Bathygobius* spp. have free upper
P fin rays and are common in
tidepools.

Guam, RM

Halfspotted goby   6.5 cm
Asterropteryx semipunctatus
Brown with tiny brilliant blue spots.
Biology: inhabits protected areas of
mixed rubble, rock and sand of lagoon
and seaward reefs, 1 to 15 m. Usually
on algae-coated dead coral rubble
and sand near shelter. Territorial
and haremic. **Range:** Red Sea to
Polynesia, n. to S. Japan, s. to
Chagos Archipelago and SE. Australia.

Redeye goby   2.5 cm
Bryaninops natans
Transparent with lavender-pink eye.
Biology: inhabits deep lagoon and
seaward reefs, 7 to 27 m. Hovers in
small groups close above branches of
Acropora corals. **Range:** Red Sea to
Cook Islands, n. to Ryukyus, s. to
GBR. **Similar:** several similar species
in region: Whip coral goby *B. yongei*
lives on wire coral *Cirrhipathes
anguina* (Red Sea to Polynesia,
2.2 cm); *B. erythrops* on surfaces of
corals, sponges, algae and seagrasses
(Red Sea to Samoa, 2.5 cm).

Sinai, HHH

Blue-striped cave goby   4.5 cm
Trimma tevegae
Pale with broad blue stripe on upper
sides. **Biology:** inhabits caves of steep
seaward reef slopes, 9 to over 45 m.
In clusters in midwater near shelter,
oriented vertically in a head-up
position. **Range:** C. Red Sea
to New Britain, n. to S. Japan, s. to
NW. Australia. **Similar:** *T. taylori* is
mostly yellow, also hovers in caves
(Red Sea to Hawaii, n. to S. Japan;
2.5 cm); other *Trimma* spp. tend to
stay on bottom, walls, or ceilings of
caves on steep reef slopes (p. 196).

Sulawesi, RM

Girdled cave goby   5 cm
Priolepis cincta
Brown with dark-edged narrow pale
bars. **Biology:** inhabits holes and
crevices of outer lagoon and seaward
reef slopes, 1 to 70 m. Usually upside-
down on roof of hole. Common, but
secretive and rarely seen. **Range:** Red
Sea and Arabian Gulf to Samoa, n. to
S. Japan, s. to S. Africa, GBR and
Tonga. **Similar:** *P. randalli* has narrow
red and white bands on D and tail
fins and conspicuous dark spot
at front base of 1st D fin (Gulf of Aqaba
to Arabian Gulf only; 5 cm); *P.
semidoliatus* has oblique pale bars on
head and front of body, also quite
secretive (Red Sea to Polynesia,
3.5 cm).

Dahab, MH

1. Diagonal shrimpgoby *Amblyeleotris diagonalis* 9 cm
Biology: inhabits open sand of clear sheltered reefs, 6 to 42 m. **Range:** Red Sea and Arabian Gulf to Solomon Is.

2. Triplespot shrimpgoby *Amblyeleotris triguttata* 8 cm
Biology: inhabits fine sand of protected areas, 1 to 17 m. **Range:** Red Sea and Arabian Gulf, s. to Tanzania.

3. Blotchy shrimpgoby *Amblyeleotris periophthalma* 9 cm
Biology: inhabits open sand of clear sheltered reefs, 8 to 30 m. **Range:** Red Sea and Arabian Gulf to Samoa.

4. Orange-striped goby *Amblygobius nocturnus* 7 cm
Biology: on silty sand of sheltered areas, 3 to 30 m. Hovers near burrow. **Range:** Red Sea and Arabian Gulf to Fr. Polynesia.

5. Clown goby *Callogobius amikami* 4 cm
Biology: secretive among rubble, 4 to 8 m. Nocturnal. Adult less colourful. **Range:** Red Sea to Gulf of Oman only.

6. Arabian goby *Cryptocentroides arabicus* 13.5 cm
Biology: on mud bottoms of shallow inner protected areas under 1 m. **Range:** Red Sea to Arabian Gulf only.

7. Zebra goby *Ego zebra* 4.2 cm
Biology: hovers in groups near caves at about 21 m. **Range:** S. Oman only, expected in upwelling areas of Yemen.

8. Spotted dwarfgoby *Eviota guttata* 2.2 cm
Biology: on dead coral rock of clear lagoon and seaward reefs, 1 to 15 m. **Range:** Red Sea and Arabian Gulf to Samoa.

9. Redspotted dwarfgoby *Eviota prasina* 3.7 cm
Biology: on coral and rock of sheltered reefs, tidepools to 6 m. **Range:** Red Sea and S. Oman to W. Pacific.

10. Sebree's dwarfgoby *Eviota sebreei* 3 cm
Biology: on massive *Porites* corals of lagoon and sheltered seaward reefs, 3 to 33 m. **Range:** Red Sea to Samoa.

11. Fan shrimpgoby *Flabelligobius latruncularius* 5.5 cm
Biology: inhabits clear sheltered reefs, 1 to 37 m. **Range:** Red Sea and Arabian Gulf to Maldives.

12. Michel's ghostgoby *Pleurosicya micheli* 3 cm
Biology: inhabits clear lagoon and sheltered seaward reefs, 12 to 38 m. Exclusively on hard corals, particularly *Platygyra sinuosa*, *Porites lobata*, *Psammocora* spp., *Pachyseris* spp., *Turbinaria* spp. and *Echinopora* spp. **Range:** Red Sea to Hawaii and Tonga.

13. Common ghostgoby *Pleurosicya mossambica* 3 cm
Biology: inhabits clear lagoon and sheltered seaward reefs, 2 to 28 m. On surface of many organisms including algae, fire corals, black corals, soft corals and hard corals. **Range:** Red Sea to Marquesas.

14. Frogface goby *Oxyurichthys papuensis* 18 cm
Biology: inhabits mud bottoms of estuaries and deep bays, 1 to 50 m. **Range:** Red Sea to New Caledonia.

15. Gulf goby *Valenciennea persica* 16 cm
Biology: inhabits large sand patches, 2 to 12 m. **Range:** C. Oman to Arabian Gulf.

16. Railway goby *Valenciennea helsdingeni* 20 cm
Biology: inhabits protected sand expanses, 2 to 30 m. Usually in pairs near hole. **Range:** S. Red Sea to GBR.

17. Spikefin goby *Discordipinna griessingeri* 2.5 cm
Biology: among live coral, rubble and sand, 3 to 27 m. Extremely cryptic. **Range:** Red Sea to Fr. Polynesia.

18. Merten's shrimpgoby *Vanderhorstia mertensi* 11 cm
Biology: inhabits protected expanses of sand, 2 to 16 m. Hovers above entrance to burrow. **Range:** Red Sea to Samoa.

19. Barall's dwarfgoby *Trimma baralli* 3.5 cm
Biology: inhabits coral-rich areas, 2 to 10 m. **Range:** Red Sea only.

20. Bloodspot dwarfgoby *Trimma flavicaudatum* 3 cm
Biology: on dead coral in coral-rich areas, 5 to 20 m. **Range:** Red Sea only.

21. Sheppard's dwarfgoby *Trimma sheppardi* 2.5 cm
Biology: secretive in caves and dropoffs in coral-rich areas, 5 to 45 m. **Range:** Red Sea to W. Pacific.

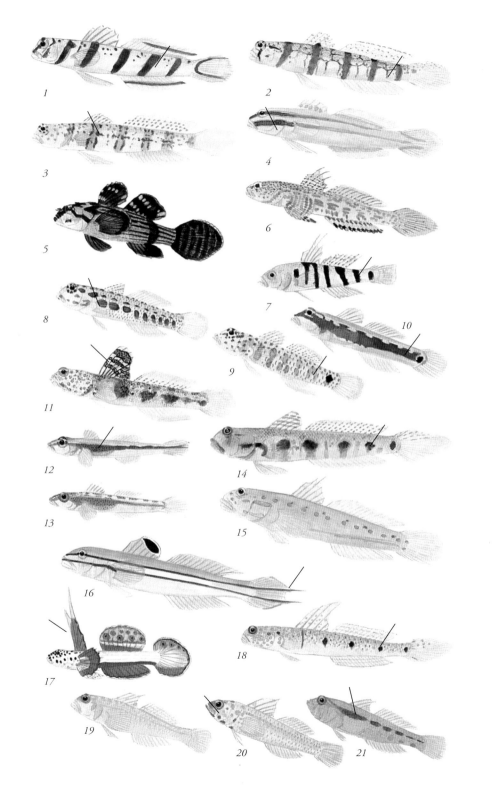

◀ Longfin spadefish   60 cm
Platax teira

Forehead steep, D and A fins broad. **Biology:** occurs singly or in schools along steep slopes or in open water of deep lagoons and bays, 1 to 20 m. Juveniles in shallow protected areas or far offshore drifting with clumps of algae. **Range:** Red Sea to Fiji, n. to Ryukyus, s. to S. Africa and GBR. **Similar:** juv. Golden spadefish *P. boersi* is nearly identical (see below). *(Photo: Maldives, MK.)*

juv.

Orbicular spadefish   57 cm
Platax orbicularis

Head smoothly curved, D, A fins short. **Biology:** occurs singly or in schools along steep slopes or in open water of deep lagoons and bays, 2 to 34 m. Juveniles in shallow protected waters and mimic a floating dead leaf. **Range:** Red Sea to Fr. Polynesia, n. to Ryukyus, s. to S. Africa and New Caledonia. **Similar:** adult Golden spadefish *P. boersi* has golden-green sheen.

juv.

Golden spadefish   41 cm
Platax boersi

Head smoothly curved, D, A fins short, has a golden sheen. **Biology:** occurs singly or schools along steep slopes or in open water of deep lagoons and bays, 1 to 15 m. Juveniles in shallow protected areas. **Range:** Red Sea to PNG, n. to S. Japan, s. to S. Africa and GBR. **Similar:** Golden spadefish is nearly identical to juv. Orbicular spadefish *P. orbicularis*, but with fewer LL scales (<54 vs >55), more D rays (≥35 vs ≤34); adult Orbicular spadefish lacks golden sheen.

SPADEFISHES   EPHIPPIDAE

Spadefishes are moderately large ovoid fishes with deep, highly compressed bodies, small terminal mouths with brush-like teeth (tricuspid in *Platax*), continuous dorsal fins, and small ctenoid scales extending over the base of the median fins. Juveniles of *Platax* have very deep bodies and greatly elevated dorsal, anal and pelvic fins that shorten with age. Spadefishes are omnivores of algae and small invertebrates. They become quite tame around dive sites where they look for handouts. In many older books, species of *Platax* are called batfishes, a name already used for the bottom-dwelling fish family Ogcocephalidae.

Marsa Shagra, RM

Eritrea, EL

RABBITFISHES

SIGANIDAE

Rabbitfishes are highly compressed fishes with venomous fin spines, a small terminal mouth with a row of small close-set incisiform teeth in the jaws, a complete lateral line, and minute cycloid scales. All species have 9 dorsal fin spines preceded by an imbedded forward-projecting spine, 7 anal fin spines, and pelvic fins with an outer and inner spine surrounding 3 soft rays. These spines can inflict an extremely painful wound, which fortunately is not normally as dangerous as those of the Stonefish (p. 64). Rabbitfishes are diurnal herbivores of algae and sea-grasses and are named for their voracious appetites. Some species may also occasionally feed on tunicates or sponges. They generally spawn on a lunar cycle with peak activity during the spring and early summer. Spawning occurs in pairs or groups on outgoing tides either at night or in the early morning. Juveniles of some species are estuarine. Rabbit-fishes are highly esteemed foodfishes. Some species are used in aquacul-ture. When alive most species are easily distinguished by colour pattern, but when stressed or dead, may be extremely difficult to distinguish due to obliterating colour patterns and identical fin-ray counts.

▲ **Stellate rabbitfish** 35 cm
Siganus stellatus laqueus
Pale with close-set black spots; nape and tail margin yellow.
Biology: inhabits coral-rich areas of lagoon and seaward reefs, 1 to 35 m. Common, adults always paired, subadults in groups. Covers large area while foraging for filamentous algae and seaweed.
Range: subspecies *laqueus* Red Sea and Gulf of Aden only; subspecies *stellatus* Mozambique to Bali. *(Photo: Mövenpick Bay, RM.)*

Forktail rabbitfish   43 cm
Siganus argenteus
Tail forked, top of head yellowish.
Biology: inhabits lagoon and seaward
reefs, 0.5 to 30 m. Common in areas
of mixed coral and rubble, also over
sand, bare rock and seagrasses.
Usually in small roving aggregations,
rarely solitary. Spawns in the Red Sea
during new moons of July to August.
Range: Red Sea to Fr. Polynesia, n. to
S. Japan, s. to Mozambique and GBR.

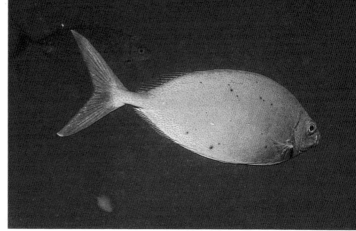

Marsa Shagra, RM

Squaretail rabbitfish   24 cm
Siganus luridus
Pale olive-green to very dark brown
often with conspicuous scratches on
skin. **Biology:** inhabits lagoon and
seaward reefs, 2 to 18 m. Common.
Solitary or in small roving groups
in sandy, rubbly or weedy areas of
upper reef slopes. Prefers areas with
little current. Feeds on benthic algae.
Range: Red Sea and Arabian Gulf to
Mauritius and Mozambique; a recent
migrant to E. Mediterranean.

Marsa Shagra, RM

Rivulated rabbitfish   30 cm
Siganus rivulatus
Pale olivaceous to mottled brown
with faint orange lines on belly.
Biology: a common inhabitant of
shallow lagoon and protected seaward
reefs, 1 to 15 m. Usually in roving
schools over weedy, sandy or dead
coral bottoms. Becomes mottled when
alarmed or asleep. **Range:** Red Sea to
Gulf of Aden; a recent migrant to E.
Mediterranean.

Marsa Shagra, RM

SURGEONFISHES AND UNICORNFISHES ACANTHURIDAE

Surgeonfishes are ovate to elongate compressed fishes with a small terminal mouth containing a single row of small close-set lanceolate or incisiform teeth, continuous dorsal and anal fins, a continuous lateral line, a tough skin with minute scales, and one or more pairs of sharp blades at the base of the tail. They have a long larval stage resulting in broad distributions for most species. Two of the three subfamilies occur in our area: the true **Surgeonfishes** (Acanthurinae) with a single scalpel-like blade which folds into a groove, and the **Unicornfishes** (Nasinae) with 1 or 2 sharp fixed keel-like bony peduncular plates. The blades are used offensively or defensively, against one another in struggles for dominance or against predators, and can inflict deep and painful wounds. Surgeonfishes are among the most conspicuous and abundant fishes of shallow coral reefs. They are diurnal herbivores or planktivores that shelter in the reef at night. Most species of *Acanthurus* graze on benthic algae while some species of *Naso* feed on fleshy brown algae and a few species of *Acanthurus* and adults of many species of *Naso* feed on zooplankton. Some species of *Acanthurus* have a thick-walled gizzard-like stomach and ingest sand as they feed. Species of *Ctenochaetus* feed primarily on detritus, and some form a key link in the ciguatera food chain and are occasionally toxic. Reproduction is typically on a lunar cycle with spawning occurring at dusk in groups or pairs. Surgeonfishes are important food-fishes on most tropical islands.

▲ **Sohal surgeonfish** 40 cm
Acanthurus sohal
Pale with narrow black stripes, belly pale; median fins black with blue edges. Juvenile with similar colour.
Biology: a common inhabitant of outer reef flats and margins of exposed seaward reefs, 0 to 10 m. Males occupy small well-defined feeding territories which include a harem of females and are aggressively defended. Males aggressively defending territories may turn dark. Spawns at new moon shortly after sunrise.
Range: Red Sea to Arabian Gulf only.
(Photo: Marsa Shagra, RM.)

Black surgeonfish 40 cm
Acanthurus gahhm
Dark brown, P fin margin yellow; tail
sometimes pale, white bar at base can
be turned off. **Biology:** occurs in
groups in open water of deep lagoons
and seaward reefs, 1 to 67 m. Feeds
on filamentous algae growing on sand
and rubble. Common, often near dive
boats where it feeds on sewage and
scraps of food. Easily approached.
Range: Red Sea and Gulf of Aden only.
Similar: *A. leucocheilus* has white band
on chin (Gulf of Aden to C. Pacific;
48 cm). ▼

Mangrove Bay, RM

Dusky surgeon 21 cm
Acanthurus nigrofuscus
Lavender-brown with orange spots on
head. **Biology:** inhabits outer reef flats
and shallow lagoon and seaward reefs,
1 to 25 m. Common, occurs in large
schools and grazes on filmentous algae
covering dead corals, rocks, or rubble.
At the bottom of the 'pecking order'
among surgeonfishes, it relies on
numerical superiority to overwhelm
the territorial defences of other
herbivores. **Range:** Red Sea to
Polynesia, n. to S. Japan, s. to
S. Africa and SE. Australia.

Elongate surgeonfish 50 cm
Acanthurus mata
Bluish-brown with thin pinstripes and
yellow eye-patch; occasionally with
white bar on tail base. **Biology:** occurs
in aggregations in open water of deep
lagoons, bays and seaward reef slopes,
5 to over 30 m. Often in turbid water,
near shelter of caves or wrecks. Feeds
primarily on zooplankton. Becomes
pale at cleaning stations. **Range:**
S. Red Sea to Fr. Polynesia, n. to S.
Japan, s. to S. Africa and GBR. Also
in the Gulf of Aden: Convict
surgeonfish *A. triostegus* which is light
grey with 5 narrow black bars (Gulf of
Aden to Panama;
27 cm). ▶

Lahami, KF

Bali, RM

Lahami Reefs, KF

Lined bristlethooth   26 cm
Ctenochaetus striatus
Yellow P fins. **Biology:** inhabits reef
flats, and lagoon and seaward reefs, 1
to 30 m. Solitary or in schools, often
with other small surgeonfishes.
Common. Feeds on blue-green algae
and diatoms growing on rubble, dead
corals, rocks and sand. **Range:** Red
Sea and S. Oman to Fr. Polynesia,
n. to S. Japan, s. to S. Africa and
GBR.

juv.

Dahab, AK

Yellowtail tang   22 cm
Zebrasoma xanthurum
Deep body with fan-like D, A fins;
deep blue with yellow tail. **Biology:**
inhabits protected and exposed coral
and rocky reefs, 0.5
to at least 20 m. Common on coral-
rich patch reefs and upper reef slopes
with holes and passages. Feeds on
filamentous algae on dead coral
surfaces. In pairs or in small groups.
Easily approached. **Range:** Red Sea
to Arabian Gulf; strays to Maldives.

Marsa Shagra, RM

Sailfin tang   40 cm
Zebrasoma desjardinii
Deep body with fan-like D, A fins;
brown with pale bars and yellow spots
and stripes. **Biology:** inhabits deep
lagoons and semi-protected seaward
reefs with moderate to rich coral
growth, 1 to 30 m. Usually in pairs or
small groups.
Juveniles solitary,
often hidden
among staghorn
corals. **Range:**
Red Sea to Java,
s. to S. Africa.

juv.

Orangespine unicornfish 45 cm
Naso elegans
D fin orange with black base;
C spines orange; ♂ with elongate
tail filaments. **Biology:** inhabits reef
flats and lagoon and seaward reefs,
1 to 90 m. Common, usually over
rubble, corals and rocks. Feeds
primarily on leafy brown algae,
including *Sargassum*. Adults in pairs,
occasionally in groups. Large ♂
occasionally territorial. Pair-spawns.
Range: Red Sea and S. Oman to Bali,
s. to S. Africa. Replaced by *N.
lituratus* (with black D fin) in Pacific.

Marsa Shagra, RM

Bluespine unicornfish 70 cm
Naso unicornis
Olivaceous with blue C spines; horn,
tail filaments elongate with age.
Biology: inhabits shallow lagoon and
seaward reefs with channels and
moats, 1 to 80 m. Common, often in
exposed surgy areas. Usually occurs in
groups and feeds on coarse leafy algae,
including *Sargassum*. **Range:** Red Sea
and S. Oman to Fr. Polynesia, n. to S.
Japan and Hawaii, s. to S. Africa and
SE. Australia.

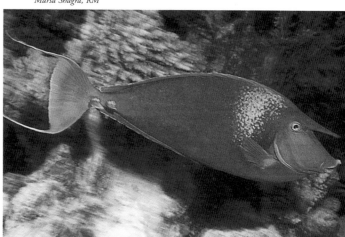

Mangrove Bay, RM

Sleek unicornfish 75 cm
Naso hexacanthus
Grey, usually tan below; preopercle
and operculum with black margin,
lips white. **Biology:** inhabits deep
lagoon and steep seaward reef slopes,
6 to 137 m. Common, usually in
groups a few metres from the reef.
Feeds on large zooplankton. Courting
♂ flashes white patch and bars on
nape and upper body. **Range:** Red Sea
to Fr. Polynesia, n. to S. Japan and
Hawaii, s. to S. Africa and SE.
Australia. **Similar:** *N. fageni*, has
bulbous snout (Oman to Philippines,
rare; 80 cm). ▼

Sipadan, KF

Maldives, RM

Spotted unicornfish   60 cm
Naso brevirostris
Brown to grey with dark vertical
lines and spots; horn elongates with
age. **Biology:** occurs in midwater
along deep lagoon and steep seaward
reef slopes, 0 to 45 m. Uncommon,
usually in groups. Feeds on
zooplankton. Courting ♂ flashes
a pale blue band behind head.
Range: Red Sea to Polynesia,
n. to S. Japan and Hawaii, s. to
S. Africa and SE. Australia.

MOORISH IDOL
ZANCLIDAE

Moorish idol   22 cm
Zanclus cornutus
Very closely related to surgeonfishes
but without a caudal blade. Has an
elongate snout and dorsal fin
produced into a long filament. A
single species with black and pale
yellow bars that resembles a
bannerfish (p. 132). **Biology:** inhabits
rocky and coral reefs, 1 to 145 m.
Uncommon off Oman and Yemen.
Feeds primarily on sponges,
occasionally on plant material.
Usually swims in small groups over
large areas. Has a long larval stage,
nearly full-grown when settling on the
reef. **Range:** Gulf of Aden and Oman
to Mexico, n. to S. Japan and Hawaii,
s. to S. Africa and SE. Australia.

Kenya, EL

BARRACUDAS   SPHYRAENIDAE

Barracudas are elongate silvery fishes characterized by a pointed head, large mouth with projecting lower jaw,
long knife-like teeth on the jaws and roof of the mouth (palatines), two widely separated dorsal fins, pelvic
fins located well behind the pectoral fins, forked tail, small cycloid scales and a well-developed lateral line.
They are voracious predators of other fishes. A large barracuda is capable of cutting a large fish in two with a
single bite and causing severe injury to a human. Some species are primarily diurnal, while others are nocturnal and occur in near-stationary schools during the day. The young of most species and adults of some are
common in estuarine areas. Attacks on man by the Great barracuda have been documented but these are
invariably the result of mistaken identity or outright provocation such as being speared. Attractants such as
metallic jewelry flashing in the sun or speared fish, particularly in murky water, are frequent causes of attacks.
The Great barracuda has also been implicated in ciguatera fish poisoning in many areas, but not Africa. Six
species of barracuda have been recorded from the Red Sea and Gulf of Aden.

Great barracuda 1.9 m; 38 kg
Sphyraena barracuda
Sides often with a few scattered
dark spots; tail emarginate, dark with
white tips. **Biology:** inhabits estuaries,
lagoons, bays and seaward reefs, 1 to
100 m. In open water above or near
reefs. Juveniles common in estuaries,
often in groups. Adults solitary.
Curious but not dangerous unless
provoked or attracted by shiny
objects in murky water. **Range:** Red
Sea and Gulf of Oman to Polynesia,
s. to S. Africa; tropical Atlantic.

Sipadan Is., KF

Blackfin barracuda 1.7 m
Sphyraena qenie
Sides with dark chevrons; tail black,
emarginate. **Biology:** occurs in
large semi-stationary schools at
current-swept promontories or in
deep lagoons or at channel entrances.
May occur at these sites for months.
Probably disperses at night to feed.
Range: Red Sea to Panama, s. to
S. Africa and Fr. Polynesia.

Nabaa Bay, EL

Pickhandle barracuda 1.5 m
Sphyraena jello
Sides with dark chevrons; tail
yellowish-green. **Biology:** occurs in
schools in deep lagoons or coastal reefs
near current-swept promontories.
More apt to occur in turbid water
than Blackfin barracuda. Uncommon.
Range: Red Sea and Arabian Gulf to
Fiji, n. to S. Japan, s. to S. Africa and
GBR.

Fiji, JB

Yellowtail barracuda 37 cm
Sphyraena flavicauda
Yellowish-green above with 2
dusky stripes; tail yellowish-green.
Biology: occurs in schools above
deep lagoon and seaward reefs
during the day. Probably disperses
at night to feed. **Range:** Red Sea to
Samoa, n. to Ryukyus, s. to S. Africa
and GBR. **Similar:** several similar
small species: *S. obtusata* with deeper
body, less yellow (Red Sea to Samoa,
30 cm).

Nabq, EL

TUNAS AND MACKERELS

SCOMBRIDAE

Tunas and mackerels are typically elongate, fusiform silvery fishes with separate dorsal fins, several detached finlets, a deeply forked tail, two or more peduncular keels, and a simple lateral line. They swim continuously, eat large quantities of food, and grow rapidly. Most species are swift predators of the open sea, but a few are closely associated with coral reefs. Many tunas maintain high core body temperatures. When near reefs, the oceanic species prey heavily on larval and early juvenile reef fishes and crustaceans. Reef-associated species prey either on large zooplankton or fishes that occupy the water above the reef. Most species are utilized as food.

▲ **Striped mackerel**   38 cm
Rastrelliger kanagurta
Biology: inhabits midwaters of deep lagoons, bays, harbours and protected seaward reef slopes, 1 to 70 m. In tightly packed schools. Feeds by straining zooplankton through gill-rakers with mouth wide open and operculum flared. Attracted to lights at night. A foodfish, but flesh soft and easily spoils. **Range:** Red Sea to Samoa, n. to Ryukyus, s. to S. Africa and GBR *(Photo: Marsa Alam, KH).* Kawakawa *Euthynnus affinis* occasionally along clear reef slopes (circumglobal; 38 cm).▼

Yellowfin tuna   2.2 m; 176 kg
Thunnus albacares
2nd D and A fins lengthen with age. **Biology:** oceanic, surface to 300 m. In large schools in open sea. Occasionally visits steep dropoffs or enters bays. Feeds on fishes, squids and crustaceans, often at surface along current lines. Among the world's most valuable food and sport fishes. **Range:** circumglobal, in all waters above 15°C.

Marsa Alam, KH

Dogtooth tuna   2.0 m; 131 kg
Gymnosarda unicolor
silver; large mouth with large canine
teeth; tips of D and A fins white.
Biology: solitary or in small groups in
midwater along steep lagoon, channel,
or seaward reef slopes, 1 to 100 m. A
voracious predator of fishes,
particularly fusiliers and other
planktivores. Large adults may be
ciguatoxic. Sometimes curious of
divers. **Range:** Red Sea and Gulf of
Oman to Fr. Polynesia, n. to S. Japan,
s. to S. Africa and New Caledonia.

Maldives, RM

Double-lined mackerel   1.0 m
Grammatorcynos bilineatus
Elongate, somewhat compressed;
second LL behind P fin sharply
curved. **Biology:** occurs near the
surface of coastal and offshore waters,
occasionally near pinnacles and
seaward reefs. Solitary. **Range:** Red Sea
and Gulf of Aden to Samoa, n.
to Ryukyus, s. to S. Africa and GBR.
Similar: Wahoo, *Acanthocybium
solandri* is less compressed and inhabits
open seas, rarely along steep seaward
slopes (circumglobal; 2.1 m). ▼

GBR, RM

**Narrow-barred Spanish
mackerel**   2.45 m; 45 kg
Scomberomorus commerson
Compressed; silver with wavy dusky
bars. **Biology:** solitary or in small
groups in midwater along deep lagoon
and seaward reef slopes,
1 to 200 m. A common coastal
pelagic species that migrates long
distances. Feeds primarily on
sardines, anchovies and fusiliers.
May approach divers, but never
lingers. A highly valued foodfish,
probably the most important one
in Oman. **Range:** Red Sea to Fiji,
n. to Korea and Japan, s. to S. Africa
and SE. Australia; migrant to
E. Mediterranean.

Palau, RM

Lahami, KF

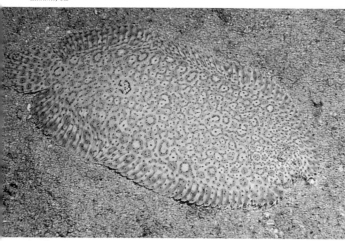

Mangrove Bay, RM

Sulawesi, LI

FLATFISHES

Order
PLEURONECTIFORMES

Highly compressed fishes that live with one side flat against the bottom. Both eyes migrate to one side before settlement and the blind side remains pale. Capable of remarkable colour changes to match the bottom. Several families, with all species carnivores of benthic invertebrates and fishes.

LEFTEYE FLOUNDERS
BOTHIDAE

Panther flounder   39 cm
Bothus pantherinus
Biology: on sand of lagoons, bays and sheltered seaward reefs, 1 to over 60 m. Often partially buried. Elongate P fin of ♂ used in courtship, territorial displays or when alarmed. **Range:** Red Sea and Arabian Gulf to Fr. Polynesia, n. to S. Japan, s. to S. Africa and SE. Australia.

SOLES   SOLEIDAE

Flatfishes with eyes on right side.

Moses sole   26 cm
Pardachirus marmoratus
Dark spots and ocelli with yellow flecks; eyes on right side, mouth below snout. **Biology:** inhabits shallow sandy areas near reefs, 1 to 15 m. Common, but usually completely buried except for eyes and nostrils. Secretes a milky toxin from pores at the base of D and A fins. **Range:** Red Sea and Arabian Gulf to Sri Lanka, s. to S. Africa.

Banded sole   16 cm
Soleichthys heterorhinos
Pale with dark bars; eyespots at rear of D and A fins. **Biology:** inhabits shallow protected sandy areas near reefs. Usually buried except for eyes and nostrils. Active and exposed only at night. Wary, capable of remarkable bursts of speed. Juvenile with blue fin margins, mimics a toxic flatworm. **Range:** Red Sea to Samoa, n. to S. Japan, s. to SE. Australia. Many similarly banded species, but these lack the eyespots.

▲ Blue triggerfish   55 cm
Pseudobalistes fuscus
Biology: inhabits lagoons, bays and
semi-protected seaward reefs, 0.5 to
50 m. Prefers areas of sand and rubble
near patch reefs or scattered coral
heads, also in seagrass beds. Spends
considerable time blowing away sand
to expose invertebrates, often with an
entourage of wrasses and other fishes
waiting to dart in for a meal. Wary,
but may be aggressive when guarding
a nest. Common. **Range:** Red Sea to
Fr. Polynesia, n. to S. Japan, s. to S.
Africa and GBR. *(Photo: Nuweiba, EL.)*

juv.

TRIGGERFISHES   BALISTIDAE

Triggerfishes are characterized by a relatively deep, compressed body
with eyes set high on the head, a small terminal mouth with large stout
incisiform teeth, a first dorsal fin with a spike-like spine followed by two
smaller ones, all of which fit into a groove, second dorsal and anal fins
with only rays, a pelvic fin reduced to a spiny knob, and a tough skin
covered with large armour-like non-overlapping scales. The first dorsal
spine may be locked upright and depressed only by lowering the second
smaller 'trigger' spine. Triggerfishes normally swim by undulating their
second dorsal and anal fins, but will use their tail for rapid bursts. When
alarmed, or at night, they wedge themselves in a hole by erecting the first
dorsal spine and pelvic girdle. They can be removed easily if one is able
to reach inside and unlock the spine. Most triggerfishes are solitary diurnal carnivores of a wide variety of benthic animals including crustaceans,
molluscs, sea urchins, other echinoderms, coral, tunicates and fishes.
Some feed largely on zooplankton and other pelagic animals such as salps
and jellyfishes, or on benthic algae. Their powerful jaws and teeth easily crush hard-shelled prey or snip off pieces of coral. Triggerfishes lay
demersal eggs in a nest which is aggressively guarded by the male. At
these times, some of the larger species will attack and bite an intruding
diver, and are capable of causing serious injury.

Egypt, KF

Yellowmargin triggerfish 60 cm
Pseudobalistes flavimarginatus
Tan with pink face, median fin margins yellow. **Biology:** inhabits deep lagoons, bays, channels and sandy notches of seaward reefs, 2 to 50 m. Builds nests of rubble (2 × 0.7 m) in sandy channels or seagrass beds. Aggressive when guarding eggs. May be ciguatoxic. **Range:** Red Sea to Fr. Polynesia, n. to S. Japan. s. to S. Africa and New Caledonia.

juv.

Titan triggerfish 75 cm
Balistoides viridescens
Olivaceous with light and dark areas, dark band above upper lip. **Biology:** inhabits lagoons and seaward reefs, 1 to 40 m. Usually solitary. Feeds on wide variety of invertebrates including corals, molluscs, crustaceans, sea urchins and worms as well as some algae. Becomes aggressive towards divers when guarding eggs and attacks by hitting or biting. May cause serious injury. Ciguatoxic in some areas. **Range:** Red Sea and Gulf of Aden to Fr. Polynesia, n. to S. Japan, s. to Mozambique and New Caledonia.

Bali, RM

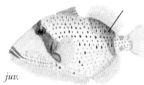

juv.

Orange-striped triggerfish
Balistapus undulatus 30 cm
Green with curved orange stripes; juv., ♀ with stripes on snout. **Biology:** inhabits coral-rich areas of lagoon and seaward reefs, 1 to 50 m. Solitary or in small groups. Feeds on fishes, corals, algae and benthic invertebrates. Excavates shallow nest in sand or rubble of channels. Probably haremic, spawns in loose groups. **Range:** Red Sea to Fr. Polynesia, n. to S. Japan, s. to S. Africa and GBR.

♀ *Marsa Shagra, RM*

Redtooth triggerfish   40 cm
Odonus niger
Dark blue-green with elongate tail
lobes and red teeth. **Biology:** inhabits
current-swept seaward reef slopes,
3 to 55 m. Abundant above exposed
terraces with little to moderate coral
growth and patch reefs. Forms large
aggregations high in the water and
feeds on drifting pelagic animals and
zooplankton. Hides in holes when
alarmed leaving tail lobes protruding.
Juveniles solitary or in groups around
isolated patch reefs. **Range:** Red Sea
and Arabian Gulf to Fr. Polynesia,
n. to S. Japan, s. to S. Africa and
GBR.

Marsa Shagra, RM

Indian triggerfish   25 cm
Melichthys indicus
Black with white tail margin and
D and A fin bases. **Biology:** inhabits
coral-rich seaward reef slopes, 5 to
53 m. Often well above the bottom.
Feeds on zooplankton. **Range:** S. Red
Sea and Oman to Sumatra, s. to S.
Africa. **Similar:** Starry triggerfish
Abalistes stellatus inhabits sandy to
muddy bottoms, occasionally near
reefs, 4 to 120 m (Red Sea to Fiji;
60 cm). ▼

Similan Is., RM

Largescale triggerfish   60 cm
Canthidermis macrolepis
Elongate; grey with dark P fins and
eyes; juv. ovoid with pale spots.
Biology: pelagic near drifting objects,
1 to 30 m. Visits steep seaward reefs or
offshore pinnacles to nest in sand
patches. Male guards eggs, possibly
also guards newly hatched larvae.
Solitary and wary. **Range:** Red Sea
to Gulf of Oman only. **Similar:** very
similar to *C. maculatus* (circumglobal,
33 cm).

Mangrove Bay, KF

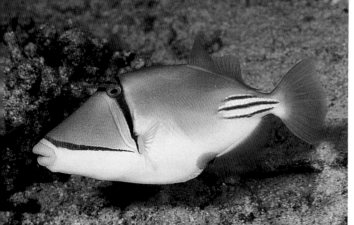

Mövenpick Bay, KF

Picassofish 30 cm
Rhinecanthus assasi
Tan above, white below; blue
stripes between and below eyes.
Biology: inhabits areas of mixed
rubble, sand and coral rock of lagoon
and shallow seaward reefs, 1 to 25 m.
Territorial and wary, always near
shelter. **Range:** Red Sea to Arabian
Gulf only. **Similar:** Wedge piccasofish
R. rectangulus (S. Red Sea to Fr.
Polynesia). ▼

Abu Galum, RM

Bluethroat triggerfish 22 cm
Sufflamen albicaudatus
Brown with white bar on tail base; ♂
with blue throat. **Biology:** inhabits
shallow lagoon and sheltered seaward
reefs, 2 to 20 m. Usually on sand and
rubble near scattered low corals. Feeds
on benthic invertebrates. Shy, retreats
to hole when alarmed. **Range:** Red
Sea only. **Very similar:** *S. chrysopterus*
lacks white bar on tail base (Oman
and Socotra to Samoa; 30 cm).

juv.

Oman, KF

Bridled triggerfish 38 cm
Sufflamen fraenatus
Grey-brown with pale stripe on
cheek. **Biology:** inhabits areas of
sand, rubble and scattered corals
of seaward reef slopes, 8 to 150 m.
Range: S. Red Sea (rare) and Oman
(common) to Fr. Polynesia, n. to S.
Japan, s. to S. Africa and SE. Australia.

juv.

▲ Honeycomb filefish 25 cm
Cantherinus pardalis
Face with pale bluish lines; sides
with polygonal brown spots, white
spot on upper caudal peduncle.
Biology: a common inhabitant
of coral-rich slopes of lagoon and
seaward reefs, 1 to 25 m. Also in
seagrass beds. Solitary. **Range:** Red
Sea and S. Oman to Fr. Polynesia,
n. to S. Japan, s. to S. Africa and SE.
Australia. *(Photo: Marsa Shagra, EL.)*
Similar: Blackvent filefish
Thamnaconus melanoproctes has stripes
instead of spots (Gulf of Aden to
Oman; 20 cm).▼

FILEFISHES   MONACANTHIDAE

Filefishes are closely related to triggerfishes. They differ by having more
compressed bodies, a longer and thinner first dorsal spine, much smaller
or no second dorsal spine, smaller and fewer teeth, and much smaller
non-overlapping scales, each with setae that give the skin a file-like texture. Most species have a brush-like patch of elongate setae in front of the
caudal peduncle which are usually better developed or even hooked in
males. Unlike triggerfishes, most filefishes are able to change their colour
to closely match their surroundings and are relatively secretive. Species of
the genus *Paraluteres* are remarkable mimics of poisonous puffers of the
genus *Canthigaster*. They share the puffer's colours and swimming behaviour and are more easily approached than other filefishes. One has to look
very hard to detect evidence of the dorsal spine which is kept depressed
or see the shape of the clear soft dorsal and anal fins which are more
broad-based than those of the puffers. Most species of filefishes feed on a
wide variety of invertebrates, but a few specialize on corals or zooplankton. In addition to reef-dwelling species, there are many others that prefer soft bottoms of turbid coastal areas or deeper continental shelves and
slopes. The species whose reproductive behaviour is known lay demersal
eggs that may be guarded by at least one of the parents.

Lahami, KF

Scrawled filefish   1.0 m
Aluterus scriptus
Olivaceous with blue spots and short
curved lines; tail large and broom-like.
Biology: inhabits lagoon and seaward
reefs, 2 to 80 m. Typically along
protected reef margins and slopes.
Juveniles pelagic around drifting
seaweed. **Range:** circumtropical; Red
Sea and Gulf of Oman in our area.

Marsa Shagra, EL

Broom filefish   20 cm
Amanses scopas
Dark brown with white lips; ♂
with patch of spikes near tail base.
Biology: inhabits clear lagoons,
bays and semi-protected seaward reefs,
1 to 20 m. Common in areas of mixed
sand, rubble and corals, and on coral-
rich reef crests. Usually in pairs.
Somewhat wary and secretive, always
near shelter. **Range:** Red Sea to Fr.
Polynesia, n. to S. Japan, s. to S.
Africa and GBR.

Nabq, RM

Harlequin filefish   7 cm
Oxymonacanthus halli
Blue-green with orange spots.
Biology: inhabits coral-rich areas
of lagoons, bays and sheltered
seaward reefs, 0.5 to 30 m. Typically
in pairs among *Acropora* corals or
among branched *Millepora* fire corals.
Occasionally in groups. Feeds
exclusively on *Acropora* coral polyps
which are extracted from their calices.
Range: Red Sea only. Replaced in
rest of Indo-Pacific by the very similar
O. longirostris with spot instead of bar
on tail (E. Africa to Samoa; 9 cm).

Spotted toby mimic 8 cm
Paraluteres arqat
Dark brown to olive with small
white spots above, pale below. 1st
D fin a spine; 2nd D and A fins
broad-based (vs. narrow-based
soft-rayed single D and A fins only
Canthigaster margaritata p. 222).
Biology: a remarkable mimic of the
poisonous Red Sea toby *Canthigaster
margaritata*. Apparently rare. Probably
territorial and haremic. **Range:** Red
Sea only. **Similar:** Indian Ocean
species (Andaman Sea, probably more
widespread) mimics the puffer
Canthigaster solandri.

Lahami, KF

Blacksaddle toby mimic 11 cm
Paraluteres prionurus
White with black saddles and smaller
black spots below; ♂ more colourful
than juveniles or ♀. 1st D fin a spine;
2nd D and A fins broad-based (vs.
narrow-based soft-rayed single D and
A fins only in *Canthigaster valentini*,
p. 222). **Biology:** inhabits clear
lagoon and seaward reefs, 1 to 25 m.
A remarkable mimic of the poisonous
blacksaddle toby *Canthigaster
valentini*. Solitary or in small groups.
Males territorial and probably
haremic. **Range:** Gulf of Aden and
Oman to Marshall Islands, n. to
S. Japan, s. to S. Africa and New
Caledonia.

♂, Sulawesi, RM

Lozenge filefish 25 cm
Stephanolepis diaspros
Snout concave, base of soft D
fin elevated; brown with pale
reticulations. **Biology:** inhabits
coastal rocky and coral reefs and soft
bottoms. Often caught by trawlers.
Juveniles associated with jellyfishes at
surface. **Range:** Red Sea to Arabian
Gulf only; a recent immigrant to the
E. Mediterannean. Randall's filefish
Pervagor randalli is a smaller
somewhat secretive inhabitant of
shallow reefs to 10 m (Red Sea to
Djibouti only; 10 cm). ▼

Juv., Marsa Alam, BE

TRUNKFISHES

OSTRACIIDAE

Trunkfishes are encased in a box-like shell of bony polygonal plates with small gaps for the mouth, eyes, gill openings, anus, and caudal peduncle. The surface of the plates is usually rough and in some species has angular ridges armed with spines. The mouth is small and low with thick lips and a single row of conical to incisiform teeth. Trunkfishes are slow-moving diurnal predators of a wide variety of small sessile invertebrates and algae. Like some triggerfishes, some trunkfishes will hang head-down and blow sand away to expose worms and other invertebrates. Those species that have been studied are haremic with males that defend a large territory containing non-territorial females and subordinate males. Spawning in pairs occurs at dusk, usually above a conspicuous outcrop. When under stress, trunkfishes secrete a highly toxic substance, ostracitoxin, which may be lethal to other fishes or even themselves if kept in a confined space. Swimming is gondolier-like, with soft dorsal and anal fins moving in a sculling motion.

▲ **Yellow boxfish** 45 cm
Ostracion cubicus
Juvs. yellow with small white ocelli, becoming mustard-yellow with growth; large ♂ s blue.
Biology: a solitary inhabitant of lagoon and semi-protected seaward reefs and offshore patch reefs with abundant coral growth, 1 to 45 m. Usually close to shelter or under corals or rocks. Juveniles in *Acropora* staghorn corals. Feeds on various benthic invertebrates and algae. Common, somewhat wary.
Range:: Red Sea and Arabian Gulf to Polynesia, n. to Gulf of Oman, Ryukyus, s. to S. Africa and N. New Zealand (*Photo: ♂, Lahami, KF*).

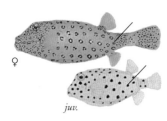

♀

juv.

Bluetail trunkfish 15 cm
Ostracion cyanurus
♀ dusky yellow with black spots;
♂ olive above with black-spotted blue
sides; juv. yellow. **Biology:** inhabits
shallow lagoon and protected seaward
reefs with moderate coral growth,
to 25 m. Uncommon, usually
solitary and always close to shelter.
Shy and secretive. **Range:** Red Sea
to Arabian Gulf only.

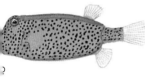

♂, Marsa Shagra, RM

Longhorn cowfish 46 cm
Lactoria cornuta
Straw-yellow with long paired spines
in front of eyes and at lower rear of
carapace. **Biology:** occurs on sand,
rubble, algae-covered rocks and
among seagrasses of shallow lagoons
and coastal reefs, 1 to 100 m.
Uncommon, solitary. Feeds on small
invertebrates by blowing away sand
to expose them. **Range:** Red Sea
and Gulf of Oman to Polynesia,
n. to S. Japan, s. to S. Africa and
SE. Australia.

Guam, RM

Thornback boxfish 30 cm
Tetrasomus gibbosus
Back with large thorn on elevated
ridge; light olive-grey with pale blue
spots. **Biology:** inhabits shallow sandy
flats of lagoon and coastal reefs with
algae and seagrasses and offshore soft
bottoms, 2 to 110 m. Occasionally
near patch reefs. Wary. **Range:** Red
Sea and Arabian Gulf to Indonesia,
n. to S. Japan, s. to S. Africa and
GBR; immigrant to E.
Mediterranean.

Dahab, JN

PUFFERS

Puffers are named for their ability to inflate themselves by drawing water into a specialized chamber near the stomach. The resulting prickly spherical ball deters many predators. Puffers have a tough, highly flexible scaleless and often prickly skin, beak-like dental plates with a median suture, a single posterior dorsal fin, and lack fin spines, pelvic fins, or ribs. The lateral line is inconspicuous or absent. Puffers harbour one of nature's most powerful toxins, tetradotoxin. The viscera, gonads and skin are usually the most toxic portions. The flesh is often safe, and in some areas is considered a delicacy, but it may occasionally be toxic and has caused many deaths. Toxicity varies greatly with species, area and season but may not always affect predatory fishes. Tobies (genus *Canthigaster*) have a repellent skin secretion that deters predation. Puffers lay demersal eggs. At least one species of *Canthigaster* is haremic. The diet of puffers varies among species. Some feed on a wide variety of plants and animals including fleshy and calcareous algae, sponges, cnidarians, worms, molluscs, crustaceans, echinoderms and tunicates. Others feed primarily on a particular item such as *Acropora* coral tips. Puffers inhabit all tropical and temperate seas and a few have invaded estuarine and fresh waters. Many species will bite if handled.

Mövenpick Bay, KF

Starry puffer   1.0 m
Arothron stellatus
White with close-set black spots; subadult with dusky patches. **Biology:** inhabits sandy stretches of deep lagoon and seaward reefs, 2 to 52 m. Often rests on sand, occasionally hangs in midwater above dropoffs. Juveniles in sandy areas of protected inner reefs. Solitary and common. May bite! **Range:** Red Sea and Arabian Gulf to Fr. Polynesia, n. to S. Japan, s. to S. Africa and N. New Zealand.

juv.

Whitespotted puffer   50 cm
Arothron hispidus
Brown with white spots sometimes forming reticulations; rings around eyes and P fin base. **Biology:** inhabits sheltered lagoons, bays and protected fringing reefs with moderate coral growth, 1 to 50 m. Common, often rests on corals, rubble or sand during the day. Solitary and slow moving. Feeds on sponges, tunicates, polychaetes, crabs, corals, molluscs, sea urchins, algae and detritus. **Range:** Red Sea and Gulf of Oman to Panama, n. to S. Japan, s. to S. Africa and SE. Australia. The related Immaculate puffer *A. immaculatus* is uniformly pale (Red Sea to Ryukyus; 28 cm).

Nuweiba, EL

Masked puffer   30 cm

Arothron diadematus

Tan with dark mask through eyes to P fin base. **Biology:** inhabits outer reef flats and coral-rich areas of lagoons, bays and seaward reefs, to 25 m. Solitary and slow moving. Often lies on bottom. Common. Courts and spawns in groups. **Range:** Red Sea only. **Similar:** replaced by Blackspotted puffer *A. nigropunctatus* from Gulf of Aden to Samoa (30 cm) with less black around P fin.

Shams Alam Reefs, EL

Yellowspotted puffer   14 cm

Torquigener flavimaculatus

Dark specks and reticulations with white spots above, pale below. **Biology:** inhabits shallow areas of coarse sand, rubble and seagrasses near coral reefs, 0.3 to 57 m. Swims close to bottom, or rests on sand and rubble. **Range:** Red Sea to Arabia Gulf, s. to Kenya and Seychelles; recent migrant to Mediterranean.

Jordan, AK

Crowned toby   13.5 cm

Canthigaster coronata

Cream with blue spots; 4 broad black saddles above. **Biology:** inhabits open areas of sand and rubble with scattered corals of lagoons and protected seaward reefs, 5 to 79 m. Remains near bottom. Solitary or in pairs. **Range:** Red Sea and S. Oman to Hawaii, n. to S. Japan, s. to S. Africa and SE. Australia; Pacific population has yellow instead of blue spots. If distinct, then the Red Sea–Indian Ocean species is undescribed. **Similar:** *C. valentini* has yellow-orange spots and middle 2 saddles extending to belly (p. 222).

Dahab, PM

Marsa Shagra, EL

Mangrove Bay, RM

Oman, KF

Red Sea toby 12 cm
Canthigaster margaritata
Dark brown to olive with small dark-edged spots. **Biology:** inhabits reef flats, lagoons, channels and protected seaward reefs, 1 to 30 m. Usually in open areas with mixed sand, rubble, seagrasses and corals. Solitary or in pairs. **Range:** Red Sea and Gulf of Aden only. **Similar:** very similar to *C. solandri* (Gulf of Aden and E. Africa to Fr. Polynesia; 11 cm). Nearly indistinguishable from the filefish mimic *Paraluteres arqat* (p. 217).

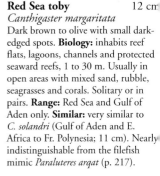

Dwarf toby 6 cm
Canthigaster pygmaea
Dark brown with blue spots and lines. **Biology:** inhabits coral-rich areas of lagoons, bays and sheltered seaward reefs, 1 to 30 m. Very secretive, usually in holes or crevices, sometimes around *Millepora* fire corals. Solitary and shy, usually on the move between holes. **Range:** Red Sea only.

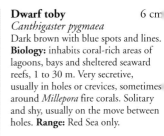

Blacksaddle toby 10 cm
Canthigaster valentini
White with 4 black saddles and numerous brown spots. **Biology:** inhabits lagoon and seaward reefs, 1 to 55 m. Feeds primarily on filamentous algae and tunicates, to a lesser extent on corals, bryozoans, polychaetes, echinoderms, molluscs and coralline red algae. Males are territorial and haremic and spawn with a different female each day. Females choose nesting sites of a tuft of algae. Hatching occurs in 3 to 5 days at sunset. The larval stage lasts 9 to 15 weeks. **Range:** Gulf of Aden and Oman to Fr. Polynesia, n. to S. Japan, s. to S. Africa and SE. Australia. Nearly indistinguishable from the filefish mimic *Paraluteres prionurus* (p. 217).

Porcupinefishes share the ability of puffers to inflate themselves, but have prominent spines over the head and body, larger eyes, an inconspicuous lateral line, and lack a median suture on the dental plates. The spines, actually highly modified scales, are either three-rooted and rigid (burrfishes, *Chilomycterus* and *Cyclichthys*), two-rooted and movable (*Diodon*), or a combination of the two. The spines of *Diodon* normally lie backwards, but stand erect when the fish inflates itself. A large inflated *Diodon* is an impenetrable fortress capable of choking a large shark to death. The beak-like jaws are well-suited for crushing the hard shells of prey such as molluscs, crustaceans, or sea urchins and may inflict a severe bite. Spawning has been observed in one species, *Diodon holocanthus*. It spawns at the surface, either at dawn or dusk, as pairs or groups of males with a single female. In many areas porcupinefishes are eaten or dried and sold as curios. In some areas they are reported to be poisonous due to either tetradotoxin or ciguatera.

Yellowspotted burrfish   34 cm
Cyclichthys spilostylus
Dusky above, spines short and rigid,
those above in pale spots, those below
in dark spots. **Biology:** inhabits
lagoon and sheltered coastal and
seaward reefs, 3 to 90 m. Prefers
seagrass beds with coral heads and
sand patches, occasionally around
wrecks. Usually inactive under ledges
or resting on the bottom by day.
Easily approached. **Range:** Red Sea
and Gulf of Oman to Galapagos
Islands. **Similar:** Blacklip burrfish
Cyphodiodon calori inhabits shallow
sandy lagoons with scattered corals,
2 to 5 m (S. Oman to Australia,
s. to E. Africa; 30 cm). ▼

Nuweiba, EL

Orbicular burrfish   15 cm
Cyclichthys orbicularis
Brown with clusters of dark
spots; spines short and rigid.
Biology: inhabits protected areas
of sand and rubble of coastal reefs,
3 to 20 m. Inactive by day under
corals, rocks, or in large sponges.
Feeds on crabs, molluscs and worms.
Uncommon. **Range:** Red Sea and
Arabian Gulf to NE. Australia, n. to
Japan, s. to S. Africa.

Salalah, KF

Mangrove Bay, RM

Porcupinefish 80 cm
Diodon hystrix
Tan with dark spots, pale below;
spines long and depressible.
Biology: inhabits lagoon and seaward
reefs, 2 to 50 m. Most often along reef
margins. During the day, usually
inactive under ledges or in caves,
occasionally hovering high in the
water. Forages at night for molluscs
and hermit crabs. Juveniles pelagic.
Range: circumtropical; all locations in
our area.

Bali, RM

Black-blotched porcupinefish
Diodon liturosus 50 cm
Tan above with large black blotches;
spines depressible, shorter than in
D. hystrix. **Biology:** an uncommon
inhabitant of coastal and seaward reefs
and offshore patch reefs, 5 to 90 m.
Usually inactive under ledges
or in crevices by day. Forages at
night on hard-shelled invertebrates.
Range: S. Red Sea and Oman to
Fr. Polynesia, n. to S. Japan, s.
to S. Africa and SE. Australia.
Similar: *D. holocanthus* but with
black blotch between eyes and
longer spines (below).

Salalah, KF

Spiny balloonfish 29 cm
Diodon holocanthus
Pale with large black blotches and
smaller black spots, the number of
spots decreasing with age. Spines long
and depressible. **Biology:** inhabits
lagoon and seaward reefs, 1 to 100 m.
More often over open bottoms than
other species. Juveniles pelagic to a
size of at least 7 cm. More common in
cooler warm-temperate waters than
in tropical seas. **Range:** circumglobal
in tropical and warm-temperate seas;
Red Sea to Gulf of Oman in our area.

REPTILES
SNAKES
Order SQUAMATA

Yellow-bellied sea snake   1.1 m
Pelamis platurus
Black back, yellow belly, flattened tail.
Biology: pelagic, occasionally washed ashore. Primarily surface-dwelling, occasionally with drifting seaweed. Gives birth to 2 to 8 live young. Feeds at the surface on small fishes which it grabs by whipping its head sideways. May mimic a floating stick to attract them. **Range:** S. Africa to Panama, n. to Arabian Gulf and Japan.
(Photo: Maldives, HV.)

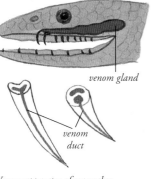

venom gland

venom duct

Venom apparatus of sea snakes

SEA SNAKES
ELAPIDAE subfamily Hydrophiinae

Sea snakes are completely aquatic, perfectly adapted to life in the sea. They have flattened paddle-like tails, nostrils that can close, and can partially respire through the skin (up to 20% O_2 and 100% CO_2) and remove salt through glands under the tongue. **True sea snakes** (Hydrophiinae; 54 species) are viviparous, that is they bear live young, while **sea kraits** (Laticaudinae; 5 species) are ovoparous, that is they lay eggs. Sea snakes can submerge for up to 2 hours to depths of 100 m. They may avoid decompression sickness by respiring quickly, then diving before nitrogen bubbles can form. Sea snakes have poor eyesight. They detect prey by sampling the water with their tongue which is slid across an olfactory organ (Jacobsen's organ) at the roof of the mouth, and by detecting the prey's vibrations with their muscles. Most species feed primarily on eels and small fishes that inhabit holes and burrows. Some feed on prey in the open by striking laterally. Sea snakes are normally non-aggressive unless molested, but a few species may be aggressive, particularly during the mating season when they may wrap themselves around divers. However, they may be unable to grip a wetsuit with their small mouths and short fangs. Fatal bites occur primarily among fishermen removing snakes from their nets. Sea snakes should never be handled since their venom is up to 10 times stronger than cobra venom. The Beaked sea snake, *Enhydrina schistosa* may inject enough venom in a single bite to kill over 50 people. Symptoms of a bite are swelling, vomiting, and paralysis of the respiratory muscles. Additional enzymes destroy muscle cells resulting in brown urine from the breakdown of myoglobin. A bite victim should be kept still and calm and receive immediate medical treatment. There are about 12 species of sea snakes in the Arabian Gulf and coast of Oman, but none reach the Red Sea, apparently unable to get past the cool water upwelling of the S. Arabian Peninsula.

SEA TURTLES

Order TESTUDINES

Sea turtles are highly adapted for life in the sea. They migrate over great distances and most return to their natal beaches to reproduce. Mating occurs in the water and only the females come to shore briefly to deposit their eggs. Breeding cycles vary, usually averaging about once every 2 years. Nesting usually occurs during summer or autumn, when a female lays several batches of eggs at intervals of 2 to 3 weeks. Incubation lasts about 2 months. Hatching usually occurs at night and mortality is extremely high with only about 1 in 100 surviving to reproduce. Hatchlings face a gauntlet of predators ranging from ghost crabs and seabirds on the beach to fishes in the water. Most sea turtles are primarily carnivores that feed on a variety of animals, ranging from sponges, jellyfishes and anemones to crustaceans and fishes. Green turtles feed increasingly on algae and seagrasses as they grow. Adult Green turtles can remain submerged for up to 5 hours by slowing their heart rate to 1 beat per minute. Sea turtles are an ancient group that has remained essentially unchanged for 150 million years. They have outlived the dinosaurs but face possible extinction at the hand of man. All 7 species are considered threatened or endangered. Their meat and eggs are used for food and their shells for jewelry. Countless numbers die each year from entanglement in fishing nets or by ingesting plastic bags. The Leatherback turtle is classified in a family of its own, **Dermochelidae**, all others are in the family **Chelonidae**.

▲ **Green turtle**   1.53 m; 150 kg
Chelonia mydas
Single pair of prefrontal scales.
Biology: inhabits all reef zones. Uncommon in the Red Sea. Young feed primarily on soft-bodied animals, adults feed primarily on plants, including mangroves and seagrasses. Nests every 2 to 3 years, laying 100 to 150 eggs per clutch at 10 to 15 day intervals. Incubation usually lasts 45 to 60 days with nests cooler than 30°C producing mostly ♂'s, and warmer nests producing mostly ♀s.
Range: all tropical and subtropical waters. *(Photo: Marsa Abu Dabab, BE.).*

prefrontal scale

Hawksbill turtle   1.14 m;
Eretmochelys imbricata   77 kg
2 pairs of prefrontal scales, shell plates overlap. **Biology:** inhabits all reef zones. The most common turtle in the Red Sea. Omnivorous, feeds primarily on invertebrates, including venomous Portuguese man-of-war. Clutch size

Mangrove Bay, RM

ranges from 53 to 206. The species most often used for 'tortoiseshell' products. **Range:** all tropical and subtropical waters.

Loggerhead turtle   2.13 m;
Caretta caretta   540 kg
2 pairs of prefrontal scales. Short bill. **Biology:** rare in Red Sea. Feeds on invertebrates, fishes, algae and seagrasses. Clutch size ranges from 64 to 200. Incubation lasts 49 to 71 days with eggs cooler than 28°C becoming ♂'s, those above 32°C becoming ♀s, those in between mixed. **Range:** all tropical and temperate seas. **Similar:** the smaller Olive ridley

Lepidochelys olivacea, inhabits primarily shallow lagoons and bays (all tropical seas; Arabia in our area).

Greece, HG

Leatherback turtle   2.44 m;
Dermochelys coriacea   867 kg
Carapace leathery with 5 length-wise ridges. **Biology:** the world's largest turtle. Pelagic, rarely near reefs. Feeds on jellyfishes, salps and other soft-bodied animals. Often killed by ingesting plastic bags or by fishing lines and drifting nets. Dives to at least 1,500 m. Adults warm-blooded with core temp. up to 18°C above water temp. Nests only on a few remote tropical beaches. Clutch size 50 to 170, but with many yolkless eggs. Incubation lasts 53 to 74 days. Numbers have declined precipitously in recent years. **Range:** all tropical to sub-polar seas.

Guayana, KF

MARINE MAMMALS
DOLPHINS AND WHALES

Order CETACEA

Dolphins and whales, known collectively as cetaceans, are highly streamlined, highly intelligent social animals that evolved from land mammals millions of years ago. They are divisible into two groups, those with teeth (suborder Odontoceti: all dolphins and toothed whales) and those with baleen for straining food from the water or from soft bottoms (suborder Mysticeti: baleen whales). Dolphins communicate by a complex language of organized sounds and use echolocation, a system of sonar utilizing clicks and whistles, to 'see' prey as well as surroundings.

DOLPHINS

DELPHINIDAE

1. Risso's dolphin *Grampus griseus* 3.83 m
Biology: inhabits steep slopes that drop below 300 m. Feeds on squid. Lives over 30 years. **Range:** all seas >10°C.

2. Pantropical spotted dolphin *Stenella attenuata* ♀ to 2.4 m; ♂ to 2.56 m, 120 kg
Beak short, body slender; juveniles without spots, adults with small white spots (density variable with age and area). **Biology:** occurs both inshore and offshore. Migrates to the Red Sea in pods of 5 to 500. Uncommon. Size at birth, 85 cm. **Range:** all tropical and subtropical seas. **Similar:** The Pacific hump-backed dolphin *Sousa chinensis* is grey to white with a small triangular dorsal fin above a basal hump. It is common in the S. Red Sea and Gulf of Aden (Red Sea to E. Australia, 2.80 m).

3. Rough-toothed dolphin *Stena bredanensis* 2.65 m; 160 kg
Biology: over deep water, usually seaward of continental and island shelves. In close-knit groups of 20 to 50.
Range: nearly all tropical and warm-temperate seas; S. Red Sea and Gulf of Aden in our area.

4. Pygmy killer whale *Feresa attenuata* 2.6 m
Biology: a rare oceanic species usually in groups of under 50. Pugnacious disposition. Feeds on squids and fishes.
Range: pantropical.

5. Short-finned pilot whale *Globicephala macrorhynchus* 7.2 m
Biology: continental shelves and outer reef slopes, in pods of 15 to 50 of mixed sex and age. Capable of diving as deep as 500 m. Feeds primarily on squid. Breeding season varies by population. Gestation lasts 15 months, lactation for at least 2 years and up to 15 years for the last calf. Males live to 46 years, females to 63 years. **Range:** all tropical and warm-temperate seas.

6. Killer whale (Orca) *Orcinus orca* 9 m; 5,600 kg
Biology: reported from the Arabian Sea. In pods of up to 60. Highly variable in behaviour and diet with pods specializing to suit their location. In tropics, feeds on sharks. Reproduces year round at intervals of about 5 years. Gestation lasts 15 to 18 months, lactation about 2 years. Males live up to 60 years, females to 90 years. **Range:** all polar and temperate seas with widely scattered occurrences in tropics. Uncommon in Red Sea.

SPERM WHALE

PHYSETERIDAE

7. Sperm whale *Physeter macrocephalus* ♂ to 18.3 m, 57,000 kg; ♀ to 11 m, 24,000 kg
Biology: rarely in deep waters near reefs, usually near deep undersea canyons. Feeds near bottom on medium to giant squids as large as 200 kg. Can dive below 3,000 m for as long as 2 hours. ♀s and immatures in small local areas for long periods, mature ♂s range over long distances. Breeds in warm seas. Calves at intervals of 4 to 6 years, with gestation as long as 18 months, and lactation of 2 to 13 years. Lives to at least 60 years. **Range:** all seas, to edges of polar ice.

BALEEN WHALES

BALAENOPTERIDAE

8. Minke whale *Balaenoptera acutirostrata* 10.7 m; 9,200 kg
Biology: in open seas in tropics. Alone or in groups with complex social structure and groups often segregated by age and sex. Feeds on schooling fishes and krill. May give birth annually, in winter after gestation estimated at 10 months. Lactation possibly as short as 6 months. Probably lives to 50 years. **Range:** nearly all warm seas in winter, migrating to edge of polar ice in summer; Gulf of Aden in our area. **Similar:** Bryde whale, *B.edeni*, has been observed in the S. Red Sea. Blue whale *B. musculus* (Arabian Sea, Gulf of Aden).

9. Humpback whale *Megaptera novaeangliae* 17 m; 40,000 kg
Biology: occasionally near coral reefs near deep water. A rare winter migrant to Red Sea. Arabian Sea population unique by remaining in the tropics year-round. Other populations spend summers at high latitudes to feed, then migrate to tropical archipelagos to breed. Has a repertoire of complex songs, unique to each population. Female gives birth to one 4 m long calf every 2 to 3 years. Calves remain with mothers 1 to 2 years. Lives to at least 50 years. Their only predator is the Killer whale. **Range:** worldwide in all major oceans. Uncommon in Red Sea.

DOLPHINS
DELPHINIDAE

◀ **Little bottlenose dolphin**
Tursiops aduncus   2.5 m
Beak short. Distinctive black lateral
stripe. **Biology:** inhabits primarily
coastal waters. Distinct from the
larger Bottlenose dolphin,
T. truncatus which ranges into
temperate waters (3.8 m, 500 kg).
Range: Red Sea to N. Australia,
n. to S. Japan. (*Photo: Nuweiba, MT.*)

Azores, RK

Striped dolphin   2.65 m ▶
Stenella coeruleoalba
Biology: inhabits open seas of outer
continental shelves, rare near shore.
Common in S. Red Sea, in dense
pods of up to 500. Feeds on fishes
and squids. Can jump 7 m. Matures at
8 to 12 years, lives to 58 years. **Range:**
all warm seas.

Spinner dolphin   2.11 m;
Stenella longirostris   78 kg
Biology: highly acrobatic, often jumps
and spins up to 3 m above the surface.
Occurs in small groups that regularly
visit offshore patch reefs. Small inshore
groups may fuse into larger groups of
thousands to feed in the open sea.
Often accompany tunas, making them
vulnerable to drowning in purse-seine
nets. Usually shy. Gestation lasts 12
months. Size at birth as small as 60
cm. Common in Red Sea. **Range:** all
tropical and subtropical seas.

Marsa Alam, BE

False killer whale   6 m;
Pseudorca crassidens   1,360 kg
Distinctive sickle fin. **Biology:** usually
in waters deeper than 1,000 m, in
groups of 10 to 20 that join into
much larger groups. Feeds on fishes,
squids, smaller dolphins and at least
the fins of large whales. Calving
interval nearly 7 years. Gestation lasts
14 to 16 months, lactation 1.5 to 2
years. Males live to 58 years, females
to 63 years. Uncommon in Red Sea.
Range: most tropical and temperate
seas.

Azores, RK

DUGONGS   Order SIRENIA: DUGONGIDAE

The dugong and 3 species of manatee (with rounded tails; W. Atlantic, Amazonia, E. Atlantic) are the only living representatives of the order Sirenia, or sea cows. The enormous 8 m long Steller's sea cow of the Bering Sea was wiped out within 27 years of its 'discovery' by seafarers in the 18th century. Dugongs are well adapted to aquatic life with a streamlined body, valves to close the nostrils and ears, and a thick skin with little hair. The upper lip has sensory bristles to detect food. They are herbivores that graze on algae and seagrasses and dig up roots. They swim slowly and use the tides to aid in migration. They live in pairs or small social groups, but occasionally occur in aggregations of up to 600. Communication seems to be by chirping sounds. Bulls may fight for females and often leave scars made by two tusk-like teeth. Gestation lasts 1 year culminating in the birth of a single metre-long pup. Weaning occurs in 2 years, but the young stay with the mother an additional 3 years. Dugongs live up to 73 years, but give birth to no more than 6 young during their lifetime. This low birth rate makes them highly susceptible to exploitation. Many indigenous people traditionally hunt dugongs for meat and oil. Others are drowned in fishing nets. In most countries it is illegal to kill them, but in a few areas, limited hunting for traditional purposes is allowed. Their future is uncertain.

Dugong
Dugong dugon   3.3 m; 400 kg (record: 4.1 m, 1,000 kg)
Massive body with flattened slightly forked tail; very old ♀s, ♂'s with tusks. **Biology:** inhabits shallow estuarine and coastal waters, usually among mangroves or in seagrass beds, occasionally on nearby coral reefs, 2 to at least 20 m. Usually shy, but occasionally curious of divers. In Egypt, consistently seen south of Qusier (Mangrove Bay, Nabaa Bay, Marsa Abu Dabab, Shams Alam Reefs, Ras Banas). Also in the Gulf of Suez, Tiran Is., Sudan and the Dahlak Archipelago. Estimates of 4,000 in the Red Sea and 6,000 in the Arabian Gulf may be too high. **Range:** Red Sea and Arabian Gulf to Vanuatu and Gilbert Is., n. to S. Japan, s. to Madagascar and Australia.
(Photo: Marsa Abu Dabab, BE.).

PART II: INVERTEBRATES

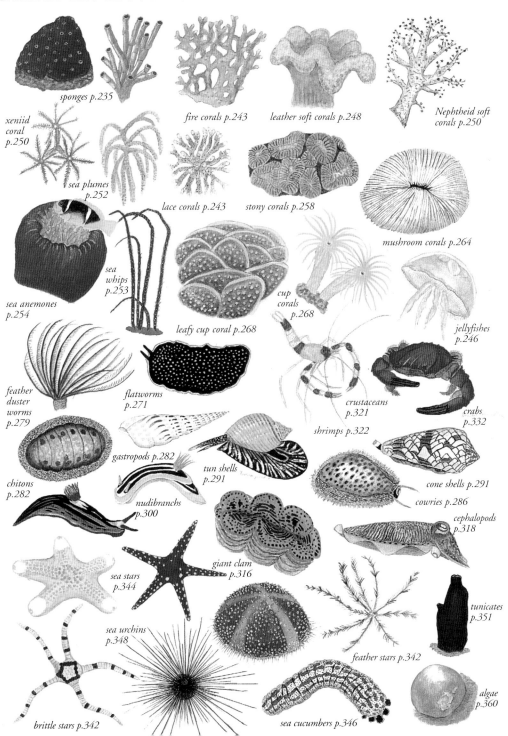

sponges p.235

fire corals p.243

leather soft corals p.248

Nephtheid soft corals p.250

xeniid coral p.250

sea plumes p.252

lace corals p.243

stony corals p.258

mushroom corals p.264

sea whips p.253

sea anemones p.254

cup corals p.268

leafy cup coral p.268

jellyfishes p.246

feather duster worms p.279

flatworms p.271

crustaceans p.321

crabs p.332

shrimps p.322

gastropods p.282

tun shells p.291

cone shells p.291

chitons p.282

cowries p.286

nudibranchs p.300

cephalopods p.318

sea stars p.344

giant clam p.316

tunicates p.351

sea urchins p.348

feather stars p.342

algae p.360

brittle stars p.342

sea cucumbers p.346

Sponges are the simplest and oldest multicellular animals. They first appeared during the Devonian, over 450 million years ago. Fossil sponge reefs occur in the Urals, the Australian Dividing Range, and other Palaeozoic mountain ranges. Sponges are sessile filter-feeders, consisting of an outer layer of tissue (cortex) and an inner layer of fibrous material (spongin) impregnated with glass-like spicules of silica or calcium carbonate. Their outer surface is riddled with small, often microscopic openings (ostia) connected to a complex network of internal canals. These are lined with specialized feeding cells, each with a thread-like cilia which projects into the canal. The combined motion of the cilia creates a current that brings in water laden with food and oxygen and removes metabolic wastes. The filtered water leaves though a larger central canal. A football-sized sponge can filter 2 tonnes of water per day.

Sponges are unique among multicellular animals by having their basic life functions carried out by cells rather than tissues or organs. Different cell types cover the outer and inner surfaces (cover cells), form the pores (pore cells), manufacture spicules for the fibrous skeleton, and create the flow of water and digest food (collar cells or choanocytes). Captured food is passed into the cytoplasm of the collar cells, engulfed there and digested in food vacuoles. Other cells roam amoeba-like throughout the sponge and are able to differentiate into any other type of cell as needed. Many sponges also host photosynthetic cyanobacteria that manufacture sugars from the sponge's metabolic wastes. These often form fans or encrustations in areas exposed to light. Sponges are hermaphrodites whose gametes mature at different times to prevent self-fertilization. Sperm is released into the water to be filtered by other sponges containing mature eggs. These then hatch into flagellated larvae which are released into the water.

Many sponges manufacture highly active chemical compounds which deter predation, or in some cases kill other animals that they overgrow. Some of these compounds have antibiotic or anticancer properties and are of great interest as a potential source of pharmaceuticals. The new herpes medicine 'Vidarabin' was synthesized from a sponge. The combination of spicules and irritating chemicals make many sponges unpleasant to handle. Predators of sponges include certain nudibranchs that use the sponge's chemicals for their own defence and angelfishes which can handle the spicules. Many sponges contain photosynthetic cyanobacteria which convert the sponges metabolic wastes into sugars and oxygen. These sponges grow as fans or encrustations in well-lit areas.

Sponges are found nearly everywhere. Boring sponges bore into the coral rock of the reef itself, eventually breaking it down. There are about 15,000 species of sponges worldwide, with perhaps 70% still undescribed. Most fall into the class **Demospongia** which have skeletons made of spongin with or without silica spicules. The skeletons of few of these are economically useful as bath sponges. The remainder of reef-dwelling species are members of the class **Calcarea** which have skeletons of calcium carbonate. Some of these, the stony-sponges (sclerosponges), resemble corals. The glass sponges (class **Hexatinellida**) have silica skeletons and are known only from deep water.

◀ **Toxic finger-sponge**
Negombata magnifica
(LATRUNCULIDAE). To 70 cm. Narrow crooked branches; reddish-brown. **Biology:** on well-lit terraces in clear water with scattered coral heads, to at least 25 m. Extremely toxic (Latrunculin). Fed upon by certain nudibranchs including *Chromodoris quadricolor*. **Range:** Red Sea and Indian Ocean. *(Photo: Zabargad, AK.)*

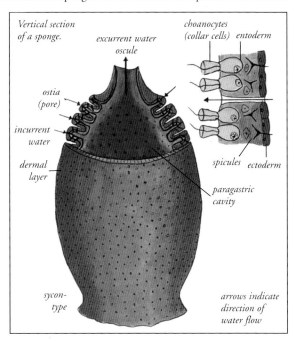

Vertical section of a sponge.

excurrent water
oscule

choanocytes (collar cells) entoderm

ostia (pore)

incurrent water

dermal layer

spicules ectoderm

paragastric cavity

sycon-type

arrows indicate direction of water flow

1. Sinai, EL

2. Marsa Shagra, EL

3. Marsa Shagra, EL

THEONELLIDAE
1 Swinhoe's sponge *Theonella swinhoe*
To *ca.* 20 cm. Robust rounded hard tubes, often encrusted with algae and epizoic organisms. **Biology:** in open areas of lagoon and seaward reefs to at least 20 m. **Range:** Red Sea to Philippines.

SUBERITIDAE
2 *Aaptos* sp. 1
To 20 cm across. Brown becoming greenish near edges of oscula due to symbiontic algae. **Biology:** on shallow lagoon and sheltered reef slopes with moderate current, to about 15 m. **Range:** Red Sea.

3 Liver sponge *Suberea* sp.
To 15 cm. Exterior olive, inside of siphons yellow. **Biology:** common on sheltered shallow reefs with little current, to 12 m. Usually on hard surfaces. **Range:** Red Sea.

4 Crusty sponge *Suberites clavata*
To about 20 cm. Encrusting on dead or living corals; creamy. **Biology:** in sedimented lagoon and seaward reefs with stretches of sand and rubble, to 20 m. A similar Mediterranean species (*S. domuncula*) contains a substance which stops the growth of blood cells in cancer tissue! **Range:** Red Sea.

CLIONIDAE
5 Red boring sponge *Pione* cf. *vastifica*
To 1 m. Encrusting with large oscula and small ostiae; red to yellow. **Biology:** primarily on shallow sheltered coral reefs, to 20 m. Bores 5 cm into coralline substrate. A dominant species in the Gulf of Aqaba. **Range:** Red Sea to W. Pacific.

AXINELLIDAE
6 Elephant-ear sponge *Acanthella carteri*
To 50 cm. Robust, usually free-standing with many spicules; yellowish-orange. **Biology:** common in shaded areas of fore-reef slopes with moderate current, to at least 60 m. Contains antibiotic substances that deter predators. **Range:** Red Sea to Australia.

7 Carter's fan-sponge *Stylissa carteri*
To 25 cm. Fan-shaped with numerous protuberances; yellow. **Biology:** on dead corals of poorly developed lagoon and seaward reef slopes with little current, to at least 15 m. **Range:** Red Sea.

HALICHONDRIIDAE
8 Arabian crust-sponge *Hemimycele arabica*
To 20 cm. Dark greenish-blue. **Biology:** on dead or living corals in well-lit areas with scattered corals and moderate current, to at least 20 m. Common. **Range:** Red Sea.

MYCALIDAE
9 Polyp sponge *Mycele fistulifera*
To 40 cm. Red, often hosts tiny white polyps of the scyphozoan *Nausithoe*. **Biology:** on shallow coral reefs, to 18 m. Encrusts dead or living corals. **Range:** Red Sea and Indian Ocean.

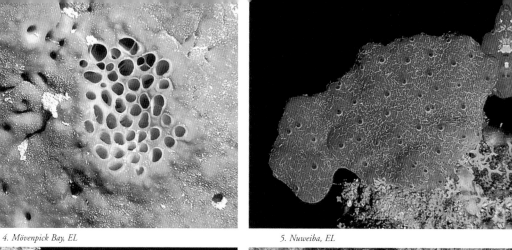

4. Mövenpick Bay, EL

5. Nuweiba, EL

6. Hurghada, JH

7. Nabq, EL

8. Marsa Bareka, EL

9. Mangrove Bay, KF

1. Hurghada, JH

2. Sinai, EL

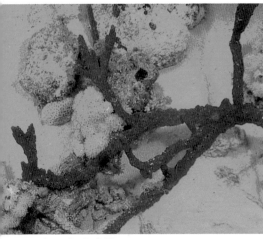

3. Shams Alam Reefs, EL

DESMACELLIDAE
1 Ehrenberg's boring sponge   *Biemna ehrenbergi*
To 10 cm. Clusters of small tubes; black. **Biology:** on reef slopes with coarse sand, rubble and living corals. Possibly toxic to humans. Do not handle! **Range:** Red Sea.

CRELLIDAE
2 Honeycomb sponge   *Crella cyathophora*
To 30 cm. Creamy body with distinctive warty oscula. **Biology:** on shaded walls and overhangs as well as open, well-lit terraces, to at least 20 m. Probably contains symbiotic algae. **Range:** Red Sea and Indian Ocean.

MYXILLIDAE
3 Iodine finger-sponge   *Iotrochota purpurea*
To 70 cm. Forms finger-like branches, dark brown. Smells strongly of iodine, hence its name. Turns dark in air. **Biology:** on open terraces with scattered coral heads, to at least 23 m. Rarely on slopes. **Range:** Red Sea.

CHALINIDAE
4 Large tube-sponge   *Haliclona fascigera*
To 80 cm. Clusters of soft tubes sometimes branched, contains spongin; rosy. **Biology:** on rocks and dead corals of fore-reef slopes, 3 to 25 m. Harbours shrimps and crinoids. **Range:** Red Sea to Australia.

CALLYSPONGIIDAE
5 Prickly tube-sponge   *Callyspongia crassa*
To 50 cm. Wide flexible tube with exterior protuberances, single or in cluster; brown. **Biology:** on coral-rich sheltered slopes or open terraces with moderate current, 5 to 30 m. Often harbours shrimps, polychaetes and gobies. Eaten by Yellowbar angelfish *Pomacanthus maculosus* (p. 135). **Range:** Red Sea to Seychelles.

6 Green sponge   *Callyspongia viridis*
To 25 cm. Encrusting with conspicuous small oscula. **Biology:** on open well-lit areas of shallow coral reefs, to about 22 m. **Range:** Red Sea.

7 Colonial tube-sponge   *Siphonochalina siphonella*
To 60 cm. Clusters of long tubes, contains spongin needles; pale lavender. **Biology:** on hard surfaces of clear lagoon and sheltered seaward reef slopes, 2 to 35 m. Common. Often harbours shrimps and polychaetes. **Range:** Red Sea only.

SPONGIIDAE
8 Leafy plate sponge   *Carteriospongia* (?) sp.
To about 30 cm. Plate-like with many oscula; grey. **Biology:** on open terraces or in deep lagoons, to about 20 m. Usually between dead or living corals. **Range:** Red Sea.

THORECTIDAE
9 Erect finger-sponge   *Hyrtios erectos*
To 50 cm. Large phallic-like rod or clusters; black to dark grey, often covered by sand. **Biology:** on shallow sheltered inshore reefs and seaward slopes, to at least 30 m. **Range:** Red Sea to Micronesia.

4. Siyul Reefs, EL

5. Mangrove Bay, RM

6. Shams Alam, EL

7. Shams Alam, EL

8. Lahami, EL

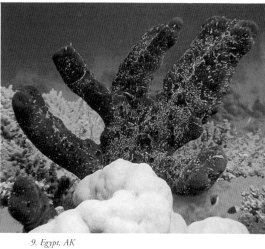

9. Egypt, AK

CNIDARIANS Phylum CNIDARIA

Cnidarians are a large and diverse group which includes the hydroids, hydrozoans, jellyfishes, anemones and corals. They are an ancient group that first appeared 550 million years ago and have radiated into thousands of species during their long history and presently include over 9,000 described species.

Cnidarians are simple multicellular animals consisting of a mouth and stomach arranged in a radially symmetric body-plan. The body tissue is divided into two layers, an outer ectoderm and an inner entoderm (gastroderm) which lines the cavity that serves as the stomach. A single opening serves as mouth, anus and sex organ. This 'mouth' is usually surrounded by stinging tentacles which provide protection and capture prey. All cnidarians have **nematocysts**, specialized cells containing a small harpoon attached to a venom sac by a coiled spring-like thread that serves as a trigger. Nematocysts fire only once, then are replaced by new ones.

Cnidarians have a life cycle that alternates between two adult body forms, the **polyp** and the **medusa**. Polyps are sessile (attached to the bottom) while medusae are free-swimming. Some cnidarians have lost the medusae stage. Most cnidarians are carnivores that feed on micro-organisms captured by the tentacles and drawn into the mouth. Many have an internal skeleton or host symbiotic single-celled algae, **zooxanthellae**, that live within their tissues and give them their colour. These algae aid their host by producing nutrients through photosynthesis, the process that converts carbon dioxide and metabolic wastes into sugars and oxygen. Stony corals are the best known zooxanthellae hosts, but many other cnidaria including anemones and soft corals as well as *Tridacna* clams, and some flatworms and tunicates, also host them. Cnidarians are classified into four classes based on morphology and life cycle.

The **hydrozoans** (class Hydrozoa) are colonies of polyps which produce medusae as a sexual generation by budding. They include reef-building fire and stylaster corals, feather- and gorgonian-like hydroids, and free-swimming or wind-driven siphonophores which are often mistaken for jellyfishes.

The **scyphozoans** (class Scyphozoa) are the pelagic jellyfishes. Their tiny polyps bud off from their top the saucer-like medusae that may grow to 2 metres in diameter.

The solitary **cubomedusae** (class Cubozoa) develop polyps and medusae in an alternation of generations. Certain cubomedusa (see p.18) nematocysts have killed humans by skin contact!

The **anthozoans** (class Anthozoa) have polyps only. Members of the subclass **Octocorallia** (leather corals, organpipe corals, soft corals, gorgonians and sea pens) have tentacles arranged in multiples of eight. Many of these host zooxanthellae. Members of the subclass **Hexacorallia** (stony corals, sea anemones, mushroom anemones and zoanthids) have tentacles arranged in multiples of six. Most of the coral reef-dwelling species host zooxanthellae. Black corals and tube anemones are included in the subclass Ceriantipatharia and do not host zooxanthellae. Many Anthozoans, particularly the stony corals, exhibit extreme variation in colour and gross morphology. Exposure to light, currents, wave action and other organisms competing for space affect the way they grow and look.

◀ Cnidarians are the dominant organisms of coral reefs with the accumulated skeletons of stony corals providing most of the reef's structure. *(Photo: Egypt, BE.)*

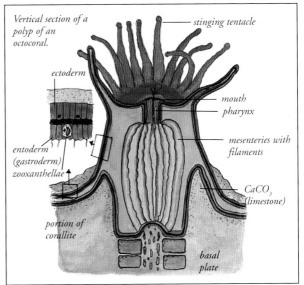

Vertical section of a polyp of an octocoral.

stinging tentacle

ectoderm

mouth

pharynx

entoderm (gastroderm) zooxanthellae

mesenteries with filaments

$CaCO_3$ (limestone)

portion of corallite

basal plate

HYDROZOANS
Class HYDROZOA

Polyp phase dominant, medusae absent or reduced. Colonize rocks and dead corals. Includes fire corals, hydroids and siphonophores.

FIRE CORALS — MILLEPORIDAE

Resemble true corals with strong calcareous skeleton and photosynthetic zooxanthellae, but lack cups. Have two forms of polyps, one for feeding by snaring zooplankton, the other with stinging cells for defence. The nematocysts contain a virulent toxin which causes painful burn-like wounds. Those with an allergic reaction may collapse.

◄ 1 Net fire coral *Millepora dichotoma*
Fans to 60 cm may form stands of up to several metres. Mustard to olive-yellow; forms branching colonies in single plane. **Biology:** along exposed upper portions of reef slopes, 0 to over 15 m. Abundant on most Red Sea reefs. Grows transverse to current for optimal exposure to plankton. Strong stinging cells may result in skin irritation for 1 to 2 weeks! **Range:** Red Sea to Samoa and Johnston Atoll, s. to S. Africa. *(Photo: Brothers Is., AK.)*

2. Mövenpick Bay, EL

2 Plate fire coral *Millepora platyphylla*
Individual plates to 60 cm form stands of up to several metres. Mustard to olive-yellow; forms knobby plates. **Biology:** inhabits upper reef slopes as well as lagoons, back-reefs and reef flats, 0 to 15 m. More common in less exposed areas than *M. dichotoma*. May overgrow *Porites* and other corals. Hosts Christmas tree worms *Spirobranchus giganteus*, vermetid molluscs and bivalve *Pteria aegyptiaca*. Sting is not severe. **Range:** Red Sea to Samoa and Johnston Atoll, s. to S. Africa.

LACE CORALS — STYLASTERIDAE

3 Violet stylaster *Distichopora violacea*
Fans to 15 cm. Violet with knobby branches containing stinging cells at tips. **Biology:** in shaded holes of upper reef slopes with fan transverse to current. Lacks zooxanthellae. Sting is feeble, usually not noticed. **Range:** Red Sea to W. Pacific.

3. Guam, RM

GORGONIAN HYDROIDS — SOLANDERIDAE

4 Gorgonian hydroid *Solanderia* sp.
Fans to 30 cm. Network of brown elastic chitinous branches in single plane with white polyps. **Biology:** primarily on reefs with moderate current, 3 to 20 m. Often in shaded areas with plane perpendicular to current. Feeds on plankton. Lacks zooxanthellae. Easily confused with gorgonians, but stings! **Range:** Red Sea to W. Pacific.

4. Dahab, AK

FEATHER HYDROIDS PLUMULARIIDAE

1 *Macrorhynchia philippina*
Tufts to 30 cm. White polyps with black central stems.
Biology: Coral reefs, 0.5 to 20 m. Grows on dead coral
and rubble. Strong sting may cause allergic reaction and
collapse. **Range:** circumtropical.

2 *Gymnagium eximium*
Tufts to 25 cm. Feather-like branches. **Biology:** on
exposed coral slopes with moderate to strong current.
Stings severely! **Range:** Red Sea to W. Pacific.

3 *Aglaophenia* sp.?
Clumps to 20 cm. Feather-like, with secondary branches
along a primary stem. **Biology:** on exposed reefs with
moderate current. **Range:** Red Sea.

HYDROIDS SERTULARIIDAE

4 *Dynamena* sp.
Tufts to 10 cm. Thecae (polyp cups) tiny, in two alternating
rows. **Biology:** mainly on dead sheltered reefs, 3 to 25 m.
A pioneer species. **Range:** Red Sea to W. Pacific.

5 *Sertularia* sp. A
Tufts to 12 cm. Thecae brownish. **Biology:** on sheltered
reefs with moderate current. **Range:** Red Sea to W. Pacific.

6 *Sertularia* sp. B
Tufts to 12 cm. Sparsely branched stalks. **Biology:** on
sheltered reefs with moderate current. **Range:** Red Sea to
W. Pacific.

FLOATING HYDROIDS VELLELLIDAE

7 Blue button *Porpita porpita*
Disc to 1 cm. Central disc with radiating blue tentacles.
Biology: pelagic, hydroid stage floats on surface,
occasionally washes ashore. Medusae stage minute. **Range:**
cosmopolitan in tropical and temperate seas.

8 By-the-wind sailor *Velella velella*
Disc to 10 cm. Clear chitinous float edged with blue
tentacles. **Biology:** pelagic, hydroid stage floats on surface,
occasionally washes ashore. Medusae stage minute. **Range:**
cosmopolitan in tropical and temperate seas.

Subclass SIPHONOPHORA: PHYSALIIDAE

Each 'animal' is a colony of small medusa-like individuals
and polyps specialized for feeding or reproduction.

9 Portuguese man-of-war *Physalia physalis*
Sail to 30 cm. A clear floating pneumatophore filled
with carbon monoxide above blue retractile tentacles
Biology: drifts on surface, occasionally over reefs. Fishing
tentacles can stretch to 30 m. Nematocysts may cause
serious injuries with intense pain and redness requiring
antihistamines and pain killers! **Range:** cosmopolitan in
tropical and warm-temperate seas.

1. Mövenpick Bay, HM

2. Sinai, AK

3. Mangrove Bay, EL

Siyul Reefs, Egypt, EL

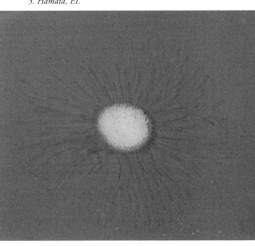

5. *Hamata, EL*

Marsa Shagra, EL

7. *Andaman Sea, RM*

. Aqaba, JN

9. *Guam, RM*

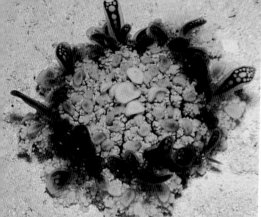
1. Nuweiba, AK

2. Nuweiba, PM

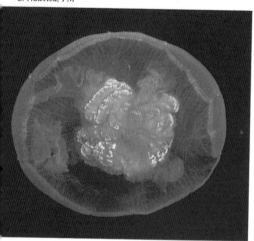
3. Shams Alam Reefs, EL

JELLYFISHES
Class SCYPHOZOA

CASSIOPEIDAE
1, 2 Upside-down jellyfish *Cassiopea andromeda*
To 15 cm across. Mouth surrounded by 8 fringed tentacles that host zooxanthellae. **Biology:** in shallow protected lagoons and seagrass beds, intertidal to 15 m. Upside-down on bottom to expose zooxanthellae to sunlight. Common. **Range:** Red Sea to Fr. Polynesia.

ULMARIIDAE
3 Moon jellyfish *Aurelia* sp. *aurita*
To 25 cm across. Transparent with four ear-like reproductive tissues, and hair-like tentacles. **Biology:** often in shallow water near coral reefs after storms. Pulsates. Stings cause a brief prickly sensation. Common. **Range:** circumtropical.

THYSANOSTOMATIDAE
4 *Thysanostoma loriferum*
To 20 cm across. Umbrella with red to violet edge; tentacles long. **Biology:** in open water over coral reefs or in lagoons after storms. **Range:** Red Sea to Polynesia.

CEPHEIDAE
5, 6 Cauliflower jellyfish *Cephea cephea*
To 15 cm across. Umbrella with about 30 tubercles on top and about 80 marginal lappets. **Biology:** in open water above coral reefs. **Range:** Red Sea to Polynesia.

NAUTISTHOEIDAE
7 Tubular sponge polyp *Nausithoe punctata* (?)
Tubes to 1 cm. Chitinous tubes with up to 70 tentacles. **Biology:** on sponges (*Desmocella, Mycale, Myxilla, Suberites*) in sheltered areas. Pelagic medusae stage up to 2 cm across. Sting severe. **Range:** circumtropical?

PELAGIIDAE
8 Luminescent jellyfish *Pelagica noctiluca*
To 10 cm across. Mouth with 4 lappets; tentacles to 1.5 m long. **Biology:** in open water near coral reefs. Luminescent. Sting severe, may cause fever. **Range:** cosmopolitan.

SEA WASPS
Class CUBOZOA

CARYBDEIDAE
9 Warty sea wasp *Tamoya gargantua*
Bell to 22cm height. Clear cuboid bell elongate with pale wart-like bumps; each corner with brown tentacle. **Biology:** polyp stage cryptic. Juvenile stage hidden by day, rises to surface at night to feed. Adult pelagic, occasionally drifts inshore. Sting severe with long-lasting effects. The large Indo-Australian *Chironex fleckeri* can kill an adult within minutes. **Range:** Indo-Pacific. **Similar:** *Carybdea alata* is entirely clear, nearly invisible (bell to 10 cm; circumtropical).

4. Marsa Alam TE

5. Oman, HF

6. Mangrove Bay, KF

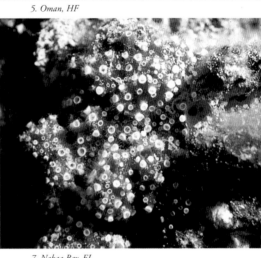

7. Nabaa Bay, EL

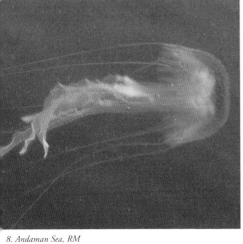

8. Andaman Sea, RM

9. Saudi Arabia, HS

1. Marsa Shagra, RM

2. Marsa Shagra, RM

3. Shams Alam, EL

ANTHOZOANS
Class ANTHOZOA
Subclass OCTOCORALLIA
SOFT CORALS and SEA FANS
Order ALCYONACEA

ORGANPIPE CORALS   TUBIPORIDAE
1 Organpipe coral   *Tubipora musica*
To 80 cm. Clumps of narrow hard red tubes housing pale grey polyps with 8 feathery tentacles. **Biology:** reef flats, lagoons and slopes, often in crevices. **Range:** Red Sea to W. Pacific.

LEATHER CORALS   ALCYONIIDAE
2 Sulphur leather coral   *Rhytisma fulvum*
To 2 m. Forms yellow encrusting sheets 1.5 cm high. **Biology:** usually on hard bottoms of sheltered dead coral reefs, 3 to at least 40 m. **Range:** Red Sea to W. Pacific.

3 Slimy leather coral   *Sarcophyton trocheliophorum*
To 80 cm. 36 species recognized, the broad-based *S. glaucum* and *S. trocheliophorum* are the most common but can't be distinguished visually. *S. trocheliophorum* is slimy to the touch, while *S. glaucum* is not. *S. glaucum* has larger spicules in its base (1–2 mm vs. 0.3–0.4 mm). **Biology:** in clear to turbid bays, harbours and sheltered seaward reefs, 1 to 30 m. **Range:** Red Sea to W. Pacific.

4 Rough leather coral   *Sarcophyton glaucum*
To 80 cm. **Biology:** on reef flats, in lagoons, and on seaward slopes. This and *S. trocheliophorum* are eaten by the egg cowrie, *Ovula ovum* (p. 288). **Range:** Red Sea to W. Pacific.

5, 6 Finger leather coral   *Sinularia polydactyla*
To 50 cm. Densely lobed encrusting colonies with long finger-like branches. **Biology:** on clear lagoon and seaward slopes, 1 to over 30 m. **Range:** Red Sea to W. Pacific.

7 Short-finger leather coral   *Sinularia leptoclados*
Clumps to 30 cm. Pale greenish-grey with finger-like branches (to 1.4 cm) forming clusters at ends of stout branches. A reef-building soft coral. **Biology:** on sheltered reef flats, in lagoons and on protected slopes, 0 to 20 m. **Range:** Red Sea to W. Pacific.

NIDALIIDAE
8 *Chironephthya variabilis*
To 50 cm. Forms large fan-like clusters with polyps at ends of branches; not fully retractile. **Biology:** under shaded overhangs or on cave ceilings exposed to currents. Uncommon in Red Sea. **Range:** Red Sea to W. Pacific.

9 *Siphonogorgia* cf. *godeffroyi*
To 60 cm. Forms fans; rusty red with white polyps. **Biology:** usually on current-swept fore-reef slopes below 15 m. Sometimes near entrance of caves. **Range:** Red Sea to W. Pacific.

4. Sinai, EL

5. Siyul Reefs, Egypt, EL

6. Mangrove Bay, KF

7. Mangrove Bay, RM

8. Sinai, AK

9. The Bells, EL

1. Sinai, EL

2. Ras Mohammed, EL

3. Egypt, HE

NEPHTHEIDAE

Have a hydroskeleton that controls stiffness of colonies by regulating internal water pressure.

1, 2 *Dendronephthya hemprichi*

To 70 cm. Pink or orange with transparent trunk; polyps in bundles of 6 to 11. **Biology:** from shallow lagoons to deep fore-reef slopes to over 100 m. A pioneer settler. Common. **Range:** Red Sea to W. Pacific.

3 *Dendronephthya klunzingeri*

To 1 m. Positive separation from *D. hemprichi* requires examination of spicules. This species usually with thicker base and red in colour with polyps in bundles of 1 to 3 on trunk and branches. Usually contracted at day. **Biology:** prefers deep water in shade of overhangs, entrances to caves, or walls with moderate current. **Range:** Red Sea to W. Pacific.

4 *Scleronephthya corymbosa*

To 20 cm. Sclerites numerous, yellow to orange; messy looking by day when retracted. **Biology:** usually on ceilings of caves, shaded walls, or at base of large gorgonians or table corals, 10 to 50 m. **Range:** Red Sea to Indian Ocean.

5, 6 *Litophyton arboreum*

To 80 cm. Pale olive-green to yellow or grey with clusters of finger-like branches on tall stalks. **Biology:** on seaward reef slopes, upright on hard bottoms. Common. **Range:** Red Sea to W. Pacific.

XENIIDAE

Possess large feathery pinnate tentacles.

7 Feathery xenid   *Anthelia glauca*

To 20 cm. Small clumps of polyps with long narrow stalk (5–10 cm) growing from basement plate, and with long pinnate tentacles. No pumping contractions. **Biology:** on upper slopes to deep fore-reefs with moderate current, 5 to at least 30 m. **Range:** Red Sea to Indian Ocean.

8 Pulsating xenid   *Heteroxenia fuscescens*

To 60 cm. Clumps of polyps with stalks *ca.* 5 cm long; pinnate tentacles shorter than in *Anthelia*. Pulsates rhythmically. Contains zooxanthellae. **Biology:** on hard bottoms of lagoons, bays and slopes with little current, 1 to 20 m. A pioneer settler. Pulsates rhythmically at about 40 times per minute to create a current to assist feeding and respiration. **Range:** Red Sea to Indian Ocean.

9 Umbrella xenia   *Xenia umbellata*

To 20 cm. Small clumps of polyps with stalks *ca.* 5 cm long; pinnate tentacles shorter than in *Anthelia*. Pulsates rhythmically, at slower rate than *Heteroxenia*. Contains zooxanthellae. In S. Red Sea, hosts the tiny pipefish *Siokunichthys bentuviai* (to 7.5 cm). **Biology:** on hard bottoms of lagoons, bays and slopes, 3 to 15 m. **Range:** Red Sea to W. Pacific.

4. Sinai, EL

5. Marsa Shagra, RM

6. Dahab, RM

7. Marsa Shagra, RM

8. Lahami, EL

9. Sinai, EL

1. Brothers I., AK

2. Sinai, EL

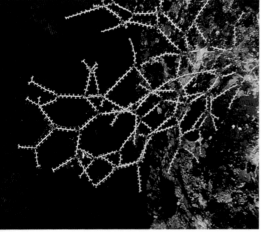

3. Egypt, HE

GIANT SEA FANS   SUBERGORGIIDAE
Large fans that lack zooxanthellae. Feed on microplankton.

1, 2 Giant sea fan   *Annella mollis*
To over 2 m. Large upright fans perpendicular to current.
Biology: on steep exposed seaward slopes and walls with
strong current, 10 to at least 50 m. **Range:** Red Sea to W.
Pacific. Red Sea population formerly known as *Subergorgia
hicksoni.*

NODED HORNY CORALS
MELITHAEIDAE
3 Noded Horny Corals   *Acabaria delicatula*
To 30 cm. Thin interlocking branches with red nodes.
Biology: in shaded areas under overhangs in bays, harbours
and on seaward reefs, 2 to 20 m. **Range:** Red Sea to Indian
Ocean.

4 *Acabaria biserialis*
To 80 cm. Interlocking branches with distinct nodes; red
with white polyps on planer side only. **Biology:** on deep
seaward slopes. Branches perpendicular to current.
Common around Elat. **Range:** Red Sea.

5 *Clathraria* cf. *rubrinodis*
To 100 cm. Bushy with dichotomous branches, red inside.
Biology: in lagoons, bays, and on slopes with moderate
current, 5 to at least 40 cm. **Range:** Red Sea only.

SEA PLUMES   PLEXAURIIDAE
6 Sulphur sea plume   *Bebryce sulphurea*
To 50 cm. Tangled mass of red branches with orange
polyps. **Biology:** seaward reef slopes, in notches.
Range: Red Sea, distribution elsewhere uncertain.

7 Sea plume   *Rumphella* cf. *aggregata*
To 50 cm. Branches elongate and drooping at ends; polyps
olive. **Biology:** in lagoons and on slopes with moderate
current, 2 to 25 m. Hosts photosynthetic zooxanthellae.
Common in S. Egypt. **Range:** Red Sea to Indian Ocean.

SEA WHIPS   ELLISELLIDAE
8 Red cluster whip   *Ellisella juncea*
To 60 cm. Dichotomously branched. **Biology:** on hard
bottoms of seaward reef slopes exposed to current, 30 to
60 m. **Range:** Red Sea to W. Pacific. **Similar:** Red whip
coral *Juncella rubra* occurs as clusters of single stalks
reaching 2 m in height.

SEA PENS   Order PENNATULACEA
PTEROEIDIDAE
9 Feather sea pen   *Pteroeides* sp.
To 10 cm. Specialized octocorals with a single stalk
anchored into sediment by a muscular peduncle.
Three families, this one broad and feather-like. **Biology:**
sandy expanses and slopes. Feeds on zooplankton.
Range: Red Sea, probably e. to W. Pacific.

4. The Bells, EL

5. Marsa Shagra, RM

6. The Bells, EL

7. Marsa Shagra, RM

8. Sinai, AK

9. Sinai, AK

1. Dahab, BE

2. Sinai, EL

3. Safaga, JH

Subclass CERIANTIPATHARIA

Order ANTIPATHARIA: ANTIPATHIDAE
1 Branching black coral　*Antipathes dichotoma*
Bushes to 2 m. Flexible skeleton with reddish-brown skin and polyps. **Biology:** on steep current-swept seaward reef slopes usually below 30 m. Slow growing. Feeds on plankton. Skeleton used for jewelry. **Range:** Red Sea to Polynesia.
2 Yellow wire coral　*Cirripathes anguina*
To 2 m. Single strand 1 cm wide. **Biology:** on steep reef faces, usually below 10 m. Feeds on microplankton. **Range:** Red Sea to Polynesia.
3 Spiral wire coral　*Cirripathes spiralis*
To 2 m. Single spiral. **Biology:** on steep reef faces, usually below 10 m. Feeds on microplankton. **Range:** Red Sea to W. Pacific.

TUBE ANEMONES Order CERIANTHARIA
CERIANTHIDAE
4 *Cerianthus* sp.
Tube width to 10 cm. Lives in tough flexible tube anchored in sediment. **Biology:** on silty sand of lagoon and seaward reefs, below surge to over 40 m. **Range:** Red Sea to Polynesia.

Subclass HEXACORALLIA

Order ZOANTHINIARIA: ZOANTHIDAE
5 Poisonous mat zoanthid　*Palythoa tuberculosa*
Polyps to 1 cm. Encrusting leathery mat. **Biology:** in shallows exposed to water motion, to 15 m. Highly poisonous, contains palytoxin produced by symbiotic bacteria. **Range:** Red Sea to Polynesia.

Order CORALLIMORPHARIA
MUSHROOM ANEMONES
ACTODISCIDAE
6 *Discosoma* sp.
Width to 10 cm. In large colonies of closely-spaced animals. **Biology:** shallow semi-protected reefs. **Range:** Oman to W. Pacific.

SEA ANEMONES　　　Order ACTINARIA
SWIMMING ANEMONES
BOLOCEROIDIDAE
7 *Boloceroides mcmurrichii*
Width to 15 cm. **Biology:** in lagoons and on upper reef slopes, 1 to 30 m. Solitary or in groups. Contains zooxanthellae. Clumps of tentacles may detach and swim off to form new colonies. **Range:** Red Sea to Polynesia.

ALICIIDAE
8 Precious anemone　*Alicia pretiosa*
Width to 30 cm. Trunk with knobs of nematocysts. **Biology:** among rubble or coralline rocks of deep lagoon and seaward slopes. Contracted by day, expanded at night. Causes severe sting. **Range:** Red Sea only.
9 Boxer crab anemone　*Triactis producta*
Width to 7.5 cm. **Biology:** in crevices, among rubble, or on fire corals, 0.3 to 30 m. Juveniles used for defence by Boxer crab *Lybia leptochelis*. **Range:** Red Sea to Fr. Polynesia.

4. Oman, KF

5. Sipadan Is., Malaysia, KF

6. Egypt, AK

7. Elat, JN

8. Mövenpick Bay, EL

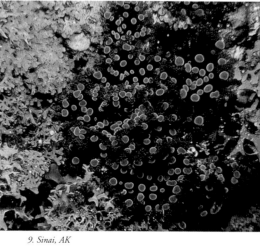

9. Sinai, AK

1. Mangrove Bay, RM

2. Mövenpick Bay, RM

3. Kenya, RM

ACTINODENDRIDAE

1, 2 Bubble anemone *Entacmaea quadricolor*
Width to 40 cm. Tentacles slightly transparent, often but not always bulbous near tip. **Biology:** anchored in crevices or among corals on lagoon and seaward reefs, 1 to 30 m. Common. Hosts anemonefishes and shrimps. **Range:** Red Sea to Fr. Polynesia.

CARPET ANEMONES
STICHODACTYLIDAE

3 Beaded anemone *Heteractis aurora*
Width to 25 cm. Tentacles with wart-like swellings. **Biology:** on sand of shallow lagoon and seaward reefs, 1 to 30 m. Uncommon in Red Sea. Hosts anemonefishes and shrimps. **Range:** Red Sea to Fr. Polynesia.

4 Magnificent anemone *Heteractis magnifica*
Width to 1 m. Column lilac to red or brown; tentacles with rounded tips. **Biology:** on exposed deep lagoon and seaward reef slopes and edges, 1 to 30 m. On hard bottoms, often lodged between dead corals. Zooxanthellate but also feeds on zooplankton. Hosts anemonefishes, shrimps and porcelain crabs. **Range:** Red Sea to W. Pacific.

5 Leathery anemone *Heteractis crispa*
Width to 30 cm. Column leathery; tentacles long but sparse. **Biology:** on sand of shallow lagoon and seaward slopes, 1 to 5 m. Often hosts *Amphiprion bicinctus* and *Periclimenes* shrimps. **Range:** Red Sea to W. Pacific.

6 Haddon's anemone *Stichodactyla haddoni*
Width to 75 cm. Trunk warty; disc carpet-like, tentacles sticky. **Biology:** on sand or fine rubble, 3 to 40 m. Hosts anemonefish *Amphiprion sebae* and juvenile *Dascyllus trimaculatus*. **Range:** Red Sea to W. Pacific.

HELL'S FIRE ANEMONES
ACTINODENDRIDAE

7 Hemprich's fire anemone *Megalactis hemprichii*
Width to 20 cm. Tentacles long with clumps of branchlets bearing clusters of nematocysts. Stings severely. **Biology:** inhabits sheltered sandy areas, particularly among seagrasses. Hosts *Periclimenes* shrimps but not anemonefishes. **Range:** Red Sea to W. Pacific.

STICKY ANEMONES
THALLASIANTHIDAE

8 Adhesive anemone *Cryptodendron adhaesivum*
Width to 35 cm. Rim raised, tentacles short, sticky and deciduous; colour variable, usually green to brown. **Biology:** on hard bottoms of lagoon and seaward reef slopes, 0.3 to 10 m. Often inhabited by the shrimp *Periclimenes brevicarpalis* (p. 324), but rarely used by anemonefishes. **Range:** Red Sea to W. Pacific.

CRAB ANEMONES   HORMANTHIIDAE

9 Hermit crab anemone *Calliactis polypus*
Width to 4 cm. Hundreds of thin tentacles. **Biology:** symbiotic on shell of hermit crabs including *Dardanus tinctor* (p. 338) in Red Sea. **Range:** Red Sea to Polynesia.

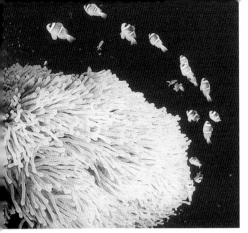

4. Sinai, AK

5. Nuweiba, EL

6. Yap, RM

7. Bali, RM

8. Mangrove Bay, EL

9. Sinai, MH

1. Marsa Shagra, RM

2. Marsa Shagra, RM

3. Marsa Shagra, RM

Order SCLERACTINIA   Stony corals

The primary reef-builders with at least 282 species known from the Red Sea and Gulf of Aden.

FINGER CORALS   POCILLOPORIDAE

1 *Pocillopora damicornis*
To 30 cm. Brown to lavender; branches with clumps of knobs, thinner and more open with increasing depth. **Biology:** sheltered to semi-exposed reef flats, lagoons and slopes to over 50 m. Abundant on reef flats and sheltered slopes. The damselfish *Dascyllus aruanus* (p. 141) and crabs *Trapezia* and *Tetralia* shelter between branches. **Range:** Red Sea to C. America.

2 *Pocillopora verrucosa*
To 40 cm. Brown to lavender; branches uniform and upright. **Biology:** exposed reef margins and slopes to 20 m. Common. **Range:** Red Sea to C. America.

3 *Seriatopora hystrix*
To 30 cm. Branches narrow and sharply pointed. **Biology:** subtidal reef flats, lagoons and protected slopes, 0.3 to 30 m. Common. **Range:** Red Sea to Fr. Polynesia. **Similar:** *S. caliendrum* has rounded tips, uncommon on upper reef slopes (Red Sea to Fiji).

4 *Stylophora pistillata*
To 20 cm. Cream to rosy; tips of branches rounded. **Biology:** reef flats and lagoon and seaward reef slopes, to 80 m. Abundant to dominant on shallow exposed reefs. Occasionally on sand. Home of many fishes and invertebrates. **Range:** Red Sea to Fr. Polynesia.

5 *Stylophora subseriata*
To 15 cm. Brown to purple; branches narrow and finely subdivided. **Biology:** lagoon and seaward reef slopes to at least 20 m. Common. **Range:** Red Sea to Fiji.

6 *Stylophora wellsi*
To 15 cm. Beige; branches short rounded knobs. **Biology:** only on exposed outer reef flats, locally common. **Range:** Red Sea and S. Oman to Madagascar.

STAGHORN CORALS   ACROPORIDAE

7 *Astreopora myriophthalma*
To 50 cm. Encrusting to hemispherical domes; calices large (to 4 mm). **Biology:** subtidal lagoon and seaward reefs. Common. **Range:** Red Sea to SE. Polynesia.

8 *Montipora stilosa*
To *ca.* 50 cm. Encrusting to submassive; beige to reddish-brown. **Biology:** shallow reefs, often at base of slopes. **Range:** Red Sea to Maldives. Many similar species.

9 *Montipora tuberculosa*
To 70 cm. Encrusting plates; brown to green to bright blue. **Biology:** in shallow sheltered reefs and slopes, to at least 30 m. **Range:** Red Sea to Polynesia.

4. Marsa Shagra, RM

5. Wadi Gimal, EL

6. Marsa Shagra, RM

7. Marsa Abu Dabab, RM

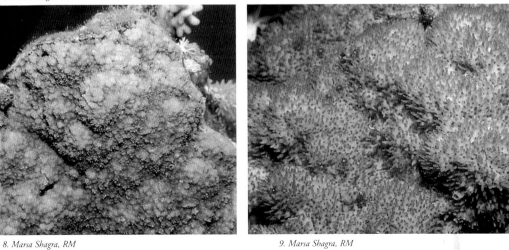

8. Marsa Shagra, RM

9. Marsa Shagra, RM

1. Marsa Shagra, RM

2. Marsa Shagra, RM

3. Marsa Shagra, EL

Acropora: the most diverse coral genus with at least 54 species in the Red Sea. Many similar, some distinctive.

1 *Acropora humilis*
To 20 cm. Calyces (polyp cups) arranged into orderly rows. **Biology:** exposed outer reef flats and upper slopes. Common. **Range:** Red Sea to Polynesia. **Similar:** *A. samoensis* with calyces less orderly (Red Sea to SE. Polynesia).

2 *Acropora microclados*
To 1 m. Irregular heads enlarging into plates. **Biology:** upper reef slopes. Usually uncommon. **Range:** Red Sea to Fiji.

3 *Acropora hemprichii*
To 1.5 m. Open bushy thickets; corallites differ in size. **Biology:** exposed reef slopes, 2 to 30 m. Common. **Range:** Red Sea to Sri Lanka. **Similar:** *A. formosa* with uniformly sized calices (Red Sea to Fr. Polynesia).

4, 5 *Acropora pharaonis*
To 1.5 m. Branches interlinked and contorted, upright when small, then spreading into plates; corallite cups pointed. Branches more open in deeper water. **Biology:** sheltered reef slopes, 2 to 30 m, exposed slopes below 15 m. A favoured home of the goby *Gobiodon citrinus* (p. 187). Common. **Range:** Red Sea to Sri Lanka. **Similar:** other large tabular species have shorter more evenly spaced branches and rounded corallite cups: *A. cytherea* with short narrow upright branches (Red Sea to Polynesia) and *A. parapharonis* with slightly longer branches (Red Sea only).

6 *Acropora lamarcki*
To 2 m. Bushy clumps that spread out into tiered table-like plates. **Biology:** clear lagoons, bays and seaward reef slopes, 1 to 25 m. **Range:** Red Sea to Sri Lanka.

7 *Acropora maryae*
To 60 cm. Branches narrow with many sub-branches and smooth rounded corallites. **Biology:** protected reef slopes. Common. **Range:** Red Sea only. **Similar:** *A. squarosa* has thicker branches, more prominent corallites (Red Sea to Chagos Arch.).

ANEMONE CORALS   PORITIDAE
8 *Goniopora columna*
To 20 cm. Large polyps with elongate trunks and 24 tentacles usually expanded by day. **Biology:** current-swept reef flats and slopes to 35 m. Forms large stands in turbid areas. Common. **Range:** Red Sea to Samoa. **Similar:** *G. ciliatus* has elongate tentacles, is common in shallow turbulent areas (Red Sea and NE. Africa).

9 *Goniopora planulata*
To 60 cm. Polyps uniformly grey. **Biology:** shallow protected reef slopes and lagoons. **Range:** Red Sea only. Other columnar species have larger brown to olive tentacles, sometimes with pink mouths.

4. Marsa Shagra, EL

5. Quseir, RM

6. Marsa Shagra, RM

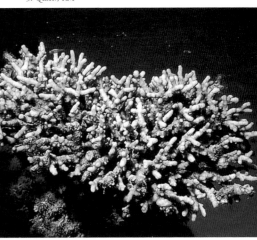

7. Gulf of Aqaba, RM

8. Hanish Is., EL

9. Abu Galum, RM

1. Nabq, EL

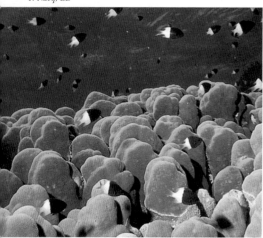

2. Mangrove Bay, RM

3. Nabq, EL

1 Mountain coral   *Porites lutea*
To 4 m across. Calyces tiny; forms microatolls on reef
flats or domes 5 to 7 m high in deeper water; pale green
to lavender or brown. **Biology:** reef flats and lagoon and
seaward reefs to at least 18 m. Among the most shoreward
of corals, dominant on many reef flats. **Range:** Red Sea to
C. America. **Similar:** *P. solida* and *P. lobata* have larger
calices, sides of *P. lobata* with more tiers.

2 Dome coral   *Porites nodifera*
To 4 m across. Rounded columns clustered into large
domes, may form patch reefs; pale olive. **Biology:** shallow
bays and seaward reefs to 18 m. Common. Shelters many
species of fishes. **Range:** Red Sea to Pakistan.

3 *Porites solida*
To 4 m across. Calyces larger than in other *Porites*; as small
mounds and columns forming large domes; brown to
yellowish-green. **Biology:** shallow clear reefs. Common.
Range: Red Sea to Polynesia. **Similar:** several species are less
common and smaller, some (*P. nigrescens*) are branching.

SIDERASTREIDAE
4 *Coscinaraea monile*
To 60 cm. Encrusting or domed; surface between polyps
fingerprint-like. **Biology:** lagoon and seaward reefs.
Common. **Range:** Red Sea to Indonesia and S. Japan.

AGARICIIDAE
5 *Gardinoseris planulata*
To 1 m. Encrusting to massive, acute ridges separate calyces;
edges sometimes laminar. **Biology:** on walls or under
overhangs in clear water. **Range:** Red Sea to Panama.

6 Porcelain coral   *Leptoseris yabei*
Stands often over 1 m. Laminar, forms tiers or whorls
when large; acute radiating ridges; calyces rectangular.
Biology: usually on flat surfaces. **Range:** Red Sea to Fiji.

7 Castle coral   *Pachyseris speciosa*
Stands to 3 m. Rounded laminar plates with concentric
ridges. **Biology:** deep lagoon and seaward reef slopes, more
common below 20 m. **Range:** Red Sea to Fr. Polynesia.

8 *Pavona varians*
To 2 m. Encrusting to laminar. **Biology:** lagoon and
seaward reefs, often tiered on lower slopes. **Range:** Red Sea
to Panama.

OCULINIDAE
9 Crystal coral   *Galaxea fascicularis*
Stands up to 5 m. Corallites with conspicuous spikes,
well-separated; cushions or irregular domes when small;
columnar to massive when large. **Biology:** as small patches
in areas exposed to surge to large stands on clear protected
reefs, 1 to over 30 m. Common. Polyps often expanded
during the day. **Range:** Red Sea to Fr. Polynesia.

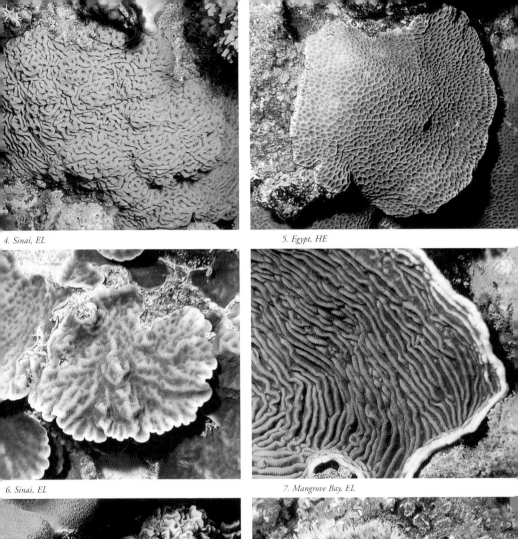

4. *Sinai, EL*

5. *Egypt, HE*

6. *Sinai, EL*

7. *Mangrove Bay, EL*

8. *Sewul, Egypt, EL*

9. *Marsa Shagra, RM*

1. Nuweiba, EL

2. Siyul, S. Egypt, EL

3. Abu Galum, RM

MUSHROOM CORALS   FUNGIIDAE

Most solitary with axial furrow and radiating septa (blade-like teeth), a few colonial. Most free-living, some attached. Juveniles may bud off adults.

1 *Ctenactis crassa*
To 48 cm. Elongate with several mouths and serrated septa. **Biology:** in deep lagoons and seaward slopes to 31 m. Often along reef edges. Common. **Range:** Red Sea to Fiji. **Similar:** *C. echinata* has a single mouth, to 60 cm.

2 *Fungia puamotensis*
To 25 cm. Thick and heavy often with central arch; usually pale brown. **Biology:** lagoons and reef slopes. Common. **Range:** Red Sea to Polynesia.

3 *Fungia scruposa*
To 24 cm. Circular to oval; some septa wavy. **Biology:** on rubble of reef flats and lagoon and seaward reef slopes, to 25 m. Common. **Range:** Red Sea to Fr. Polynesia.

4 *Herpolitha limax*
To 62 cm. Elongate with rounded ends and several mouths; often forked. **Biology:** lagoons and seaward slopes, 2 to 54 m. **Range:** Red Sea to Fr. Polynesia.

5 *Podobacia crustacea*
Plates to 50 cm. Large attached encrusting plates with numerous small mouths and pale free edges; may form tiered stands 2 or more metres across. **Biology:** lagoon and seaward reef slopes. **Range:** Red Sea to Fr. Polynesia.

LEAF CORALS   PECTINIIDAE

6 *Echinophylla aspera*
To over 2 m. Encrusting plates with smooth edges and contorted surface; may form tiers or concentric whorls; brown, green or red with red or green oral discs. **Biology:** lagoon and seaward reefs, usually on lower slopes, 1 to 100 m. **Range:** Red Sea to Fr. Polynesia. **Similar:** *E. echinata* and *Oxypora* spp. have toothed edges; *Echinopora lamellosa* has more closely packed calyces.

7 *Oxypora convoluta*
Stands up to 2 m. Highly contorted laminae with toothed edges. **Biology:** protected reefs slopes. Uncommon. **Range:** Red Sea only.

8 Elephant ear coral   *Mycedium elephantotus*
To over 1 m. Encrusting to laminar growing into large tiered plates. Brown, grey or green with green, white or red oral discs. **Biology:** on steep slopes with moderate current below surge. **Range:** Red Sea to Fr. Polynesia.

9 *Mycedium umbra*
To 60 cm. More contorted with smoother edges than *M. elephantotus*; colour distinct: purple-grey centres becoming yellowish peripherally and pink oral discs. **Biology:** on slopes with moderate current below wave action. **Range:** Red Sea to S. Oman only.

4. El Burqa, RM

5. Marsa Shagra, RM

6. Indonesia, HE

7. Siyul, S. Egypt, EL

8. Mövenpick Bay, EL

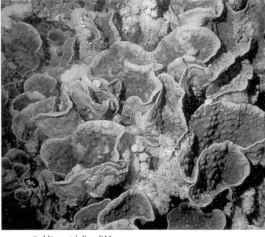

9. Mövenpick Bay, RM

1. Mangrove Bay, RM

2. Ras Mohammed, EL

3. Guam, RM

FAVIIDAE

Corallites usually large with ridged rims either well-separated or fused to neighbours; polyps fused into meandering channels in meandroid forms ('brain' corals). Many growth forms, particularly massive and meandroid but some finely branched or laminar.

1 *Echinopora fruticulosa*
To 2 m. Domes of interlocking branches, usually under 20 cm. **Biology:** lagoon and seaward reef slopes. Common. A favoured shelter for *Pseudochromis springeri* (p. 79). **Range:** Red Sea to Chagos Arch. Many other species resemble various pectinids (p. 266).

2 Honeycomb coral *Favia favus*
Over 1 m. Corallites large (> 12 mm), well separated; green to brown, centres often bright green. **Biology:** primarily in turbid water of reef flats and sheltered slopes. Common to dominant on back-reef margins. Polyps expanded at night. **Range:** Red Sea to Fr. Polynesia.

3 *Favia stelligera*
Domes of columns to 80 cm. Corallites small (< 8 mm) and conical; as encrusting patches, mounds or columns, may form huge domes; pale brown to green. **Biology:** reef flats and upper lagoon and seaward slopes. Common. **Range:** Red Sea to SE. Polynesia. Single column shown in photo.

4 *Favia lacuna*
To 1 m. Corallites nearly fused, deep with vertical inner walls; massive; olive with pale centres. **Biology:** shallow exposed slopes. Common. **Range:** Red Sea only.

5 *Goniastrea edwardsi*
To 1 m. Corallites fused and slightly angular, calyces deep; massive to columnar; pale tan to pink. **Biology:** shallow slopes. Common. **Range:** Red Sea to Samoa.

6 *Goniastrea peresi*
To 20 cm. Corallites fused and polygonal, calyces deep; encrusting to domed; pale grey to pink. Some forms resemble a *Favites*. **Biology:** shallow sheltered slopes. Common. **Range:** Red Sea to Maldives.

7 Brain coral *Platygyra daedalea*
To 2 m. Encrusting to massive; meandroid with thick walls; ridges across walls ragged. **Biology:** reef flats and slopes to 30 m, common on back-reef margins. **Range:** Red Sea to Fr. Polynesia.

8 Neat brain coral *Platygyra lamellina*
To 2 m. Encrusting to massive; meandroid with thick walls; ridges across walls neat and rounded. **Biology:** reef flats and upper reef slopes, especially back-reef margins. Uncommon. Night active. **Range:** Red Sea to Samoa.

9 *Plesiastrea versipora*
To 3 m. Encrusting, usually flat to lobed; corallites small (≤ 4 mm) and round; colour highly variable. **Biology:** most reef environments, usually in shaded areas. Uncommon. **Range:** Red Sea to SE. Polynesia.

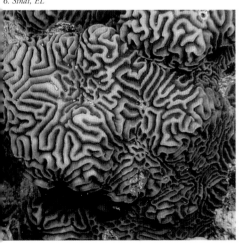

4. Mövenpick Bay, RM

5. Nabq, EL

6. Sinai, EL

7. Aqaba, EL

8. Nuweiba, EL

9. Marsa Shagra, EL

1. Marsa Shagra, RM

2. El Burqa, RM

3. Indonesia, HE

MUSSIDAE

Corallites large and attached only at base; deep valleys usually obscured by fleshy polyps; septa toothed.

1 *Acanthastrea hillae*
To 1 m. Encrusting to massive; polyps with concentric folds; colour variable, centres often pale. **Biology:** most reef zones, usually uncommon. **Range:** Red Sea to Fr. Polynesia.

2 *Lobophyllia corymbosa*
To over 2 m. Flat to hemispherical; corallites round to meandroid; brown to mustard with pale centres, polyps usually appear spiny. **Biology:** lagoon and sheltered reef slopes. Common. **Range:** Red Sea to Fr. Polynesia.

3 *Lobophyllia hemprichii*
To 5 m. Flat to hemispherical; corallites usually meandroid; colour variable with pale centres, polyps appear fuzzy. **Biology:** lagoon and sheltered slopes to 40 m. Common to dominant. **Range:** Red Sea to SE. Polynesia.

TRACHYPHYLLIIDAE
4 *Trachyphyllia geoffroyi*
To 12 cm. Flabello-meandroid with large fleshy polyps. Green to brown or blue. **Biology:** free-living on soft bottoms of protected slopes. **Range:** Red Sea to New Caledonia.

CARYOPHYLLIIDAE
5 *Euphyllia glabrescens*
To 1 m. Corallites 2–3 cm and 1.5–3 cm apart; polyps with long tentacles. Always expanded, resembles an anemone. **Biology:** deep slopes, 20 to 80 m. Conspicuous but rare in Red Sea. **Range:** Red Sea to Fiji.

6 Bubble coral *Plerogyra sinuosa*
To 3 m. Flabello-meandroid with deep, broad valleys; polyps with large bulbous tentacles that may obscure valleys. Creamy-grey. **Biology:** open, often turbid areas of protected slopes, 10 to 45 m. **Range:** Red Sea to Fr. Polynesia. **Similar:** *Physogyra lichtensteini* has smaller, more elongate bubbles (Red Sea to W. Pacific).

CUP CORALS   DENDROPHYLLIDAE
7 Red cup coral *Tubastrea coccinea*
To 30 cm. Orange with yellow tentacles. **Biology:** in caves and under overhangs, to 25 m. Lacks zooxanthellae, feeds on zooplankton. **Range:** Red Sea to Panama.

8 Dark cup coral *Tubastrea micrantha*
To 1 m. Dark brown to green bushes exposed to current. The only reef-building coral without zooxanthellae. **Biology:** along reef edges and under overhangs. Feeds at night on large plankton, particularly salps. **Range:** Red Sea to Fiji(?).

9 Leafy cup coral *Turbinaria reniformis*
To 3 m. Foliaceous to tiered; corallites conical; yellow-green. **Biology:** clear, semi-sheltered slopes with moderate current. Common. **Range:** Red Sea to Fr. Polynesia. **Similar:** *T. mesenterina* is usually grey-green to brown.

4. Bali, RM

5. Bali, RM

6. Mangrove Bay, EL

7. Nuweiba, RM

8. Mangrove Bay, RM

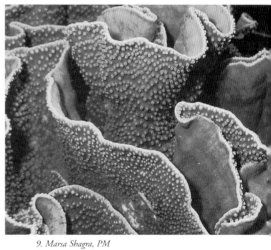

9. Marsa Shagra, PM

WORMS

'Worm' is a generalized term commonly applied to a diverse array of elongate or flattened legless invertebrates. There are at least 17 phyla of worm-like marine animals. Most live hidden within the reef or within other animals as symbionts or parasites. We include a few forms likely to be seen by divers.

Flatworms (phylum PLATYHELMINTHES) are the first animals to develop bilateral symmetry, anterior-posterior differentiation with a 'head', and three tissue layers. They are the likely ancestors of all higher animals. Most are symbiotic or parasitic (including tapeworms and flukes). Those seen by divers are the free-living marine polyclads (class Turbellaria, order Polycladida). These leaf-like creatures seem to flow over the bottom. Their mucus-covered bodies are driven by coordinated strokes of cilia. Surprisingly rapid locomotion or swimming occurs by undulations of well developed mesodermal muscles. Orientation occurs through information gathered by two eyespots and two sensory tentacles on top of the head which is fed to a tiny brain. Most polyclads feed on living and dead material sucked in through a ventral muscular pharynx that also serves as the anus. Digestion occurs in a primitive ciliated gastrovascular cavity connected to branched canals that reach every corner of the body. At least one recently discovered species is a voracious predator of small cowries. It is able to sense their chemical trail and outrun them as they flee. In a scene out of a bad horror film, it injects its victim with a paralysing substance. Within the span of a minute it everts its pharynx into the shell, separates tissue from shell, and ingests the soft portions intact and perhaps still alive, leaving an empty shell. Turbellarinas are hermaphrodites. During copulation, each partner transfers sperm into the other. It is stored until needed by yolk-laden eggs ready to hatch into planktonic larvae. The vivid colours of the polyclads most noticed by divers serves to warn predators that they are distasteful or toxic. There are also many species that are partially transparent or cryptic, such as the cowrie predator, which are likely non-toxic.

Ribbon worms (phylum NEMERTINA) are long highly contractile animals with a long proboscis used for impaling prey, and a complete digestive tract with a true mouth and anus.

Fireworms, Feather duster worms and their relatives (phylum ANNELIDA, class Polychaeta) have ringed bodies consisting of a sequence of anatomically similar segments enclosing an intestine, closed circulatory system with muscular arteries instead of a heart, and ventral nerve cords. Each side of each segment has a parapode, originally used for respiration and locomotion. Adaptive radiation of parapodes has resulted in the great diversity of annelids seen today. Parapodes of fireworms are armed with horny bristles that may inflict painful stings. Parapodes on the heads of nereids have evolved into jaws and sensory projections used for finding, catching and killing prey. Circular and longitudinal muscles power the crawling and swimming of free-living annelids. The sedentary forms (sedentaria) evolved from free-swimming forms by segregating segmental body functions to different regions of the body and building tubes in which to live. Feathered tentacles used for breathing and filtering plankton evolved from head parapodes. Feather duster worms (Sabellidae) secrete mucus used as mortar for building tubes of sand while Christmas tree worms (Serpulidae), with their double tentacle crowns, secrete calcareous tubes which can be sealed with an operculum (lid). Terebellids live hidden in crevices and holes with only long sticky mucus-coated tentacles exposed. These sedentary forms have retained the sensory equipment of their free-living ancestors which enables them to react rapidly to vibrations and changes in light. Dispersal and colonization is by planktonic trochophora larvae.

Bryozoans (phylum BRYOZOA) are tiny (*c.* 1 mm) colonial animals that live in external tubes connected into encrusting or fan-like networks. They have a crown of ciliated tentacles and one-way digestive tract. Some individuals (zooids) within colonies may be specialized for defence, reproduction or other functions. Most of the 5,000 known species are marine and are small and inconspicuous. A few that form fan-like networks resemble miniature corals and are quite beautiful.

◀ **Gold-dotted flatworm** *Thysanozoon* sp. (Platyhelminthes: Pseudocerotidae) To 5 cm. Black with yellow papillae; marginal band white; 2 pseudotentacles as folds along edge of body. **Biology:** on sand and rubble of lagoons and sheltered seaward slopes, 1 to 25 m. **Range:** Red Sea to W. Pacific. (*Photo: Eritrea, AK.*)

1. Jeddah, JK

2. Jeddah, JK

3. Jeddah, JK

PSEUDOCEROTIDAE

1 Bedford's flatworm
Pseudobiceros bedfordi

To 10 cm. Brown to black with yellow crossbands and dots and wavy black marginal band; 2 pseudotentacles and paired penis. **Biology:** on protected reefs, 2 to at least 20 m. An elegant swimmer and swift crawler. Feeds on ascidians and crustaceans. **Range:** Red Sea to W. Pacific.

2 Pin-striped flatworm
Pseudobiceros fulgor

To 2 cm. Brown with yellow pin-stripes and black marginal band; pseudotentacles dark. **Biology:** shallow rocky and coral reefs. Under stones and in crevices by day. Secretes large amounts of mucus. **Range:** Red Sea to W. Pacific.

3 Glorious flatworm
Pseudobiceros gloriosus

To 9 cm. Black with burgundy, pink and orange marginal band. **Biology:** on reef slopes to at least 20 m. On encrusted hard bottoms under shaded ledges. Nocturnal, swims by undulating. **Range:** Red Sea to Polynesia.

4 Pleasing flatworm
Pseudobiceros cf. *gratus*

To 5 cm. Creamy with 3 dark bands and curly black marginal band. **Biology:** on sand or among rubble of shallow lagoons and reef margins, to at least 10 m. Under stones by day, active by night. Fragile. Feeds on bryozoans and ascidians. **Range:** Red Sea to Polynesia.

5 Cryptic flatworm
Pseudobiceros murinus

To *ca.* 2 cm. Cryptically coloured with numerous black dots and a yellow marginal band. **Biology:** primarily on rubble or dead corals of shallow reefs. **Range:** Red Sea.

6 Starry flatworm
Pseudobiceros stellatus

To *ca.* 2 cm. Dark blue with a few white starry dots. **Biology:** intertidal and shallow coral reefs. Under rocks by day. **Range:** Red Sea.

7 Gold-margin flatworm

Pseudobiceros uniarborensis

To *ca.* 2 cm. Brown with distinctive golden-black marginal band. **Biology:** on encrusted hard or sandy bottoms of shallow protected reefs. Nocturnal. **Range:** Red Sea to Philippines.

8 Devil's flatworm *Thysanozoon (Acanthozoon)* sp. 2
To *ca.* 3 cm. Orange-red with numerous thorny black-tipped papillae and creamy marginal band. **Biology:** on sand or fine rubble of protected shallow reefs. **Range:** Red Sea.

9 Pinky flatworm *Thysanozoon (Acanthozoon)* sp. 3
To *ca.* 2 cm. Creamy with brownish-pink papillae, frilly margin and 2 white pseudotentacles. **Biology:** shallow coral reefs, under stones by day. A fast crawler. **Range:** Red Sea.

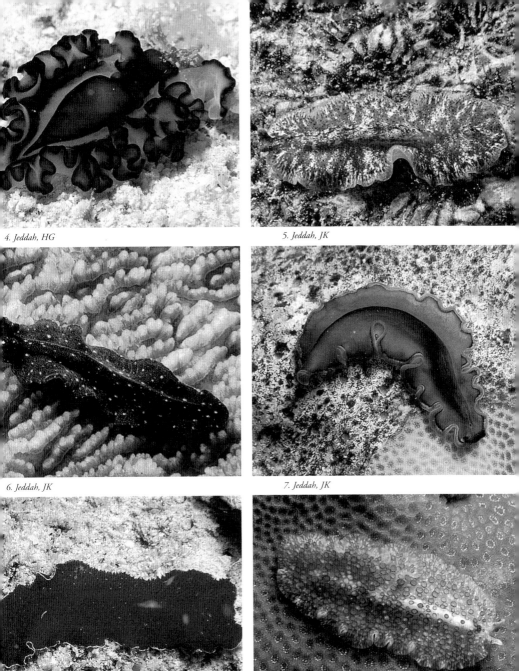

4. Jeddah, HG

5. Jeddah, JK

6. Jeddah, JK

7. Jeddah, JK

8. Jeddah, JK

9. Jeddah, JK

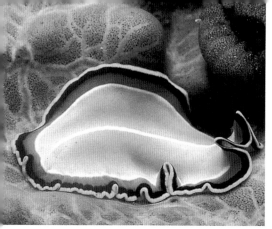

1. Jeddah, JK

2. Jeddah, JK

3. Jeddah, JK

1 Twoband flatworm *Pseudoceros bimarginatus*
To 3 cm. Creamy with white median line; marginal band
yellow, black, orange and white. **Biology:** on reef flats and
in lagoons, often under coral heads on sand or rubble.
Range: Red Sea to Micronesia.

2 Tiger flatworm *Pseudoceros* cf. *dimidiatus*
To 8 cm. Colour highly variable: black or yellow with yellow
or black stripes; marginal band orange. **Biology:** on hard
bottoms of protected bays and lagoons to at least 18 m. Day-
active and probably toxic or distasteful. Feeds on tunicates,
mucus and soft corals. **Range:** Red Sea to Polynesia.

3 Rusty flatworm *Pseudoceros* cf. *ferrugineus*
To 5 cm. Rusty-red to scarlet with tiny creamy dots;
marginal band yellow and purple; looks deep blue at
depth. **Biology:** on rocky and coral reefs, to at least 25 m.
Active by day and night. Probably toxic. Feeds on colonial
tunicates. **Range:** Red Sea to Polynesia.

4 Gosliner's flatworm *Pseudoceros goslineri*
To 7 cm. Creamy with orange and red spots; marginal
band brilliant purple. **Biology:** on reef flats and sheltered
upper reef margins. Active by night on well-encrusted hard
bottoms. Feeds on tunicates and probably also on sponges.
Range: Red Sea to New Guinea.

5 Dotted flatworm *Pseudoceros leptostichus*
To 3 cm. Creamy with black and white dots; marginal
band yellow with black spots. **Biology:** under rubble or
stones of shallow reefs by day. Feeds on colonial tunicates.
Range: Red Sea to PNG and Australia.

6 Burgundy flatworm *Pseudoceros* cf. *rubronanus*
To 3 cm. Burgundy with irregular white spots. **Biology:** on
hard bottoms of shallow protected coral reefs, 0.5 to at
least 8 m. Nocturnal. **Range:** Red Sea only.

7 Harlequin flatworm *Pseudoceros scintillatus*
To 2 cm. Black with irregular greenish spots encircled by
white ring; marginal band yellow. **Biology:** under or near
coral heads by day. Feeds on tunicates. **Range:** Red Sea to
Philippines and Australia.

8 Scriptus flatworm *Pseudoceros scriptus*
To 1.5 cm. White with black stripes and spots; marginal
band yellow. **Biology:** shallow coral reefs, often under
boulders. Feeds on ascidians. **Range:** Red Sea to Philippines.

9 Orange-dotted flatworm *Pseudoceros* sp. 1
To *ca.* 3 cm. White with orange dots; marginal band green
and black. **Biology:** inhabits reef flats, shallow lagoons and
protected seaward reefs. Feeds on tunicates. **Range:** Red
Sea. Undescribed.

4. Jeddah, JK

5. Jeddah, JK

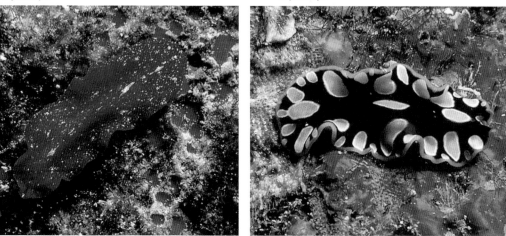

6. Jeddah, JK

7. Jeddah, JK

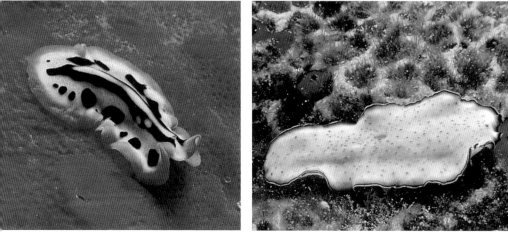

8. Jeddah, JK

9. Jeddah, JK

1. Jeddah, JK

2. Jeddah, JK

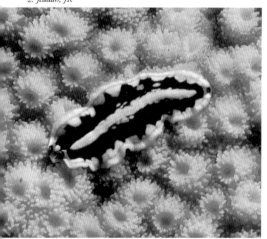

3. Jeddah, JK

1 Black flatworm   *Pseudoceros* sp. 2
To *ca.* 1 cm. Black with blue and yellow marginal band.
Biology: inhabits reef flats, shallow lagoons and protected
seaward reefs, to about 14 m. **Range:** Red Sea.
Undescribed.

2 Velvet flatworm   *Pseudoceros* sp. 3
To *ca.* 2 cm. Black with yellow and green marginal band.
Biology: shallow coastal reefs, to at least 5 m. Hidden
under stones by day. **Range:** Red Sea. Possibly the same as
P. depiliktabub (PNG).

3 Mourning flatworm   *Pseudoceros* sp. 4
To *ca.* 1 cm. Black with broad yellow and white marginal
band. **Biology:** inhabits living and dead corals of shallow
reefs, to at least 7 m. **Range:** Red Sea. Undescribed.

4 Carpet flatworm   *Pseudoceros* sp. 5
To *ca.* 3 cm. Creamy with white spots; marginal band
black with a narrow yellow edge. **Biology:** inhabits shallow
coastal bays and lagoons. On richly encrusted hard
bottoms by night. **Range:** Red Sea. Undescribed.

5 Red-edged flatworm   *Pseudoceros* sp. 6
To *ca.* 4 cm. Irregular white and black spots; marginal
band red. **Biology:** usually on well-encrusted rubble of
poorly developed shallow reefs. Uncommon. **Range:** Red
Sea only. Undescribed.

EURYLEPTIDAE
6 Marbled flatworm   *Eurylepta* (?) sp.
To *ca.* 4 cm. White with a few dark spots; marginal band
black with alternating brown and white spots. **Biology:** on
sand or fine rubble of protected bays and lagoons. Feeds
on ascidians by day. **Range:** Red Sea.

7 Leopard flatworm   *Maritigrella* (?) sp.
To *ca.* 3 cm. White with leopard-like black spots; marginal
band with black and brown stripes. **Biology:** on shallow
poorly developed coral reefs. In crevices and under stones
by day. **Range:** Red Sea.

PERICELIDAE
8 Deer flatworm   *Pericelis* sp.
To 7 cm. Pale brown with irregular white spots; marginal
band rosy. **Biology:** on sand under stones and rubble of
shallow coral reefs. A fast crawler that feeds on
invertebrates. Locally common. **Range:** Red Sea.

PLANOCERIDAE
9 Translucent flatworm *Paraplanocera* cf. *oligoglena*
To 4 cm. Translucent with pale and dark marks; internal
organs visible. **Biology:** inhabits reef flats and shallow
lagoon and protected seaward reefs. Under stones by day.
A swift crawler over hard bottoms by night. Probably feeds
on crustaceans. **Range:** Red Sea to Hawaii.

4. Jeddah, JK

5. Jeddah, JK

6. Marsa Bareka, EL

7. Jeddah, JK

8. Jeddah, JK

9. Jeddah, JK

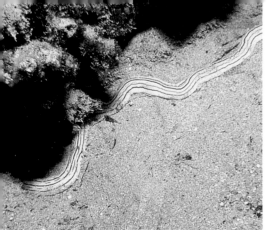

1. Marsa Shagra, RM

2. Quseir, JH

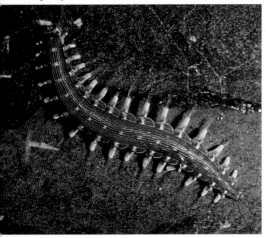

3. Egypt, JH

RIBBON WORMS   Phylum NEMERTINA

1 Five-striped ribbon worm
Baseodiscus cf. *quinquelineatus*
To 2 m. Highly contractile; white with 5 narrow dark lines. **Biology:** most reef zones, primarily over sand. Actively hunts by night. **Range:** Red Sea to W. Pacific(?) **Similar:** *B. hemprichii* has single longitudinal line.

Phylum ANNELIDA, Class POLYCHAETA
FIREWORMS   AMPHINOMIDAE

2 Orange fireworm   *Eurythoe complanata*
To 15 cm. Sides with clumps of toxic bristles. **Biology:** all reef zones. Hidden by day, active by night. A scavenger and predator of live corals and other invertebrates. Bristles cause a burning irritating wound or rash. **Range:** circumtropical.

FIREWORMS   HESIONIDAE

3 Splendid fireworm   *Hesione splendida*
To *ca.* 6 cm. Sides with clumps of toxic bristles, head with 8 pairs of tentacular cirri; pale with brown striations and spots. **Biology:** under coral rubble in shallow protected areas. A carnivore, probably hunts primarily at night. **Range:** circumtropical.

FEATHER DUSTER WORMS SABELLIDAE
4 Colonial frost worm   *Filogranella elatensis*
Crowns to 1 cm. Colonial masses of fine calcareous tubes. **Biology:** in rubble or sand along reef edges. **Range:** Red Sea to W. Pacific.

5 Feather duster worm   *Sabellastarte indica*
Crown to 10 cm. Tentacular crown horseshoe-shaped to circular. **Biology:** most reef habitats, in flexible tube embedded in sand or cracks. **Range:** circumtropical? **Similar:** *S. sanctijosephi* has two-tiered tentacular crown.

6 Unid. sp.
Crown to 5 cm. Tentacular crown circular; tentacles banded. **Biology:** in sand pockets near reefs. **Range:** Red Sea to W. Pacific.

CHRISTMAS TREE WORMS SERPULIDAE
7 Christmas tree worm   *Spirobranchus giganteus*
Crowns to 1.5 cm. Tentacular crown as twin spirals; colour highly variable. **Biology:** lives in calcareous tube embedded in living corals, intertidal to over 30 m. May live up to 20 years. **Range:** circumtropical.

BRYOZOANS   Phylum BRYOZOA
Class GYMNOLAEMATA
Order CHEILOSTOMIDA

8 *Canda pecten*
Fans to 4 cm. Fine reddish-tan lattice. **Biology:** on well-encrusted surfaces, primarily in caves or under overhangs. **Range:** Red Sea. Many similar species.

9 *Rhynchozoon larreyi*
Patches to *ca.* 10 cm. Pale purple to red encrustation. **Biology:** on well-encrusted surfaces, primarily in caves or under overhangs. **Range:** circumtropical.

4. Dahab, MH

5. Sinai, EL

6. Mangrove Bay, RM

7. Sinai, SMö

8. Egypt, JH

9. Egypt, JH

MOLLUSCS: SNAILS, BIVALVES AND CEPHALOPODS

Phylum MOLLUSCA

Molluscs are highly developed animals without an inner skeleton. Over the past 540 million years they evolved a wide variety of basic body plans united by the same basic larval development. Some groups such as the ammonites went extinct long ago, while others survived to flourish into the second largest animal phylum which includes 110,000 species of snails, 20,000 species of bivalves (clams and their relatives) and 700 species of cephalopods (squids and octopuses). Presently 1,765 species are known from the Red Sea, 245 of them are endemic.

The basic molluscan body consists of a muscular foot below a visceral mass and mantle which in some groups secretes a shell made of calcium carbonate with an outer protective protein layer. The mouth, anus, gills and chemical sensors are located in a cavity under the mantle. A unique feature of many molluscs is the radula, a rasp-like feeding organ packed with chitinous teeth that is pressed upward onto the food to tear off pieces by moving back and forth. The radula has been modified into a drill used to puncture the shells of bivalves and into a venomous harpoon by the cone shells. The sexes in molluscs may be separate or hermaphroditic with either one sex changing to the other (sequential) or with both at the same time (simultaneous). Most molluscs reproduce by releasing sperm and eggs direcctly into the water, but some deposit eggs on the bottom, often as plates, chains, lumps or bands under rocks and in cavities. Most species have planktonic veliger larvae which facilitates dispersal over long distances. There are at least 5 classes of molluscs with 4 inhabiting coral reefs.

Chitons (**class Polyplacophora**) are the most primitive molluscs with shells composed of eight narrow plates surrounded by a tough fleshy girdle. Most coral reef species occur in rocky intertidal areas.

Snails and slugs (**class Gastropoda**) have a distinct head with mouth, radula and sensory organs. Torsion of the larval visceral mass of marine snails (Prosobranchia) creates the whorls of snails and places the gills at the front of the mantle cavity. In the sea slugs (Opisthobranchia, p. 296) the gills are positioned at the rear and the shell is vestigial or absent. Terrestrial gastropods and slugs (Pulmonata) lack gills but their mantle is modified into a primitive lung.

Bivalves (**class Bivalvia**) lack a head and live inside a pair of shells (valves) joined by an interlocking hinge and controlled by two powerful muscles. Perforated cilia-covered gills in the mantle cavity produce a steady current through an incurrent siphon which supplies oxygen and food particles. An inner mucus transport system brings food to the mouth. Metabolic wastes and gametes are expelled through an excurrent siphon. Multiple eyes and chemical sensors are located along the rim of the mantle. In many species both siphons are the only visible soft parts.

Cephalopods (**class Cephalopoda:** octopuses, cuttlefishes, squids and nautiluses) have a crown of tentacles surrounding a beaked mouth connected to a tubular or bulbous body that houses a pair of highly developed eyes, internal organs, mantle and mantle cavity (see p. 318).

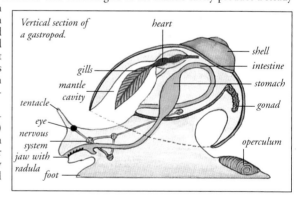

Vertical section of a gastropod.

heart
shell
intestine
stomach
gonad
operculum
gills
mantle cavity
tentacle
eye
nervous system
jaw with radula
foot

◀ **Spanish dancer** *Hexabranchus sanguineus* (OPISTHOBRANCHIA suborder DORIDACEA, family HEXABRANCHIDAE). To 30 cm. Red to purple or brown; 6 gills. **Biology:** on coral reefs, 1 to 50 m. Feeds on sponges and tunicates. Primarily nocturnal. Swims with undulating motion when disturbed. Eggs pink, attached to bottom in veil-like spiral. Often hosts a pair of *Periclimenes imperator* shrimp (p. 324), usually near gills. **Range:** Circumtropical. **Similar:** *H.* sp., with only 4 gills (Djibouti 52 cm). *(Photo: Abu Galum, EL.)*

1. Marsa Bareka, EL

2. Hurghada, JH

3. Marsa Bareka, EL

Class POLYPLACOPHORA
CHITONS   CHITONIDAE
Primitive flattened molluscs with 8 overlapping plates
surrounded by fleshy girdle. In intertidal zone clinging to
rocks. Feed on algae. 10 species in Red Sea.

1 Large chiton   _Acanthopleura vaillianti_
To 6 cm. Edge with bristles. **Biology:** rocky intertidal
zone exposed to waves. Grazes on algae at night. May
survive in dry crevices for days. **Range:** Red Sea to Arabian
Sea. **Similar:** _A. testudo_ (S. Red Sea only).

2 Red Sea chiton   _Chiton affinis_
To 4 cm. Edge without bristles. **Biology:** on reef flats and
crests exposed to waves. Algae feeder. **Range:** Red Sea only.

Class GASTROPODA
MARINE SNAILS   PROSOBRANCHIA
Complete head with sensory organs, radula, and broad
foot; 1–2 gills above head.

LIMPETS   PATELLIDAE
3 Rock limpet   _Patella_ sp.
To _ca_. 2 cm. Shell domed with radiating ridges. **Biology:**
firmly attached to intertidal rocks or rubble exposed to
surge. Feeds on algae. **Range:** Red Sea only?

ABALONES   HALIOTIDAE
4 Variable abalone   _Haliotis_ cf. _varia_
To 5 cm. Flattened oval shell with 4 to 5 holes along edge;
eyes at base of tentacles. **Biology:** on reef flats. Under
rocks during day, feeds on algae at night. **Range:** Red Sea
to W. Pacific.

KEYHOLE LIMPETS   FISSURELLIDAE
5 Black Keyhole Limpet   _Scutus unguis_
To 3 cm. Mantle covers shell. **Biology:** under rocks or rubble,
intertidal. Feeds on tunicates. **Range:** Red Sea to Australia.

TOPSHELLS   TROCHIDAE
6 Red Sea topshell   _Tectus dentatus_
To 8 cm. Pyramid shell often densely encrusted.
Biology: on hard substrate of reef flats and upper slopes.
Scrapes algae from surface. **Range:** Red Sea only.

TURBANS   TURBINIDAE
7 Tapestry turban   _Turbo petholatus_
To 8 cm. Red with black and white spirals. **Biology:** on
hard bottoms of shallow lagoon and fore-reef slopes to
10 m. Feeds on algae. **Range:** Red Sea to Fr. Polynesia.

8 Pharaoh turban   _Clanculus pharaonius_
To 2 cm. Red with black and white spirals. **Biology:** on
rubble or dead coral of reef flats and fore-reefs to 12 m.
Algae feeder. **Range:** Red Sea to Australia.

SLIPPER WINKLES   NERITIDAE
9 Wave slipper winkle   _Nerita albicilla_
To 2 cm. Colour variable, mostly black with yellow
spirals. **Biology:** on coralline rock of intertidal splash
zone. Feeds on algae. **Range:** Red Sea to Pacific.

4. Egypt, JH

5. Egypt, JH

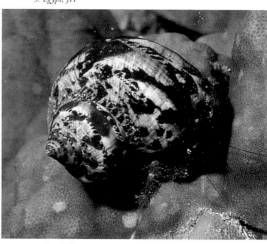

6. Gulf of Aqaba, RM

7. Abu Galum, EL

8. Hurghada, VW

9. Marsa Shagra, RM

1. Ras Mohammed, EL

2. Safaga, JN

3. Dahab, EL

HORN SHELLS — CERITHIDAE
1 Rock horn shell — *Cerithium* sp.
To 2 cm. **Biology:** in shallow lagoons and protected upper slopes, to 2 m. Solitary or in groups. **Range:** Red Sea.

CONCHES — STROMBIDAE
2 Lineated conch — *Strombus fasciatus*
To 4 cm. Creamy with dark rings. **Biology:** sandy to muddy areas of inner reef flats and shallow lagoons with seagrasses. Often eaten by octopuses. **Range:** Red Sea only.

3 Seba's spider conch — *Lambis truncata sebae*
To 25 cm. 6 to 7 elongate spines. **Biology:** usually on reef flats and in lagoons with sand and coral rubble. Feeds on filamentous algae. **Range:** Red Sea only (ssp.).

4 Jumping conch — *Strombus terebellatus*
To 5 cm. Shell with rusty brown spots, inside of aperture with fine brown lines. **Biology:** on reef flats and in shallow lagoons with coarse sand. Uses operculum for defence and locomotion. **Range:** Red Sea to Fiji.

5 Arabian tibia — *Tibia insulaechorab*
To 16.5 cm. Front of shell with long spike. **Biology:** migrates to shallows in spring to deposit eggs in spirals on sand hills made by a polychaete worm. Worm then covers eggs. **Range:** Red Sea to Philippines.

WORM SHELLS — VERMETIDAE
6 Large worm shell — *Dendropoma maxima*
To 20 cm, operculum to 15 mm across. Embedded in corals. **Biology:** in large coral heads of reef flats and upper slopes. Lives in a calcareous tube which it secretes. Produces a fine mucus veil which filters plankton and is hauled in about every 15 minutes by day and night. Juveniles grow within tube! **Range:** Red Sea to Fr. Polynesia.

MUREX SHELLS — MURICIDAE
A large family. Often with spiny projections. Can paralyse prey by drilling holes into shells and injecting a toxic purple ink 'purpurin' (the royal dye of the Roman emperors). Mostly carnivorous or feed on carrion.

7 Ramose murex — *Chicoreus ramosus*
To 20 cm. Robust shell with short blunt spines. **Biology:** in shallow protected areas of bays and lagoons. Produces purple ink. **Range:** Red Sea to W. Pacific.

8 Spotted drupe — *Drupa hadari*
To 3.8 cm. Short blunt spines. **Biology:** between or under rocks of the intertidal zone. Often overgrown by rosy calcareous algae. **Range:** Red Sea to Pacific.

9 Comb murex — *Murex forskoehlii*
To 11 cm. Several pointed spines. Long siphonal canal. **Biology:** on mud or fine sand of deep lagoon and seaward slopes, 20 to 250 m. Enters shallows in Spring to lay eggs. **Range:** Red Sea to Fiji.

4. Marsa Shagra, RM

5. Jeddah, HS

6. Marsa Shagra, RM

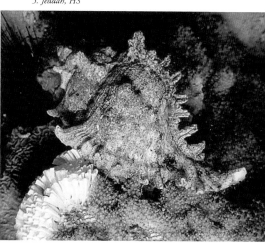

7. Hurghada, AK

8. Marsa Bareka, EL

9. Jordan, AK

1. Jeddah, HSj

2. Safaga, JH

3. Safaga, JH

COWRIES CYPRAEIDAE

When active, bilobed mantle covers and polishes the shell. Most nocturnal, hidden under rocks and in crevices by day with mantle withdrawn. Herbivores of filamentous algae or carnivores of sponges or cnidarian polyps. Lay hundreds of eggs under rubble or in crevices. *Ca.* 200 species worldwide, 48 in Red Sea.

1 Panther cowrie *Cypraea pantherina*
To 11.8 cm. Creamy with numerous large brown spots; mantle with black lines. **Biology:** under dead corals or rocks near sand of outer reef flats to deep offshore patch reefs. **Range:** Red Sea to Gulf of Aden. Replaced by **similar** *C. tigris* in rest of Indo-Pacific.

2 Gold-ringed cowrie *Erosaria annulus*
To 3.4 cm. Creamy white with narrow golden ring; mantle grey with white branched tassles. **Biology:** between and under coral heads, rubble and rocks of intertidal and shallow reefs. **Range:** Red Sea to Fr. Polynesia.

3 Money cowrie *Erosaria moneta*
To 4.4 cm. Creamy white with 3 indistinct greyish bands; mantle white with black lines. **Biology:** between and under coral heads, rubble and rocks of intertidal and shallow reefs. Often found in warm tidepools. Once traded as money in some areas. **Range:** Red Sea to Polynesia.

4 Snowflake cowrie *Erosaria nebrites*
To 5 cm. Creamy brown with numerous white and some red spots; mantle white with brush-like appendages. **Biology:** intertidal, on sand or algae-covered rocks; relatively common. **Range:** Red Sea only.

5 Thrush cowrie *Erosaria turdus*
To 5.7 cm. White with orange spots; mantle with short appendages. **Biology:** among rocks, dead coral or seagrasses of shallow protected reefs. **Range:** Red Sea to S. Africa.

6 Caurica cowrie *Erronea caurica*
To 7 cm. Colour variable. Creamy with brown dots and spots; mantle with tree-like appendages. **Biology:** on rubble or dead coral of shallow reefs. **Range:** Red Sea to Pacific.

7 Carnelian cowrie *Lyncina carneola*
To 9.4 cm. Colour variable. Creamy with dark bands, mantle with many appendages. **Biology:** under rocks and boulders of shallow reefs. Common, feeds on algae, sponges and detritus. **Range:** Red Sea to Fr. Polynesia. **Similar:** *L. leviathantitan* often gets larger (Red Sea; 9 cm).

8 Isabelle cowrie *Lyncina isabella*
To 5.4 cm. Light brown with interrupted dark lines; mantle dull black. **Biology:** in most coral reef habitats, 1 to 100 m. Usually under rocks by day. Feeds on detritus and sponges. **Range:** Red Sea to Panama.

9 Lynx cowrie *Lyncina lynx*
To 8.5 cm. Creamy with reddish spots; mantle translucent with anemone-like appendages. **Biology:** on reef flats, protected bays and fore-reefs. Locally common. **Range:** Red Sea to Fr. Polynesia.

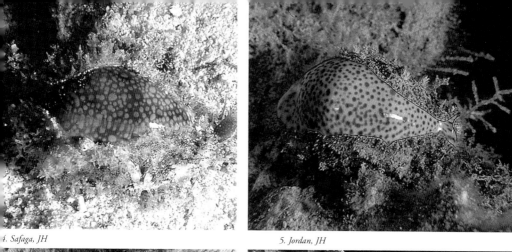

4. Safaga, JH

5. Jordan, JH

5. Sharm El Sheik, JH

7. Sharm El Sheik, JN

8. Sharm El Naga, HG

9. Jeddah, HSj

1. Guam, RM

2. Nuweiba, MT

3. Dahab, HHH

1 Indo-Pacific deer cowrie *Lyncina vitellus*
To 10 cm. Brown with white spots; mantle pale with dark lines and bulbous branched papillae. **Biology:** on reef flats and shallow protected areas. Under rocks and coral by day. **Range:** Red Sea to Fr. Polynesia.

2 Grey's cowrie *Mauritia grayana*
To 8 cm. Creamy brown network; mantle transparent with thorn-like papillae, edge with dark blotches. **Biology:** on hard bottoms of shallow fringing reefs and offshore patch reefs, 1 to 25 m. Locally common. **Range:** Red Sea to Gulf of Aden. **Similar:** Arabian cowrie *M. arabica immanis* (Red Sea to S. Africa); other similar species throughout Indo-Pacific.

3 Beauty cowrie *Luria pulchra*
To 7.6 cm. Creamy with orange-brown bands and 2 black squares at front; mantle black. **Biology:** on dead or living corals of shallow reefs. Rarely seen. **Range:** Red Sea to Gulf of Aden only.

4 Mole cowrie *Talparia talpa*
To 10.4 cm. Creamy with 4 reddish-brown bands; mantle black with white dots and finger-like papillae. **Biology:** under rocks and dead corals, 0.3 to 20 m. Nocturnal. Feeds on carrion and algae. **Range:** Red Sea to Panama. **Similar:** *T. exusta* is more pear-shaped (Red Sea to Gulf of Aden; 9 cm).

EGG COWRIES OVULIDAE
Close relatives of cowries that live primarily on octocorals and sponges. About 100 species worldwide.

5 Umbilical ovula *Calpurnus verrucosus*
To 4 cm. White; mantle white with reddish-brown spots. **Biology:** shallow coral reefs, 3 to 20 m. On and under *Sarcophyton* and *Lobophytum* soft corals on which it feeds. **Range:** Red Sea to Fiji.

6 Egg cowrie *Ovula ovum*
To 10 cm. Shell white; mantle black with yellow dots. **Biology:** shallow reefs, 0.3 to 20 m. On or near *Sarcophyton* and *Sinularia* soft corals on which it feeds. **Range:** Red Sea to Fr. Polynesia.

7 Spindle ovula *Phenacovolva* sp
To *ca.* 4 cm. Mantle yellow with wavy bands and yellow papillae. **Biology:** on protected coral reefs. **Range:** Red Sea only?

8 Honeycomb ovula *Pseudosimnia* sp
To 1.2 cm. Shell white; mantle with numerous close-set brown or red spots and white papillae. **Biology:** on coral reefs, 10 to at least 30 m. Lives and feeds on soft corals. **Range:** Red Sea only.

LAMELLARIDS LAMELLARIIDAE
9 Black lamella *Coriocella nigra*
To 10 cm. Internal shell completely covered by brown, red or black mantle with finger-like appendages; eyes at base of antennae. **Biology:** on coral debris of shallow reefs. Nocturnal. Feeds on tunicates. **Range:** Red Sea to Solomon Is. **Similar:** *C. safagae* (Red Sea only).

4. Guam, RM

5. Jeddah, JK

6. Kenya, KF

7. Jeddah, JK

8. Safaga, AK

9. Jeddah, JK

TRITONS — RANELLIDAE (CYMATIDAE)

▲ **1 Triton's trumpet** *Charonia tritonis* To 50 cm. Largest gastropod in Red Sea. Animal creamy with brown spots. **Biology:** on lagoon and seaward reefs, 2 to 40 m. Feeds on echinoderms including Crown-of-thorns (*Acanthaster planci*). During the day it hides in pockets with operculum closed. Nocturnal. At night it hunts for prey located by smell and sight. Pries proboscis and shell under prey to turn it over. May take up to 3 hours to subdue and consume prey. Female may sit on mass of red eggs. Populations in most areas greatly reduced by collecting. Traditionally used as a horn by seafarers. **Range:** Red Sea to Polynesia, s. to S. Africa. *(Photo: Jeddah, KM.)*

HORSE CONCHES — FASCIOLARIIDAE

2 Filament conch *Pleuroploca filamentosa*
To 14 cm. Siphon long. Shell with white knobs and red rings on whorls. **Biology:** on shallow reefs, often under sand and fine rubble. **Range:** Red Sea to Pacific.

3 Trapeze horse conch *Pleuroploca trapezium*
To 14 cm. Shell thick. Foot wine-red with white dots. **Biology:** on lagoon and seaward reefs, 3 to 25 m. Usually on sand or rubble near reefs. May sit on eggs. **Range:** Red Sea to W. Pacific.

Jeddah, SH

Hurghada, JN

CLUSTER WINKLES   PLANAXIDAE
4 *Planaxis griseus*
To 1.5 cm. Similar to Littorinidae but notch at bottom of columella. **Biology:** in large groups on rocky bottoms of shallow protected reefs to 1 m. **Range:** Red Sea only.

TUN SHELLS   TONNIDAE
Shell thin with parallel ridges along whorls, foot enormous and plate-like. Nocturnal predators of invertebrates that stay buried in sand by day.

5 Partridge tun   *Tonna perdix*
To 20 cm, foot enormous, to 50 cm. Each ridge of shell brown with curved white bands; animal brown with white streaks. **Biology:** on sand and rubble of shallow protected coral reefs, 1 to at least 22 m. Nocturnal, buried in sand during day. Feeds on sea cucumbers and molluscs. Moves rapidly with long siphon extended to sample water for scent of prey which is run down. Sea cucumbers are rapidly engulfed lengthwise by highly expandable proboscis. Feeding on molluscan prey may take hours. Digestive fluids include sulphuric (H_2SO_4) and hydrochloric acid (HCl) which dissolve calcareous spicules and shells. **Range:** Red Sea to Polynesia.

HELMET SHELLS   CASSIDAE
Thick stout shells that live in shallow sandy areas. Nocturnal predators of sea urchins and sand dollars. The smaller species stay buried by day. Surface structure of shell is sexually dimorphic.

6 Heavy bonnet   *Casmaria ponderosa*
To 7 cm. Shell creamy with brown dashes around aperture and distal edge of rings. **Biology:** seagrass beds, sandy stretches and dead coral rock or rubble of shallow protected coral reefs, to at least 15 m. Feeds on sea urchins. Not uncommon in Gulf of Aqaba. **Range:** Red Sea (ssp. *unicolor*) to W. Pacific.

CONE SHELLS   CONIDAE
Large family of over 500 species, with 46 in Red Sea. All are cone-shaped with a narrow lengthwise aperture and radula modified into a venomous barbed harpoon used to paralyse and kill prey (p. 18). Most species are carnivores specializing on various invertebrates or fishes. The fish-eating species are dangerous and can kill humans.

> **CAUTION:** the animal may withdraw deep into its shell. The harpoon is shot from the tip of a worm-like proboscis that can reach all regions of the shell. It is a separate structure from the more frequently visible hollow siphon. Stings produce a puncture wound. The neurotoxic venom may cause numbness, muscular paralysis and respiratory or cardiac failure by blocking neurotransmitters. Treatment is similar to that for a snake bite.

4. Marsa Shagra, RM

5. Sharm El Naga, HG

6. Nuweiba, MT

1. The Bells, AK

2. Sinai, HE

3. Hurghada, JH

1 Geography cone   *Conus geographus*
To 15 cm. Shell creamy with brown markings; animal
creamy with black spots; siphon with brown rings. **Biology:**
on protected reef flats and shallow lagoon and seaward reefs.
Common in areas with patches of sand and rubble. A
nocturnal predator of reef fishes (photo eating
a *Chromis* sp.). Among the **most dangerous** of cone
shells, responsible for many human fatalities. Stings produce
a puncture wound, followed by numbness, loss of
coordination, muscular paralysis and respiratory failure.
Quick medical treatment is essential. **Range:** Red Sea to
Marshall Is.

2 Sand-dusted cone   *Conus arenatus*
To 8.3 cm. Creamy with dark brown spots. **Biology:**
shallow protected reefs, 0.3 to 30 m. Primarily in intertidal
and on sandy stretches close to coral reefs. Feeds on
polychaetes. **Range:** Red Sea (ssp. *aequipunctatus*) to
Pacific.

3 Nussatella cone   *Conus nussatella*
To 8.3 cm. Similar to *C. arenatus* but more elongate with
fewer brown dots and indistinct pale brown spots.
Biology: on algae-covered hard bottoms of seaward coral
reefs, often below 30 m. **Range:** Red Sea to Pacific.

4 Striated cone   *Conus striatus*
To 12.5 cm. Shell creamy with irregular lengthwise bands of
perpendicular dark pinstripes; animal creamy with dark
cross-streaks. **Biology:** on sand and hard bottoms of reef flats
and lagoon and seaward reefs to at least 40 m. Nocturnal
predator of fishes. **Dangerous**, has caused human fatalities.
Range: Red Sea to Polynesia.

5 Ringed cone   *Conus taeniatus*
To 6 cm. Shell brown and white with brown lines.
Biology: under rocks or buried in sand, reef flats to at least
28 m. **Range:** Red Sea to Polynesia.

6 Tessellate cone   *Conus tessulatus*
To 6.5 cm. White with elongate yellowish spots; operculum
narrow. **Biology:** in lagoons and protected bays including
mangroves. Buried in sand or under stones by day. Feeds on
worms. **Range:** Red Sea to Pacific.

7 Textile cone   *Conus textile neovicarius*
To 13 cm. Numerous white tent-like markings; siphon with
red tip and black ring. **Biology:** in most coral reef habitats,
from reef flats and lagoon and seaward reefs to at least 50 m.
Nocturnal, usually under rocks, rubble or buried in sand by
day. Feeds on gastropods including other cone shells. Lives
up to 9 years in captivity. **Very dangerous**, has caused human
fatalities. **Range:** Red Sea, other subspecies to Polynesia.

8 Dashed cone   *Conus vexillum*
To 15 cm. Shell creamy, fading to purplish towards base.
Biology: on shallow and deep reefs with sandy and seagrass
areas. **Range:** Red Sea to Fr. Polynesia.

9 General cone   *Conus generalis*
To 9 cm. Shell yellow with ochre rings, tip dark. **Biology:** on
sheltered shallow rocky and coral reefs. Buried in sand by day,
feeds at night on polychaetes. **Range:** Red Sea and Indian
Ocean (ssp. *maldivus*) to W. Pacific.

4. Sinai, AK

5. Egypt, HE

6. Marsa Shagra, RM

7. Dahab, HHH

8. Sinai, HE

9. Sinai, AK

VIOLET SNAILS — JANTHINIDAE
1 Violet snail *Janthina janthina*
To 4 cm. Shell white with violet hue. **Biology:** drifts with raft of bubbles at surface. Feeds on siphonophores including Man-of-war and By-the-wind sailor (p. 244). **Range:** circumtropical.

HARP SHELLS — HARPIDAE
2 Love harp *Harpa amouretta*
To 7.5 cm. Shell pale with intricate wavy lines; animal white with yellow dots. Siphon long. **Biology:** in sand of lagoon and seaward reefs, 3 to 200 m. Nocturnal predator of small crabs which are immobilized with a mucus sack then killed. Buried by day. May cast off foot when captured. **Range:** Red Sea to Fr. Polynesia.

AUGERS — TEREBRIDAE
Columella elongate into tapered spike. Carnivores of invertebrates. Some species with venomous radula used to paralyse prey, none dangerous to humans.

3 Crenulate auger *Terebra crenulata*
To 15 cm. Columella ridged, pale brown. **Biology:** in sand of shallow lagoons and protected seaward reefs, 0 to over 10 m. Usually buried. **Range:** Red Sea to Polynesia.

CYPRAEIDAE
4 **Fringe cowrie**, *Purpuradusta fimbriata* to 2 cm
5 **Sieve cowrie**, *Cribrarula cribraria* to 2.5 cm
6 **Honey cowrie**, *Erosaria helvola* to 3 cm
7 **Limacine cowrie**, *Staphylaea limacina* to 2.5 cm

CONIDAE
8 **Princely cone**, *Conus aulicus* to 10 cm
9 **Tented cone**, *Conus pennaceus* to 6 cm
10 **Striatellus cone**, *Conus striatellus* to 5.5 cm
11 **Virgin cone**, *Conus virgo* to 15 cm

TROCHIDAE
12 **Maculated top**, *Trochus maculatus* to 5 cm

RANELLIDAE
13 **Pear triton**, *Cymatium pyrum* to 10 cm

BUCCINIDAE
14 **Common phos**, *Phos senticosus* to 4 cm

NASSARIIDAE
15 **Jewel mudsnail**, *Nassarius gemmulatus* to 2 cm

OLIVIDAE
16 **Inflated olive**, *Oliva bulbosa* to 4.5 cm

MITRIDAE
17 **Reticulated mitre**, *Scabricola fissurata* to 5 cm

TEREBRIDAE
18 **Marlin spike**, *Terebra maculata* to 25 cm

TURRITELLIDAE
19 **Screw turritella**, *Turritella terebra* to 16 cm
(Photo: Cismar, Germany, V. Wiese.) ▶

1. Shadwan Is., FV

2. Safaga, JH

3. Jeddah, JK

SEA SLUGS OPISTHOBRANCHIA
Order CEPHALASPIDEA
BUBBLE SHELLS   HYDATINIDAE
1 Paper bubble shell   *Hydatina physis*
Shell to 6.5 cm. Shell thin, creamy with narrow brown
lines. **Biology:** in protected shallows, 0.5 to 10 m. Usually
on sand or seagrasses with scattered coral heads. Feeds on
polychaetes, incorporates their toxins. **Range:** Red Sea to
Polynesia. **Similar:** *H. amplustre* and *H. zonata* (Red Sea).

BUBBLE SHELLS   BULLIDAE
2 Rosy bubble shell   *Bulla ampulla*
Shell to 6 cm. Shell thin, rosy; mantle orange-red with
white dots; two side lobes. **Biology:** buried in sand of reef
flats and seagrass beds with coral heads. A nocturnal
carnivore. Shells often on beaches. **Range:** Red Sea to
Pacific.

HEAD SHIELD SLUGS   AGLAJIIDAE
3 Blue-spotted shield slug   *Chelidonura livida*
To 5 cm. Black with blue markings; body with 4 lobes.
Biology: on mud or sand of shallow protected reefs. Feeds
on worms and molluscs. **Range:** Red Sea only.

4 Yellow lip shield slug   *Chelidonura flavolobata*
To 5 cm. Dark blue with two yellow head lobes. **Biology:**
Feeds on large brain corals. Lifts front of body when
disturbed. **Range:** Red Sea only. **Similar:** *C. sandrana* is black
with white spots and dots (Red Sea, W. Indian Ocean; 2 cm).

Order ANASPIDEA
SEA HARES   APLYSIIDAE
Sea hares have a vestigial internal shell. When disturbed,
they release a toxic purple ink metabolized from their diet
of Blue-green algae.

5 Ringed sea hare   *Aplysia dactylomela*
To 40 cm. Colour variable with distinctive green or black
rings. **Biology:** on sand and algae-covered rubble of
protected shallows. Produces up to 180 million eggs per
year. **Range:** circumtropical. **Similar:** *A. parvula* is green
with small white spots (Red Sea to Australia; 6 cm).

6 Giant sea hare   *Dolabella auricularia*
To 30 cm. Greenish-grey mottled in brown with white
speckles. **Biology:** protected silty areas with scattered algae
and seagrasses to at least 15 m. **Range:** Circumtropical.

7 Long-tailed sea hare   *Stylocheilus longicauda*
To 6 cm. Yellow or brown. **Biology:** on algae-covered hard
bottoms of shallow reefs. **Range:** Red Sea to Mexico.

8 Striated sea hare   *Stylocheilus striatus*
To 7.5 cm. Creamy with fine wavy brown lines. **Biology:** on
algae or seagrasses of shallow protected reefs. Feeds on plants.
Range: circumtropical.

9 Indian sea hare   *Notarchus indicus*
To 6 cm. Balloon-like with fine papillae. **Biology:** on algae
and *Halophila* seagrasses in protected shallows. Swims in
somersaults by jet propulsion. Aggregates in spring for
breeding. **Range:** Red Sea to Australia.

4. Safaga, JH

5. Jeddah, JK

6. Guam, RM

7. Jeddah, JK

8. Nuweiba, MT

9. Sharm El Naga, HG

1. Shab Sham, HG

2. Jeddah, JK

3. Jeddah, JK

SACOGLOSSANS Order SACOGLOSSA

With or without vestigial shell. Highly specialized herbivores with a well-developed mouth and radula for piercing the cells and sucking out chloroplasts. Feed primarily on the green algae *Caulerpa, Halimeda, Vaucheria* and *Cladophora.*

SAP-SUCKING SLUGS ELYSIIDAE
1 Ornate elysia *Elysia ornata*
To 3 cm. Pale green with black dots; parapodial margins black and orange. **Biology:** on algae-covered bottoms, in goups when mating. Feeds on chloroplasts of the green algae *Caulerpa, Codium* and *Halimeda.* Incorporates the chloroplasts which become symbiotic and continue photosynthesis within the slug! Eggs green. **Range:** Red Sea to Polynesia.
2 Decorated thuridilla *Thuridilla decorata*
To 2 cm. White with distinctive undulating brown band. **Biology:** on shallow algae-covered silty to gravel bottoms. Nocturnal, hides in crevices of boulders or under coral heads by day. **Range:** Red Sea only.

ORNATE SAP-SUCKING SLUGS CALIPHYLLIDAE
3 Elegant cyerce *Cyerce elegans*
To 3 cm. Cerata transparent with fine white lines and blue spot, often held upward. **Biology:** on hard bottoms encrusted with algae or sponges. Cerata contain a noxious mucus. **Range:** Red Sea, Arabian Sea to Polynesia.
4 Oriental polybranch *Polybranchia orientalis*
To 7 cm. Creamy with leaf-like rhinophores and gills. **Biology:** on algae-covered dead corals or rubble. Probably feeds on *Caulerpa.* **Range:** Red Sea to Fr. Polynesia.

SIDE-GILLED SLUGS Order NOTASPIDEA

Gills on right side between mantle and foot. Nocturnal carnivores. May secrete sulphuric acid when disturbed.

PLEUROBRANCHS PLEUROBRANCHIDAE
5 Orange gumdrop *Berthellina citrina*
To 3 cm. Orange to yellow. **Biology:** on sand or rubble with scattered rocks, 0.5 to 25 m. Nocturnal. Feeds on sponges and tunicates. **Range:** Red Sea to Fr. Polynesia.
6 Forskal's pleurobranch *Pleurobranchus forskalii*
To 20 cm. Papillae not as round as in *P. grandis.* **Biology:** on sand, mud, seagrassses and corals. Nocturnal, feeds on sponges and tunicates. **Range:** Red Sea to Polynesia.
7 Grand pleurobranch *Pleurobranchus grandis*
To 10 cm. Whitish, often with three red bands; papillae as rounded polygons. **Biology:** on sand, mud, seagrasses and corals. Buried by day. Feeds on sponges, anemones and tunicates. **Range:** Red Sea to Polynesia.
8 Pale pleurobranch *Pleurobranchus* cf. *albiguttatus*
To 3 cm. Creamy with fine scattered brown spots. **Biology:** on sand, rubble or corals. **Range:** Red Sea to S. Africa, to New Caledonia and Guam.
9 Peppered pleurobranch *Pleurobranchus* sp.
To *ca.* 5 cm. Creamy with fine brown spots. **Biology:** on reefs subject to sedimentation. **Range:** Red Sea. Undescribed.

4. *Jeddah, JK*

5. *Jeddah, JK*

6. *Abu Galum, RM*

7. *Marsa Shagra, RM*

8. *Jeddah, JK*

9. *Jeddah, JK*

1. Jeddah, JK

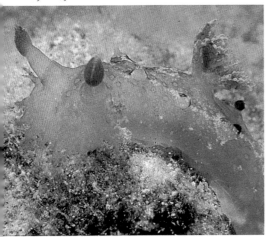

2. Jeddah, JK

3. Jeddah, JK

NUDIBRANCHS Order NUDIBRANCHIA

Often brightly coloured slugs with 1 or 2 pairs of antennae-like rhinophores used for orientation. Most have a patch of naked feathery gills on their back, but some respire through ventral gill leaflets (*Phyllidia*). Most are highly specialized nocturnal carnivores of specific sessile animals. Those that feed on noxious or venomous animals store the toxins or nematocysts of their prey for their own defence.

Suborder DORIDACEA
POLYCERIDAE

1 Red Sea nembrotha　　　*Nembrotha megalocera*
To 8 cm. 3 gills. **Biology:** on sponges, rubble or corals, 5 to 20 m. Feeds on sponges and hydroids. Swims with undulating movements. **Range:** Red Sea only.

2 *Plocamopherus* sp. 1
To *ca.* 2 cm. Cream with orange dots. **Biology:** on densly encrusted hard bottoms of shallow reefs. Feeds on bryozoans. **Range:** Red Sea only?

3 *Plocamopherus* cf *ocellatus*
To *ca.* 2 cm. Pale with indistinct spots. **Biology:** on sand or rubble of shallow protected reefs. Feeds on bryozoans. Lays a yellow egg ribbon. **Range:** Red Sea only?

4 Orange-lined tambja　　　*Tambja affinis*
To 7 cm. Black with green and orange stripes. **Biology:** on hard bottoms of reef slopes, 5 to at least 20 m. Feeds on bryozoans. **Range:** Red Sea, s. to Tanzania and Zanzibar.

5 Pacific thecacera　　　*Thecacera pacifica*
To 6 cm. Orange with distinctive blue markings.
Biology: intertidal and shallow flats, near seagrasses. Usually under coral or rubble. Feeds on bryozoans. Lays a yellow egg ribbon. **Range:** Red Sea to Indonesia, s. to S. Africa.

GYMNODORIDIDAE

6 Ceylon gymnodorid　　　*Gymnodoris ceylonica*
To 4 cm. White with red spots, edge of mantle yellow. Eggs orange. **Biology:** on sand and rubble of seagrass beds, 5 to 20 m. A voracious nocturnal predator of the sea hare *Stylocheilus longicauda* (p. 296). **Range:** Red Sea to Marshall Is., s. to Aldabra Atoll.

7 Citrus gymnodorid　　　*Gymnodoris citrina*
To 3 cm. Creamy with yellow spots, gills yellow.
Biology: on coral rubble of poorly developed reefs, to at least 10 m. Feeds on opisthobranchs. **Range:** Red Sea to Fr. Polynesia, Indonesia and Japan.

8 White gymnodorid　　　*Gymnodoris* sp.
To 6 cm. White with white bulbous protuberances.
Biology: on rubble or sponges of shallow reefs.
Feeds on opisthobranchs. **Range:** Red Sea to S. Africa, Australia.

9 Striated gymnodorid　　　*Analogium striatum*
To 5.5 cm. White or yellow with distinctive orange stripes. **Biology:** on sand or fine rubble, often partly buried. Feeds on *Placobranchus ocellatus*. **Range:** Red Sea to Micronesia.

4. Jeddah, JK

5. Sharm El Naga, HG

6. s. Egypt, AK

7. Jeddah, JK

8. Jeddah, JK

9. Jeddah, JK

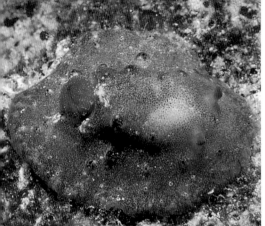

1. Jeddah, JK

2. Jeddah, JK

3. Jeddah, JK

HALGERDIDAE

1 Warty dorid *Actinocyclus verrucosus*
To 10 cm. Greyish or olive green with small tubercles.
Biology: among encrusted rocks or rubble of shallow reefs.
Nocturnal, under rocks by day. Feeds on sponges of the
family Halisarcidae. **Range:** Red Sea.

2 Clumpy dorid *Asteronotus cespitosus*
To 20 cm. Brown with large tubercles arranged in concentric
rings. **Biology:** reef flats and shallow reef slopes. Primarily
nocturnal. Usually beneath rocks and coral heads by day,
on sand or rubble by night. Feeds on sponges. Hosts the
shrimp *Periclimenes imperator* which is usually on gills.
Range: Red Sea to Polynesia, s. to Mauritius.

3 Willey's dorid *Halgerda willeyi*
To 9 cm. Yellow ridges with black and yellow lines between.
Biology: on exposed coral reefs, in open or hidden under
rocks. Feeds on sponges. **Range:** Red Sea to Marshall Is., s. to
Tanzania.

DISCODORIDIDAE

4 Fragile discodorid *Discodoris fragilis*
To 9 cm. Creamy with irregular greyish patches. **Biology:** on
reef flats and shallow reefs exposed to sedimentation. Usually
under rocks. Feeds on sponges. **Range:** Red Sea to W. Pacific.

5 Black-pitted discodorid *Sclerodoris apiculata*
To 5 cm. Greenish or blackish with large black ocelli; often
covered with sediment. **Biology:** on encrusted coralline
rocks or rubble of tidepools and shallow reefs. Feeds on
sponges. **Range:** Red Sea to Australia, s. to S. Africa.

DORIDIDAE

6 *Hoplodoris pustulata*
To 5 cm. Covered with round nodules; reddish-brown
with diffuse pale areas. **Biology:** inhabits shallow coral
reefs. **Range:** Red Sea only.

7 Pustulose dorid *Hoplodoris grandiflora*
To 3.5 cm. Covered with round nodules; edges of nodules
white except centrally. **Biology:** Found on shallow
coralline and coral reef. Feeds probably on sponges.
Range: Red Sea to Hawaii.

8 Rabbit dorid *Trippa intecta*
To 6 cm. Brown with white mid-dorsal stripe. **Biology:** on
sand or rubble of shallow reefs, 3 to 10 m. Feeds on sponges.
Range: Red Sea to Indonesia.

PLATE DORIDS   PLATYDORIDIDAE

9 Carpet dorid *Platydoris striata*
To 10 cm. Creamy with fine brown dorsal lines.
Biology: on hard bottoms of shallow reefs. Usually
hidden under rocks or coral heads. Feeds on sponges.
Range: Red Sea to Christmas Is., Indian Ocean.

4. *Jeddah, JK*

5. *Jeddah, JK*

6. *Sharm El Naga, HG*

7. *Jeddah, JK*

8. *Jeddah, JK*

9. *Jeddah, JK*

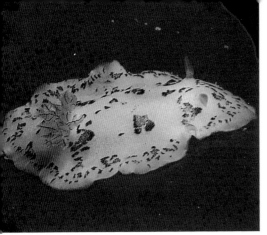

1. Jeddah, JK

2. Jeddah, JK

3. Jeddah, JK

1 Dusky dorid *Platydoris scabra*
To 8 cm. Greyish blotches, rhinophores and mantle edge orange. **Biology:** under rocks or coral heads of shallow reefs, 0.3 to 5 m. Nocturnal. Feeds on sponges. **Range:** Red Sea to Marshall Is.

KENTRODORIDS KENTRODORIDIDAE
2 Mourning dorid *Jorunna funebris*
To 6 cm. White with black rings. **Biology:** on shallow reefs, 4 to at least 18 m. Prefers shady overhangs or caves. Feeds on sponges. **Range:** Red Sea to Marshall Is., s. to S. Africa.

3 Panther dorid *Jorunna pantherina*
To 3 cm. Grey with irregular light brown spots.
Biology: turbid reefs subject to sedimentation, 3 to at least 20 m. Feeds on sponges. **Range:** Red Sea to W. Pacific.

4 Tiger dorid *Jorunna* sp.
To *ca.* 3 cm. Yellow with dark markings. **Biology:** in rubble of turbid protected reef flats and slopes subject to sedimentation. Under rocks by day. **Range:** Red Sea only?

CHROMODORIDS
CHROMODORIDIDAE
Small brightly coloured nudibranchs that feed on sponges and incorporate their toxins into glands in the mantle. 34 species in Red Sea.

5 Rosy-tipped nudibranch *Cadlinella* sp.
To 2 cm. Yellowish-white; cerata with red tips.
Biology: primarily on subtidal flats and in surge channels and tunnels. Feeds on transparent sponges.
Range: Red Sea only? **Similar:** *C. ornatissima* lacks yellow network on back (Red Sea to S. Japan; 2 cm).

6 Kangaroo nudibranch *Ceratosoma trilobatum*
To 10 cm. Body with long tail; mantle with lateral protuberances, yellow spots and purple spots along margin. **Biology:** on soft and hard bottoms, 5 to at least 18 m. Feeds on toxic sponges. **Range:** Red Sea to Polynesia, s. to S. Africa. **Similar:** *C. tenue* has smaller lateral projections and a secondary pair (Indo-West Pacific, 10 cm).

7 Miamira ceratosoma *Ceratosoma miamirana*
To 8 cm. Distinctive green with yellow protuberances.
Biology: usually on poorly developed coral reefs with rubble and algae-covered coralline rocks, 2 to 30 m.
Range: Red Sea to Polynesia.

8 African chromodorid *Chromodoris africana*
To 6.5 cm. Broad black stripes with narrow white stripes; gills and edge of mantle yellow. **Biology:** in most reef habitats, 1 to 15 m. **Range:** Red Sea to Tanzania.

9 Pyjama chromodorid *Chromodoris quadricolor*
To 4.5 cm. Similar to *C. africana* but more slender with blue and white stripes on top; rhinophores and gills deep orange. **Biology:** reef flats and lagoon and seaward reefs. Often in groups. Common. Feeds on the red sponges *Negombata* sp. and *Pione* sp. **Range:** Red Sea to Tanzania.

4. *Jeddah, JK*

5. *Jeddah, JK*

6. *Arabian Gulf, AK*

7. *Jeddah, JK*

8. *El Nabq, EL*

9. *Jeddah, JK*

1 Ringed chromodorid *Chromodoris annulata*
To 6 cm. White with orange spots and deep purple rings; gills and rhinophores purple. **Biology:** on hard bottoms of caves and shady walls, to 15 m. Usually on encrusting sponges. **Range:** Red Sea, Arabian Sea to E. Africa and Thailand (Indian Ocean).

2 Sprinkled chromodorid *Chromodoris aspersa*
To 4 cm. Dark purple dots, edge of mantle, gills and rhinophores yellow. **Biology:** on coral rubble of reef flats. Nocturnal, hidden under rocks by day. Uncommon. **Range:** Red Sea to Indonesia and W. Australia.

3 Decorated chromodorid *Chromodoris* cf. *decora*
To 2 cm. Mantle with wavy white lines and magenta dots. **Biology:** on shallow reefs, often on algae-covered coralline rocks. **Range:** Red Sea only.

4 Faithful chromodorid *Chromodoris fidelis*
To 2 cm. White, mantle with wavy red edge; gills and rhinophores dark brown. **Biology:** on richly encrusted rubble and rocks of shallow reefs. Rare. Feeds on the sponge *Aplysilla violacea*. **Range:** Red Sea to Marshall Is.

5 Twin chromodorid *Chromodoris geminus*
To 5.5 cm. Mantle with dark purple spots with blue rings, margin dark grey and blue. **Biology:** on poorly developed reefs, down to 50 m. Feeds on the red sponge *Negombata* sp. **Range:** Red Sea to Indonesia. **Similar:** *C. kunei* (Indo-Pacific) and *C. tritos* (western Indian Ocean).

6 Obsolete chromodorid *Chromodoris obsoleta*
To 6 cm. Irregular red net-like spots, edge of mantle orange and purple. **Biology:** on encrusted rubble and rocks of coral reefs, 5 to 15 m. **Range:** Red Sea to Arabian Sea only. **Similar:** *C. tinctoria* has white mantle with rusty reticulations (Red Sea to Australia; 7 cm).

7 Laboute's chromodorid *Noumea* cf. *laboutei*
To 2 cm. Transparent white with yellow gills and rhinophores. **Biology:** on encrusted, algae-covered, well-sedimented bottoms. Rare. **Range:** Red Sea to New Caledonia.

8 Pinky chromodorid *Durvilledoris lemniscata*
To 1.5 cm. Three red dorsal lines, edge of mantle yellow-purplish. **Biology:** on algae-covered rocks and corals of shallow reefs, 0.5 to 12 m. **Range:** Red Sea to Indonesia, s. to S. Africa.

9 Two-eyed chromodorid *Durvilledoris pusilla*
To 1.5 cm. Reddish mantle with two white eye-like dorsal spots. **Biology:** on rubble, coarse sand and algae-covered rocks of shallow reefs. **Range:** Red Sea to Japan and Solomon Islands.

1. Jeddah, JK

2. Jeddah, JK

3. Jeddah, JK

4. Jeddah, JK

5. Jeddah, JK

6. Jeddah, JK

7. Jeddah, JK

8. Jeddah, JK

9. Jeddah, JK

1 Black-edged glossodoris *Glossodoris atromarginata*
To 10 cm. Creamy; edge of mantle an undulating black and blue line. **Biology:** usually under coral heads and sponges of outer reefs. Feeds on sponges. Discharges fluid when disturbed. **Range:** Red Sea to Fr. Polynesia, s. to S. Africa.

2 Charlotta's glossodoris *Glossodoris charlottae*
To 8 cm. Mantle reddish with pale round spots, margin light and dark blue. **Biology:** on algae-covered bottoms of poorly developed reefs, rare. **Range:** Red Sea only.

3 Blue-edged glossodorid *Glossodoris cincta*
To 5.5 cm. Mottled brown and white; edge of mantle with distinctive light blue, dark and ochre bands. **Biology:** on shallow patch reef of sand slopes, usually under rubble. Discharges white fluid from mantle when disturbed. **Range:** Red Sea to Fiji, s. to Seychelles.

4 Brown-edged glossodorid *Glossodoris hikuerensis*
To 8 cm. Cream suffused in red; edge of mantle greyish-brown. **Biology:** on coral rubble of poorly developed reefs, to at least 16 m. **Range:** Red Sea to Fr. Polynesia. **Similar:** *G. rufomarginata* has white mantle with orange-red speckles (Red Sea to Fiji; 4.5 cm).

5 Pale glossodorid *Glossodoris pallida*
To 2 cm. Translucent milky white with yellow margin. **Biology:** coral reef slopes and patch reefs, 0.5 to 30 m. Often active by day. Feeds on sponges. **Range:** Red Sea to Fiji.

6 Fire glossodorid *Hypselodoris infucata*
To 5 cm. Colour variable with black and yellow spots; gills and rhinophores orange-red. **Biology:** shallow reefs, on hard bottoms densely covered with sponges, hydroids and algae. Feeds on sponges. **Range:** Red Sea and Oman to Fiji, s. to S. Africa.

7 *Hypselodoris* sp.
To 2 cm. Linear patches of white between longitudinal purple lines on body, foot and tail. Rhinophores and gills brown. **Biology:** beneath coral rubble and sponges. Feeds on sponges. **Range:** Red Sea and Indian Ocean?

8 Candy hypselodorid *Hypselodoris maridadilus*
To 5 cm. Distinctive red to magenta stripes. Named for Swahili word for beautiful, *maridadi*. **Biology:** silty reefs, 2 to at least 10 m. In groups on or under rubble and dead corals. Feeds on sponges. **Range:** Red Sea to Polynesia.

9 Beautiful risbecia *Risbecia pulchella*
To 12 cm. Body with orange spots, edge of mantle purple. **Biology:** on coarse sand and rubble of protected coral reefs, 3 to at least 25 m. Often in pairs. Feeds on sponges. **Range:** Red Sea and W. Indian Ocean. **Very similar:** *R. ghardaqana* has fewer and larger orange spots (Red Sea; 5 cm).

1. Jeddah, JK

2. Jeddah, JK

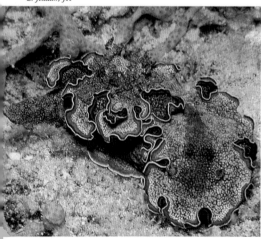

3. Jeddah, JK

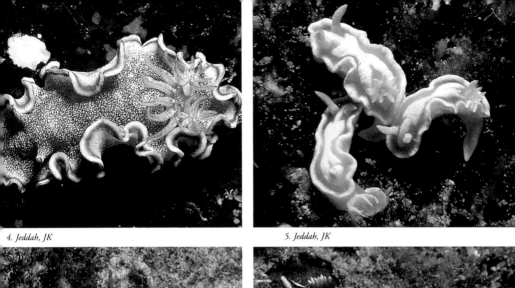

4. Jeddah, JK

5. Jeddah, JK

6. Nuweiba, PM

7. Jeddah, JK

8. Jeddah, JK

9. Marsa Bareka, EL

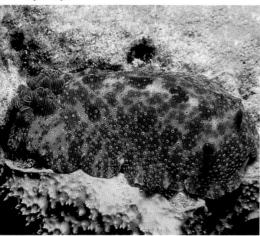

2. Jeddah, JK

3. Jeddah, JK

DENDRODORIDS
DENDRODORIDIDAE
11 species in Red Sea.

1 Crown dendrodorid   *Dendrodoris coronata*
To 4 cm. Greenish-yellow with tiny black dots.
Biology: on encrusted rubble or sand of shallow reefs.
Feeds on sponges and tunicates. **Range:** Red Sea to Japan.

2 Peppered dendrodorid   *Dendrodoris jousseaumi*
To 2 cm. Irregular brown blotches with numerous tiny dark
dots. **Biology:** on coarse sand and rubble of poorly developed
reefs. Feeds on sponges. **Range:** Red Sea to W. Pacific.

3 Black dendrodorid   *Dendrodoris nigra*
To 8 cm. Black; mantle of juv. with orange or pink margin.
Biology: on sand and encrusted hard bottoms of poorly
developed reefs. Occasionally under rocks in tidepools.
Range: Red Sea to Polynesia.

4 Panther dendrodorid   *Dendrodoris* cf. *nigropunctata*
To 3 cm. Grey with irregular black spots. **Biology:** on poorly
developed shallow reefs, to at least 8 m. **Range:** Red Sea.

WART SLUGS   PHYLLIDIIDAE
Respire through leaf-like gills located ventrally between
foot and mantle. Species feed on sponges and produce a
bitter toxic mucus with a strong chemical scent. 10 species
in Red Sea.

5 Rueppell's wart slug   *Fryeria rueppelii*
To 4 cm. Distinctive yellow and black markings, orange
margin; anus below posterior end of mantle. **Biology:** on
poorly developed lagoon and seaward reefs with abundant
sponges, on which it feeds. **Range:** Red Sea only.

6 Wave wart slug   *Phyllidia undula*
To 5 cm. No dorsal gills. Distinctive black, orange
and white markings. Possibly a colour variant of *P. ocellata*.
Biology: on lagoon and seaward reefs, 1 to 22 m. On sand,
rubble and sponges on which it feeds and lays its eggs.
Range: Red Sea and S. Africa.

7 Varicose wart slug   *Phyllidia varicosa*
To 11.5 cm. Tubercles and rhinophores yellow, a black
stripes between ridges. **Biology:** on lagoon and seaward
reefs, 1 to at least 30 m. Common, in open over sand,
rubble or coral rock by day and night. Feeds on toxic
sponges and incorporates the toxins into its mucus.
Range: Red Sea to Polynesia, s. to S. Africa.

8 Pustulose wart slug   *Phyllidiella pustulosa*
To 6 cm. Black with rosy or white tubercles arranged in
clusters. **Biology:** on lagoon and seaward reefs, 2 to 40 m.
Common. **Range:** Red Sea to Polynesia.

Suborder DENDRONOTACEA
TREE SLUGS   BORNELLIDAE
9 *Bornella stellifer*
To 4 cm. Creamy, rhinophores with red rings. 5 pairs of
gills. **Biology:** among rocks and rubble in shallow water.
Feeds on hydroids. **Range:** Red Sea and S. Africa to
Taiwan.

4. Jeddah, JK

5. Jeddah, JK

6. Marsa Bareka, EL

7. Oman, KF

8. Hurghada, AK

9. Jeddah, JK

1. Jeddah, JK

2. Aqaba, AK

3. Jeddah, JK

TRITON TREE SLUGS   TRITONIIDAE

Have distinctive oral veil; 8 species in Red Sea.

1 Bushy triton   *Marionia glama*
To 6 cm. Pale green or brown sheathed rhinophores.
Biology: on shallow sheltered reefs. Feeds on gorgonians
and soft corals. **Range:** Red Sea to S. Africa.

2 Green triton   *Marionia viridescens*
To 3 cm. 2 rows of tufted rhinophores along back; marbled
brown. **Biology:** in lagoon and seaward reefs below 6 m.
Usually on encrusted hard bottoms. Feeds on gorgonians
and soft corals. **Range:** Red Sea to Polynesia. **Similar:** *M.
rubra* is pinky red (Red Sea to Mauritius; 10 cm).

3 Blue triton   *Marioniopsis cyanobrachiata*
To 5.5 cm. Yellow to blue-green; tips of gills and rhinophores
blue-green. **Biology:** in sheltered shallows. Feeds on polyps
of *Heteroxenia* soft corals which it also mimics. **Range:** Red
Sea to Philippines.

LEAF SLUGS   TETHYIDAE

4 Leaf slug   *Melibe bucephala*
To 12 cm. Rhinophores translucent and tree-like. **Biology:**
on shallow algae-covered reefs. Locally common during algae
bloom in spring. Feeds on detritus and benthic crustaceans
by sweeping filamentous algae with its oral veil. Swims when
disturbed. **Range:** Red Sea to W. Thailand.

Suborder ARMINACEA   ARMINIDAE

5 Gonad dermatobranch   *Dermatobranchus
gonatophorus*
To 2 cm. Has oral veil, contractile rhinophores and dorsal
ridges with black and white lines. **Biology:** on deep turbid
reefs to at least 22 m. Feeds on soft corals and gorgonians.
Range: Red Sea to Indonesia and S. Japan.

6 Olive dermatobranch   *Dermatobranchus* sp.
To *ca.* 6 cm. Pale olive-green with longitudinal ridges.
Biology: on turbid reefs, 0.5 to 10 m. **Range:** Red Sea.

AEOLIDS   Suborder AEOLIDACEA

Aeolids incorporate the nematocysts of cnidarian prey for
use in their own defence. These are stored in cnido-sacs of
the tentacle-like cerata which are also used for respiration.
Named after the Greek god of the wind 'Aeolus'.

FLABELLINIDS   FLABELLINIDAE

7 Ringed flabellina   *Flabellina bilas*
To 2.5 cm. Body white with orange patches, cerata with
red rings. **Biology:** on *Eudendrium* hydroids on reef slopes
and walls, 10 to 30 m. **Range:** Red Sea to W. Pacific.

8 Purple flabellina   *Flabellina rubrolineata*
To 4 cm. Colour variable: usually pale with 3 violet stripes;
cerata with orange-purplish tips. **Biology:** reef slopes and
walls. Feeds on hydroids and horny corals, common.
Range: Red Sea to Polynesia, s. to Aldabra Atoll.

9 Rosy-ringed facelina   *Facelina rhodopos*
To 2 cm. Creamy to yellow, swollen cerata with rosy rings.
Biology: intertidal zones with rocks and sand. Rare.
Range: Red Sea, maybe Polynesia.

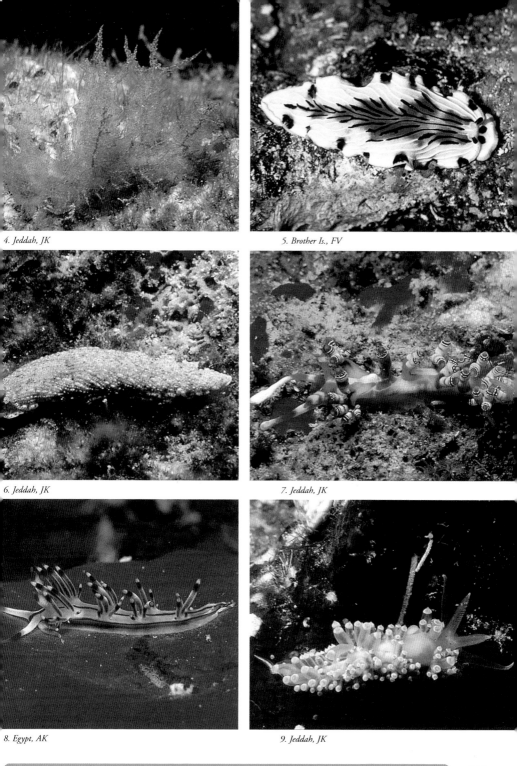

4. Jeddah, JK

5. Brother Is., FV

6. Jeddah, JK

7. Jeddah, JK

8. Egypt, AK

9. Jeddah, JK

1. Jeddah, JK

2. Jeddah, JK

3. Jeddah, JK

AEOLIDS   AEOLIDIIDAE

1 Creamy aeolid   *Baeolidia fusiformis*
To 2.5 cm. Translucent creamy with several pale dots and
rings. **Biology:** in rubble. Cryptic, associated with
unidentified sea anemones on which it feeds. **Range:** Red Sea
and Japan.

2 Mottled aeolid   *Limenandra* cf. *nodosa*
To 3.5 cm. Mottled light brown with 2 long banded oral
tentacles. **Biology:** on coral rubble or coarse sand. Possibly
two species involved. **Range:** circumtropical.

3 Translucent aeolid   *Spurilla major*
To 5 cm. Translucent creamy with dark speckles. **Biology:**
on shallow reefs with encrusted rubble, to 5 m. Feeds on
soft corals and anemones. **Range:** Red Sea to E. Africa.
Similar: *S. australis* with distinctive blue and orange rings
on inflated cerata (Red Sea and Australia, 3 cm).

DRAGON SLUGS   GLAUCIDAE

4 Indian dragon   *Caloria (Phidiana) indica*
To 2 cm. Cerata with yellowish, blue and red bands.
Biology: on hard bottoms, 3 to 15 m. Common. Feeds on
hydroids and stores nematocysts as defence against
predators. **Range:** Red Sea to Polynesia, s. to S. Africa.

5 Man-o-war dragon   *Glaucilla marginata*
To 2 cm. Shape distinctive; blue and white. **Biology:** pelagic.
Inhales air for buoyancy and hangs head-down at surface.
Feeds on siphonophores including *Physalia* and *Velella*, and
uses their nematocysts for defence. Often blown ashore.
Range: circumglobal. **Similar:** *Glaucus atlanticus* has cerata
in single rows (circumglobal, 4 cm).

6 Coral dragon   *Phyllodesmium hyalinum*
To 3 cm. Cryptic, colour variable to match host; parapodial
tentacles recurved. **Biology:** on shallow reefs. In groups of
up to 8 on the soft corals *Heteroxenia fuscescens* and *Xenia
umbellata* (Sinai). Feeds on their polyps. **Range:** Red Sea to
Pacific.

7 Iridescent dragon   *Phyllodesmium magnum*
To 13 cm. Tips of cerata yellow. **Biology:** on shallow reefs
to at least 10 m. Feeds on *Sinularia* soft corals. Its mucus
protects it from nematocysts. **Range:** Red Sea to Australia
and Marshall Is.

8 Large dragon   *Pteraeolidia ianthina*
To 10 cm. Colour variable depending on zooxanthellae in
tissues. **Biology:** on shallow coral reefs. Feeds on hydroids
and leather corals and incorporates their zooxanthellae as
symbionts. **Range:** Red Sea to Polynesia.

9 Cup-coral slug   *Phestilla melanobranchia*
To 3.5 cm. Cerata orange to dark brown, depending
on coral prey. **Biology:** on shallow reefs, usually under
Tubastrea corals on which it feeds. Also feeds
on hydroids. Deposits white egg masses directly on these
corals. **Range:** Red Sea to Polynesia. **Similar:** *P. lugubris*
feeds on *Porites* corals (Red Sea to Hawaii).

4. Jeddah, JK

5. Jeddah, JK

6. Jeddah, JK

7. Jeddah, JK

8. Jeddah, JK

9. Jeddah, JK

Class BIVALVIA   Bivalves

Most species are filter feeders and live buried in sediments with only siphon openings exposed. Others are attached to hard objects, while some burrow in corals.

GIANT CLAMS   TRIDACNIDAE

The largest bivalves. Sit with opening pointed up to expose brightly coloured mantle to sunlight. Symbiotic zooxanthellae synthesize food from metabolic wastes. Also filter zooplankton. 7 species, 2 in Red Sea.

1 Common giant clam   *Tridacna maxima*
To 35 cm. Colour of mantle highly variable; shell with narrow scutes. **Biology:** attached to live or dead corals on shallow reef tops and slopes, 0 to at least 20 m. Mantle often flush with surface due to growth of surrounding coral or secretions from clam that erode limestone. Common. **Range:** Red Sea to Fr. Polynesia.

2, 3 Fluted giant clam   *Tridacna squamosa*
To 40 cm. Colour of mantle highly variable, often with conspicuous spots; shell with broad scutes. **Biology:** on lagoon and seaward reefs, 0.3 to over 20 m. Juveniles attached to hard object, large individuals free, often on sand near corals. Uncommon. **Range:** Red Sea to Samoa.

PEARL OYSTERS PTERIDAE: Filter feeders
attached by strong byssal threads. 10 species in Red Sea.

4 Egyptian wing oyster   *Pteria aegyptiaca*
To 10 cm. Shell with long flange. **Biology:** attached to gorgonians and corals, 2 to over 22 m. Solitary or in groups. **Range:** Red Sea only.

5 Common pearl oyster   *Pinctada margaritifera*
To 30 cm. Lips with interlocking teeth. **Biology:** attached to rubble or limestone, reef flats to over 30 m. Produces gem-quality pearls. **Range:** Red Sea to E. Pacific.

FILE CLAMS LIMIDAE: 10 species in Red Sea.

6 Fragile file clam   *Limaria fragilis*
To 4 cm. Long sticky orange tentacles easily shed. **Biology:** on lagoon and seaward reefs. Usually tucked in crevices or on sand under overhangs. May swim by rapidly opening and closing shell. **Range:** Red Sea to Pacific. **Similar:** *Lima vulgaris* has shorter tentacles. (Red Sea to Polynesia; 5 cm).

HONEYCOMB OYSTERS GRAPHAEIDAE:

7 Honeycomb oyster   *Hyotissa hyotis*
To 10 cm. Opening zigzag-shaped. **Biology:** inhabits lagoon and seaward reefs to 35 m. Attached to black corals, piers or mangrove roots. **Range:** Red Sea to Pacific.

SCALLOPS PECTINIIDAE: 20 species in Red Sea.

8 Coral scallop   *Pedum spondyloideum*
To 3 cm. Mantle brightly coloured; eyes red. **Biology:** deeply embedded in *Porites* corals with only opening visible. Feeds on plankton. **Range:** Red Sea to Fiji.

THORNY OYSTERS SPONDYLIDAE:
8 species in Red Sea.

9 *Spondylus varius*
To 20 cm. Mantle colourful; shell with short thorns. **Biology:** current-swept reefs, 5 to 40 m. Feeds on plankton. **Range:** Red Sea to Marshall Is. **Similar:** *S. spinosus* has longer thorns (Red Sea).

1. Marsa Shagra, RM

2. Marsa Shagra, KF

3. Mangrove Bay, RM

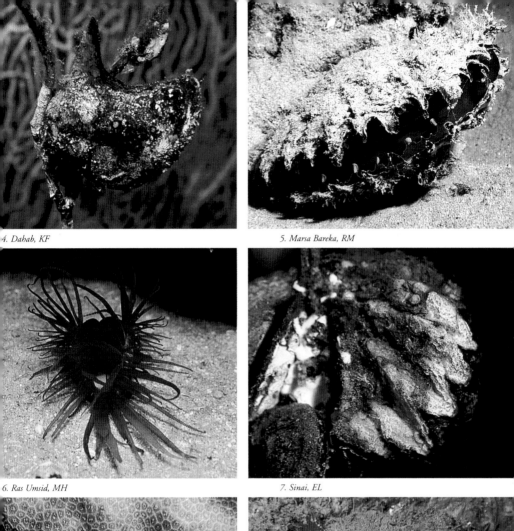

4. Dahab, KF

5. Marsa Bareka, RM

6. Ras Umsid, MH

7. Sinai, EL

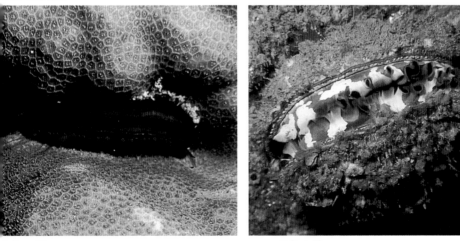

8. Egypt, EL

9. Sinai, EL

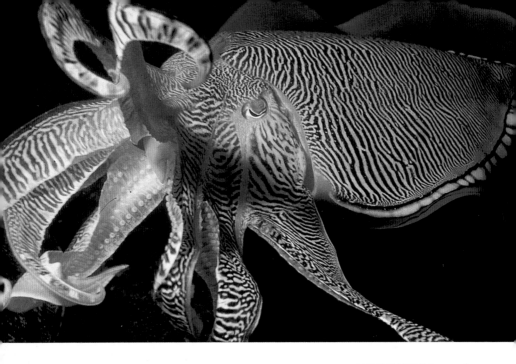

OCTOPUSES, CUTTLEFISHES, SQUID AND NAUTILUSES
Class CEPHALOPODA

Cephalopods inhabit all seas from the shoreline to the deepest abyss and range greatly in size, from 1 cm to 20 m. They are the most intelligent invertebrates and have highly developed eyes able to see colours. They are capable of amazing and instantaneous changes in colour and skin texture. They swim backwards by jet-propulsion, powered by water ejected from the mantle cavity through a tubular funnel. Octopuses and squid capture prey with tentacles lined with suction cups. They escape predators by ejecting a cloud of irritating black ink. Males have a specialized arm used for passing a packet of sperm to the female. Packets of eggs are attached to the bottom where they hatch into tiny larvae or juveniles. Most cephalopods breed only once after which they are programmed to die.

2. Mangrove Bay, RM

3. Marsa Bareka, EL

Order SEPIOIDEA
CUTTLEFISHES
SEPIIDAE

Mantle broad and slightly compressed, 8 arms and 2 retractile tentacles. 10 species in Red Sea.

◀ 1, 2 Pharao cuttlefish *Sepia pharaonis*
To 40 cm (mantle). Mantle with distinctive blue edge, striped during courtship. **Biology:** lagoon and seaward reefs, 0.3 to 110 m. Typically hangs motionless just above bottom, camouflaged to match surroundings. Easily approached, may instantly flash waves of colours down body or change skin texture. Often in shallow water by night. Feeds on crustaceans and fishes. **Range:** Red Sea to Japan. *(Photo 1: courtship, Oman, KF.)*

3 Hooded cuttlefish *Sepia prashadi*
To 30 cm. Mantle usually with skin flaps. **Biology:** over sand, fine rubble and seagrass of shallow lagoons and protected bays, to 30 m. Feeds over sand at night on crustaceans and fishes. **Range:** Red Sea to S. Africa.

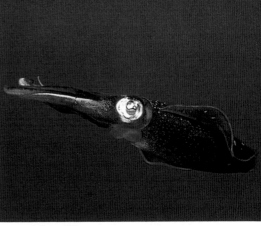

4. Egypt, PH

SQUID   Order TEUTHIOIDEA
COASTAL SQUID
LOLIGINIDAE

4 Common reef squid *Sepiotheutis lessoniana*
To 40 cm (mantle). Mantle cylindrical, arms nearly as long, eyes large. **Biology:** in midwater of lagoon and seaward reefs, shoreline to 100 m. In small schools, often near surface. Uses countershading to become nearly invisible. May flash iridescent colours. Short lifespan, matures in 10 months. **Range:** Red Sea to Polynesia.

Order OCTOPODA
OCTOPUSES
OCTOPODIDAE

Subdue prey with venomous saliva, but the Red Sea species harmless to man. Female broods over strings of eggs, protecting them from predators and blowing water across them until hatching, but starves herself in the process. Lifespan 1 to 5 years. 8 species in Red Sea.

5 Reef octopus *Octopus cyaneus*
To 1.4 m. Often with large oval spots on each side above base of arms. **Biology:** inhabits shallow protected to exposed reefs, from shoreline to over 25 m. Territorial, entrance of den often marked by midden pile of mollusc and crustacean shells. Usually hidden in crevices and holes by day, but may also sit in open with eyes held high to survey surroundings. Often reddish-brown with smooth skin, but may instantly change colour and skin texture. Often mimics dead coral. Feeds primarily on crustaceans, occasionally on molluscs and fishes. **Range:** Red Sea to Polynesia. **Similar:** nocturnal *O. macropus* has white spots (circumtropical; 1.5 m); *O. aegina* has fine papillae (Red Sea; 30 cm).

6 Red-spotted octopus *Octopus* sp.
To *ca.* 5 cm. Pale with small red spots. **Biology:** on sand and rubble of shallow reefs. Nocturnal. **Range:** Red Sea. **Similar:** many similar small species on Indo-Pacific reefs, some undescribed.

5. Marsa Shagra, RM

6. Marsa Bareka, EL

CRUSTACEANS

Crustaceans along with spiders, millipedes and insects are members of the most successful phylum of animals on earth, Arthropoda ('jointed legs'). Arthropods originated in marine waters 350 million years ago and have radiated into millions of species of insects and an estimated 50,000 others, mostly crustaceans. The secret of their success is the evolution of chitin, a three-layered carbohydrate that is lightweight, waterproof, flexible as well as rigid and highly adaptable. As the exoskeleton, it covers the entire exterior, including jointed legs, eyes, antennas and specialized mouthparts. Many crustaceans reinforce the layers of chitin with lime to form thick rigid segments that provide greater protection but may limit mobility. Since the exoskeleton cannot grow with the body, it periodically ruptures and is cast off in a process known as moulting. For several hours after moulting the new larger exoskeleton remains soft leaving the animal highly vulnerable to damage or predation. This is also when copulation occurs by being the only time when a female is able to accept a spermatophore (sperm sac) from the male.

The adaptability of each segment's paired appendages to different requirements has resulted in the arthropods' great diversity. In crustaceans, the first two segments are fused into a cephalothorax which covers the major organs and is connected to eyes, antennas, mouthparts (maxillas, jaws), claws (chelipeds) and legs. The abdomen is usually a series of similar rings connected to gills, swimming appendages and reproductive organs. In many species, the last segments contain the animal's largest muscles specialized for rapid escape strokes. In crabs, the abdomen is folded under a massive carapace. The joints of all appendages have sensory hairs, connected to a central nervous system, which are used to gather information on their surroundings. Shrimps, lobsters and crabs carry densely packed leaf-like gills under their carapace, attached to and moved by the walking legs. Many small shrimps have small pincers on their legs. Crustaceans have compound eyes formed by thousands of two-lensed cylinders. The eyes of crabs are located on erectile stalks that can fold sideways into grooves in the carapace. Many nocturnal crustaceans orient themselves with long antennae and thousands of sensory cells distributed throughout their exoskeleton. Many tiny crabs and shrimps are specialized commensals of sponges, cnidarians and echinoderms. Crustaceans range in size from tiny microscopic forms to crabs spanning over 3 m. They are divisible into several classes, but only two have species large enough to be noticed by divers.

Members of the class MAXILLOPODA have appendages modified into filter-feeding strainers. **Barnacles** (subclass Cirripedia) are attached. **Copepods** (subclass Copepoda) are tiny. Most are free-living zooplankton that feed on phytoplankton and are a major food source for planktivorous fishes. Others are benthic consumers of diatoms and bacteria, or external parasites of fishes.

The huge class MALACOSTRACA includes most large crustaceans as well as isopods, amphipods and mysids. The heavily armoured **Mantis shrimps** (order STOMATOPODA) have powerful raptoral appendages designed for either smashing or spearing prey. **Shrimps**, **lobsters** and **crabs** are among the 10,000 species of decapods (order DECAPODA) that may bear pincers on the first 1 to 3 pairs of appendages. Penaeid shrimps have branched gills and release their eggs into the sea. All others have unbranched gills and brood their eggs under their tails. The newly hatched larvae are released into the sea where they spend a period of days to as long as a year as pelagic plankton.

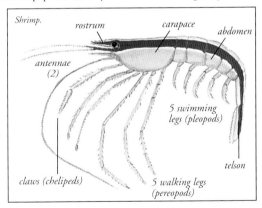

Shrimp.

rostrum · carapace · *abdomen*

antennae (2)

5 swimming legs (pleopods)

telson

claws (chelipeds)

5 walking legs (pereopods)

XANTHIDAE
◄ **Splendid spooner**   *Etisus splendidus*
To 20 cm. Red with black claws. **Biology:** nocturnal, feeds on algae, gastropods and urchins. Often eaten, but similar species may be poisonous. **Range:** Red Sea to Polynesia. *(Photo: Sinai, MH.)*

1. Safaga, JH

Class MAXILLOPODA
Barnacles Subclass CIRRIPEDIA
Adult stage permanently attached to other objects and encased in a thick shell of fused plates. Sweeps water for food with feather-like jointed appendages.
1 Goose barnacle *Lepas anatifera*
To 20 cm. Five white plates; attached by stalk. **Biology:** in clumps on piers, boats and drifting material. **Range:** circumglobal in tropical and temperate seas. **Similar:** *L. anserifera* has longer stalks.
2 Giant barnacle *Tetraclita squamosa*
To 3 cm. Four thick plates. **Biology:** intertidal rocky shores exposed to waves. Closer to water than *Tetrachthalmulus oblitteratus*. **Range:** Red Sea to W. Pacific.
3 Star barnacle *Tetrachthalmulus oblitteratus*
To 1 cm. Plates form star-like pattern. **Biology:** rocky intertidal zone. **Range:** Red Sea to Seychelles.

Class MALACOSTRACA
Mantis shrimps Order STOMATOPODA
Heavily armoured mantis-like creatures with pair of powerful raptorial appendages designed for either smashing or spearing prey. Can cause nasty wounds.
4 Club mantis *Gonodactylus chiragra*
To 10 cm. **Biology:** intertidal and shallow reefs. In hole on hard bottoms, identified by midden pile of shell fragments. Smashes gastropod shells to get to the soft parts or hermit crab inside. **Range:** Red Sea to Fr. Polynesia.
5 Falcon mantis *Gonodactylus falcatus*
To 7 cm. **Biology:** shallow reefs, intertidal to 9 m. A smasher. **Range:** Red Sea to C. Pacific.

2. Marsa Bareka, EL

Shrimps, lobsters, crabs Order DECAPODA
Commercial shrimps, prawns PENAEIDAE
Primitive decapods with branched gills, that release eggs into the sea. Many species on soft bottoms.
6 *Metapenaeopsis* sp.
To *ca.* 9 cm. Body and legs with red reticulations and white spots. **Biology:** primarily in sandy areas. Nocturnal, probably buried by day. **Range:** Red Sea (to Pacific?).

Coral shrimps STENOPODIDAE
7 Boxer shrimp *Stenopus hispidus*
To 5 cm. White and red bands on body and legs. **Biology:** coral reefs, intertidal to over 35 m. Usually in pairs in crevices. Waves long antennae to attract fishes to be cleaned. Common. **Range:** circumtropical.

Dancing shrimps RHYNCHOCINETIDAE
8 Green-eye dancing shrimp *Cinetorhynchus reticulatus*
To 6.5 cm. Translucent with red reticulations; legs banded. **Biology:** coral slopes. Solitary or in groups in crevices or among corals. **Range:** Red Sea to Polynesia.
9 Durban dancing shrimp *Rhynchocinetes durbanensis*
To 4 cm. Wine-red with white lines and spots; rostrum very long. **Biology:** on rocky and coral reefs, to at least 5 m. Usually hidden by day in caves and crevices, sometimes in groups. Common. **Range:** Red Sea to W. Pacific.

3. Hurghada, JH

4. Jeddah, JK

5. Jeddah, JK

6. Dahab, AK

7. Nuweiba, RM

8. Jeddah, JK

9. Dahab, PH

1. Guam, RM

2. Burma, MK

3. Jeddah, JK

GNATHOPHYLLIDAE
Small family of shrimps associated with echinoderms.

1 Bumblebee shrimp   *Gnathophyllum americanum*
To 1.2 cm. Creamy with narrow dark bars. **Biology:** intertidal to over 15 m. Associated with sea urchins, sea stars and sea cucumbers. Feeds on their tubed feet. **Range:** circumtropical.

2 Harlequin shrimp   *Hymenocera elegans*
To 5 cm. Cream with blue-edged brown spots. **Biology:** lagoon and seaward reefs. Territorial, in pairs. Turns over sea stars to feed on tubed feet and internal tissue. **Range:** Red Sea to Indonesia. Replaced by *H. picta* in Pacific.

PARTNER SHRIMPS   PALAEMONIDAE
Second pair of legs with small pincers. Includes many species commensal on invertebrates and many cleaners.

3 Superb coral shrimp   *Coralliocaris superba*
To 1.5 cm. Brown with broad white dorsal band. **Biology:** inhabit *Acropora* and *Pocillopora* corals. **Range:** Red Sea to Fr. Polynesia. **Similar:** *C. graminea* is green with white pinstripes (Red Sea to Samoa; 1 cm).

4 Popcorn shrimp   *Periclimenes brevicarpalis*
To 4 cm. Transparent with large white spots; tail with 5 orange ocelli. ♂ smaller than ♀. **Biology:** associated primarily with sea anemones, most often *Cryptodendrum adhaesivum*. In pairs or groups. Feeds on tentacles, mucus and detritus. **Range:** Red Sea to Fr. Polynesia.

5 Long-arm cleaner shrimp   *Periclimenes longicarpus*
To 2.5 cm. Transparent with white and purple spots on tail and white line between eyes. **Biology:** a cleaner associated with corals and anemones including *Xenia*, *Pleurogyra*, *Entacmaea quadricolor* and *Heteractis crispa*. Often in groups. **Range:** Red Sea to W. Pacific.

6 Eyed cave shrimp   *Periclimenes pholeter*
To 2.5 cm. Translucent pale orange. **Biology:** only in subterranean crevices and caves. Occurs with the blind shrimp *Calliasmata pholidota*. **Range:** Red Sea only?

7 Imperial shrimp   *Periclimenes imperator*
To 2 cm. Head resembles a duck's bill; dorsal band or patches of fine pale dots; colour matches host; legs and cheliped tips purple. **Biology:** commensal on large nudibranchs (particularly *Hexabranchus sanguineus*, but also *Dendrodoris tuberculosa* and *Asteronotus cespitosus*) and sea cucumbers (*Bohadschia*, *Stichopus* and *Synapta*). When on nudibranchs, usually near gills. Feeds on detritus and mucus. **Range:** Red Sea to Polynesia.

8 Sea star shrimp   *Periclimenes soror*
To 2 cm. Head resembles a duck's bill; broad pale stripe along entire dorsal midline; colour matches host. **Biology:** commensal on sea stars including *Acanthaster*, *Choriaster*, *Culcita* and *Linckia* with up to 24 per host. **Range:** Red Sea to Polynesia.

9 Red Sea wire coral shrimp   *Pontonides unciger*
To 1.5 cm. Translucent yellow with opaque bands. **Biology:** coral reefs, 3 to at least 50 m. Only on the wire coral *Cirripathes anguina*. Mimics its polyps. **Range:** Red Sea only; a **similar** species in Indian and Pacific Oceans.

4. Guam, RM

5. Marsa Shagra, RM

6. Sinai, JN

7. Sinai, MH

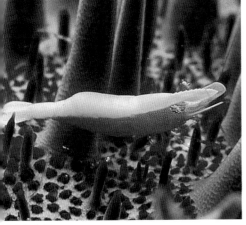

8. Jeddah, JK

9. Sharm el Naga, HG

1. Egypt, PN

2. Sinai, PH

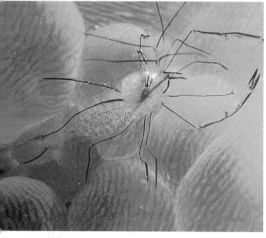

3. Jeddah, JK

1 Urchin shrimp *Stegopontonia commensalis*
To 3.7 cm. Elongate; purplish-black with narrow white
lateral stripe. **Biology:** commensal on spines of sea urchins
(*Diadema*, *Astropygia* and *Echinothrix*). Paired or alone,
facing centre of urchin. **Range:** Red Sea to Polynesia.

2 Cave cleaner shrimp *Urocaridella* sp.
To 3 cm. Tranparent with distinctive markings on tail.
Biology: coral reefs, below 10 m. Usually in groups in
caves. A cleaner that feeds on ectoparasites of fishes,
including sleeping ones at night. **Range:** Red Sea to W.
Pacific. Often misidentified as *Leandrites cyrtorhynchus*.

3 Bubble coral shrimp *Vir philippinensis*
To 1.5 cm. Tranparent with purple antennae and thin
purple line along dorsal midline and through appendages.
Biology: inhabits the bubble coral *Plerogyra sinuosa*.
Range: Red Sea to W. Pacific.

SNAPPING SHRIMPS   ALPHEIDAE

Many species make a cracking sound produced by an
enlarged claw and stay hidden in holes or under rubble.
Some maintain burrows used by certain commensal gobies.
The goby acts as lookout and warns the shrimp of
approaching danger with a flick of its tail.

4 Beautiful goby shrimp *Alpheus bellulus*
To 3.5 cm. Beige with contrasting dark green bands and
blotches and two pale crossbands. **Biology:** maintains
burrow in sand near coral reefs used by the gobies
Amblyeleotris steinitzi, *Cryptocentrus fasciatus* and
Ctenogobiops maculosus. **Range:** Red Sea to W. Pacific.

5 Djibouti goby shrimp *Alpheus djiboutensis*
To 3.5 cm. Olive with contrasting dark green bands and
blotches and two pale crossbands. **Biology:** maintains
burrow in coarse sand near coral reefs used by the gobies
Amblyeleotris steinitzi, *Cryptocentrus caeruleopunctatus*,
C. cryptocentrus, *C. lutheri* and *Flabelligobius latruncularius*.
Range: Red Sea to Indonesia.

6 Red-striped goby shrimp *Alpheus ochrostriatus*
To 4 cm. Pale grey to yellow with ochre pinstripes; two
pale crossbands. **Biology:** maintains burrow in coarse sand
near coral reefs used by the gobies *Amblyeleotris steinitzi*
and *A. fasciata*. **Range:** Red Sea to Fiji.

7 Olive goby shrimp *Alpheus rapax*
To 4 cm. Mottled olive. **Biology:** maintains burrow in fine
sand used by the gobies *Cryoptocentrus* spp., *Ctenogobiops
maculosus*, and sometimes *Amblyeleotris steinitzi*. **Range:**
Red Sea to Polynesia. **Similar:** *A. djeddensis* has dark blotch
on sides, pale crossband, and yellow antennae (Red Sea to
W. Pacific; 4 cm).

8 Red-spotted goby shrimp *Alpheus
rubromaculatus*
To 4 cm. Pale with small red marks. **Biology:** burrows in
pockets of sand along the base of coral slopes, 1 to over
20 m. The only shrimp commensal with the goby *Lotilia
graciliosa*. Burrows also used by *Cryptocentrus* spp. and
Ctenogobiops maculosus. **Range:** Red Sea to Fiji.

9 Banded snapping shrimp *Alpheus cyanoteles*
To 3 cm. Pale with dark crossbands. **Biology:** hidden
among rubble. **Range:** Red Sea. Many similar species.

4. Sharm Abhur, JR

5. Nuweiba, WT

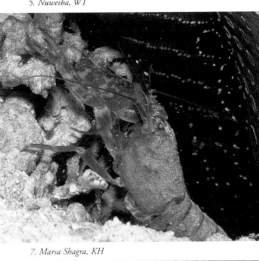

6. Red Sea, JR

7. Marsa Shagra, KH

8. Mangrove Bay, PN

9. Jeddah, JK

1. Jeddah, JK

2. Jeddah, JK

3. Sinai, JN

1 Spotted-claw snapping shrimp *Alpheus lottini*
To 4.5 cm. Reddish-brown with dark spots on chelipeds.
Biology: exclusivley in *Pocillopora* coral heads. Feeds on
tiny invertebrates, algae and mucus. **Range:** Red Sea to
Polynesia.

2 Coral snapping shrimp *Synalpheus charon*
To 2 cm. Red or creamy. **Biology:** exclusively in the corals
Pocillopora meandrina and *Stylophora pistillata*, often in
groups. **Range:** Red Sea to Panama.

HUMP-BACKED SHRIMPS
HIPPOLYTIDAE
Tail with distinctive hump, some species with pincers.
Includes a few commensals of anemones or sponges and a
few cleaners.

3 Blind cave shrimp *Calliasmata pholidota*
To 4 cm. Pale orange with reddish crossbands; blind.
Biology: only in subterranean crevices and caves without
a surface connection to the sea. Often occurs with
Periclimenes pholeter which has eyes. **Range:** Red Sea to
Polynesia.

4 Seagrass shrimp *Hippolyte ventricosa*
To 2.4 cm. Brown to green with discoloured blotches.
Biology: among seagrasses and algae. **Range:** Red Sea to
W. Pacific.

5 White-banded cleaner shrimp *Lysmata
amboinensis*
To 5 cm. **Biology:** coral reefs, 1 to 30 m. Maintains
cleaning station near shelter of crevice, most often on
isolated patch reefs. Waves its antennae to attract client
fishes, feeds on their parasites. Usually in pairs. Retreats
into crevice when alarmed. **Range:** circumtropical.

6 Marbled shrimp *Saron marmoratus*
To 7 cm. Complex floral pattern, green by day, red at
night; ♂ with elongate pincers, ♀ with longer bristles.
Biology: intertidal to over 30 m. In crevices of corals and
rocks. Nocturnal. Often associated with anemones and
Diadema sea urchins. **Range:** Red Sea to Polynesia.

7 *Saron* sp.
To 4 cm. **Biology:** among rubble. Nocturnal, may associate
with *Diadema* urchins. **Range:** Red Sea.

8 Squat shrimp *Thor amboinensis*
To 2 cm. Three white saddles, abdomen held up; ♂ larger
than ♀. **Biology:** commensal on hydroids, anemones and
corals. In pairs or groups. **Range:** circumtropical.

REEF LOBSTERS ENOPLOMETOPIDAE
First pair of legs with enlarged pincers. Closely related to
clawed lobsters of N. Atlantic, but much smaller.

9 Red reef lobster *Enoplometopus occidentalis*
To 12 cm. Prominent claws; long sensory hairs on claws,
tail, and along back. **Biology:** coral reefs to at least 15 m.
Nocturnal. Far into crevices by day, near entrance by night.
Range: Red Sea to Polynesia.

4. Egypt, JK

5. Sinai, MH

6. Jeddah, JK

7. Jeddah, JK

8. Mangrove Bay, EL

9. Hawaii, RM

1. Jeddah, JK

SHRIMP LOBSTERS   AXIIDAE
1 *Axiopsis* **sp. (?)**
To *ca.* 3 cm. Dark brown. **Biology:** in burrows in coarse sand of shallow sheltered reefs, 1 to at least 10 m. Leaves burrow at night to forage. **Range:** Red Sea only?

GHOST SHRIMPS   CALLIANASSIDAE
2 *Callianassa* **sp.**
To *ca.* 4 cm. Creamy yellow with orange bands on chelipeds. **Biology:** inhabits reef flats and shallow silty lagoons. Lives in burrows with side chambers and two entrances. Rarely leaves burrow. **Range:** Red Sea only?

MUD LOBSTERS   THALASSINIDAE
3 *Urogebia* **sp.**
To *ca.* 1.5 cm. Creamy with orange chelipeds. **Biology:** muddy bottoms of estuaries and tidal flats. Lives in and rarely leaves burrows. **Range:** Red Sea only?

SPINY LOBSTERS   PALINURIDAE
Two pairs of antennae, the outermost greatly elongate. No pincers. Shell hard and spiny. Primarily nocturnal omnivores and scavengers. All commercially valuable.

4 Painted spiny lobster   *Panulirus versicolor*
To 40 cm. Antennae white, pinkish at base; legs with white stripes. **Biology:** coral or rocky reefs, 1 to at least 50 m. Nocturnal, in crevices and holes during the day. Sometimes in small groups. A scavenger of invertebrates and dead fishes. **Range:** Red Sea to Fr. Polynesia.

5 Scalloped spiny lobster   *Panulirus homarus*
To 33 cm. Brilliant blue markings around eyes, legs dark. **Biology:** rocky and coral reefs, 1 to 90 m. Caught in traps, commercially important. **Range:** Oman to New Caledonia; Marquesas, s. to E. Africa.

6 Pronghorn spiny lobster   *Panulirus penicillatus*
To 40 cm. Bases of antennae blue, legs striped. **Biology:** exposed clear rocky and coral reefs with abundant holes, 1 to 16 m. Migrates into shallow water during summer. **Range:** Red Sea to C. America.

7 Ornate spiny lobster   *Panulirus ornatus*
To 60 cm. Legs distinctively marked in white and black. **Biology:** shallow protected reefs, often in turbid areas, usually above 10 m, rarely to 200 m. **Range:** Red Sea to Fiji; a recent migrant to E. Mediterranean.

SLIPPER LOBSTERS   SCYLLARIDAE
Outer antennae and body flattened and shield-like.

8 Ridgeback slipper lobster   *Scyllarides haanii*
To 51 cm. Series of humps along back. **Biology:** rocky and coral reefs, 10 to 135 m. Nocturnal. Uncommon. **Range:** Red Sea to Hawaii.

9 Clam digger   *Scyllarides tricdacnophaga*
To 30 cm. Abdomen spined and without hump. **Biology:** coral reefs, 1 to 112 m. Nocturnal. Observed opening *Tridacna* clams. **Range:** Red Sea to W. Thailand.

2. Jeddah, JK

3. Jeddah, JK

4. *Indonesia, RM*

5. *Oman, KF*

6. *Marshall Is., RM*

7. *Guam, IE*

8. *Marsa Alam, BE*

9. *Egypt, MK*

1. Jeddah, JK

2. Jeddah, JK

3. Jeddah, JK

SPIDER CRABS   MAJIDAE

Carapace drawn out to long pointed rostrum, legs slender. Many species attach dead material or living organisms to themselves for camouflage. Size given for all crabs is for carapace width.

1 Soft coral spider crab   *Achaeus spinosus*

To 4 cm. Creamy with red eyes and extremely long bristly legs. **Biology:** typically inhabits fire corals and *Dendronephthya* soft corals. Nocturnal, often attaches hydroids (Zancleidae) to its legs for defence and camouflage. **Range:** Red Sea to W. Pacific.

2 Invisible decorator crab   *Camposcia retusa*

To 3 cm. Rostrum short and forked. Typically completely covered with various organisms (algae, hydoids, sponges). **Biology:** on algae-covered rocky or sedimented coral reefs, to at least 15 m. Feeds on small organic particles. **Range:** Red Sea to Philippines, Japan.

3 Tuberculate spider crab   *Cyclocoeloma tuberculata*

To 5 cm. Carapace with densely packed tubercles for attaching anemones, tunicates and soft corals. **Biology:** on shallow reefs. Identification tentative, not previously recorded from the Red Sea. **Range:** Red Sea to W. Pacific.

4 Longhorned spider crab   *Lahaina* cf. *ovatus*

To 3 cm. Pair of elongate spines between eyes; appendages elongate; sparsely decorated. **Biology:** usually on soft corals. **Range:** Red Sea to W. Pacific.

5 Eyelash spider crab   *Schizophrys aspera*

To 10 cm. Reddish-brown; long spines on edge of carapace covered with small tubercles. **Biology:** coral reefs, 5 to 40 m. Nocturnal. Often covered with algae, sponges and soft corals. **Range:** Red Sea to Polynesia.

6 Sponge carrier   *Stilbognathus longispinosus* (?)

To 3.5 cm. Front of carapace with two pairs of very long spines. **Biology:** often on branching *Acropora* corals. Heavily decorated with sponges and debris including stinging polyps (*Nausithoe*?). **Range:** Red Sea only.

7 Gorgonian spider crab   *Xenocarcinus conicus* (?)

To 1.5 cm. Body a knobby elongate triangle. Red, grey or olive, often with pale dorsal band. **Biology:** on black corals, hydroids and gorgonians. Identification tentative. **Range:** Red Sea to W. Pacific.

ELBOW CRABS   PARTHENOPIDAE

8 Thorny elbow crab   *Lambrachaeus ramifer*

To 2 cm. Rostrum elongate with eyes on sides; chelipeds elongate with long spines. **Biology:** on coral reefs with abundant soft corals to over 24 m. Often covered with filamentous algae. **Range:** Red Sea to Polynesia.

ECHINODERM CRABS   EUMEDONIDAE

Symbionts of echinoderms. Often colourful.

9 Feather star crab   *Tiaramedon spinosus*

To 2.5 cm. Carapace with large blunt spines; white with red to brown stripes. **Biology:** shallow protected coral reefs, to 10 m. Lives primarily on crinoids. **Range:** Red Sea to W. Pacific.

4. Jeddah, JK

5. Aqaba, AK

6. Nuweiba, RM

7. Jeddah, JK

8. Jeddah, JK

9. Jeddah, JK

1. Marsa Alam, BE

2. Nuweiba, MT

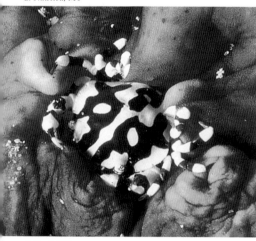

3. Sinai, MH

SWIMMING CRABS   PORTUNIDAE
Carapace wide, often with spine; last segment of rear leg flattened into a paddle for swimming or burrowing.

1 Rainbow swimming crab   *Charybdis paucidentata*
To 10 cm. Orange with blue marks. **Biology:** reef flats and slopes to 10 m. Feeds on crabs and worms. **Range:** Red Sea to Polynesia.

2 Claw-spot swimming crab   *Portunus* sp.
To *ca*. 5 cm. Tan with dark brown spot on upper claw. **Biology:** shallow sandy bottoms of seagrass beds. Nocturnal. **Range:** Red Sea only. Many similar species.

3 Harlequin crab   *Lissocarcinus orbicularis*
To 4 cm. Brown with rounded white blotches, sometimes reversed. **Biology:** commensal on tube anemones and sea cucumbers. **Range:** Red Sea to Fiji.

4 Spike-shelled swimming crab   *Thalamita picta*
To *ca*. 4 cm. Lateral spine a long spike; carapace sculpted. **Biology:** on coarse sand and rubble of shallow coral reefs. **Range:** Red Sea to Polynesia.

CORAL GUARD CRABS   TRAPEZIIDAE
5 Red-spotted guard crab   *Trapezia tigrina*
To 1.5 cm. Creamy with evenly spaced red spots. **Biology:** only among the branches of *Stylophora* and *Pocillopora* corals. Solitary or in small groups. Feeds on detritus and mucus. Defends host corals from Crown-of-thorns starfishes. **Range:** Red Sea to Polynesia.

STONE AND MUD CRABS   XANTHIDAE
Stubby crabs with heavy sculpted shells. The most speciose crab family on coral reefs with hundreds of species. Some species often poisonous.

6 Warty reef crab   *Actaea semblatae*
To 3 cm. Highly sculpted; purplish with pale tubercles. **Biology:** among rocks of shallow protected coral reefs. **Range:** Red Sea; distribution elsewhere uncertain.

7 Poison crab   *Zosimus aeneus*
To 4 cm. Highly sculpted with dark protuberances. **Biology:** among rocks of shallow reef flats and slopes. Extremely poisonous to eat. **Range:** Red Sea to Polynesia.

HAIRY CRABS   PILUMNIDAE
Recently split from Xanthidae; many species hairy.

8 Hairy reef crab   *Pilumnus* cf. *vespertilio*
To 3 cm. Creamy with long bristles resembling a mop. **Biology:** shallow protected reef flats and slopes. **Range:** Red Sea to W. Pacific?

REEF CRABS   CARPILIIDAE
Recently split from Xanthidae; carapace smooth.

9 Convex reef crab   *Carpilius convexus*
To 7 cm. Carapace smoothly rounded; orange-red often with brown and white mottling. **Biology:** reef flats and slopes with abundant shelter, to at least 20 m. Feeds on gastropods and urchins. **Range:** Red Sea to Polynesia.

4. Jeddah, JK

5. Jeddah, JK

6. Jeddah, JK

7. Sinai, HM

8. Jeddah, JK

9. Egypt, MB

1. Jeddah, JK

2. Maesa Alam, BE

3. Marsa Shagra, RM

BOX CRABS — CALAPPIDAE

Chelipeds fold tightly against carapace; knobbed claw works like can-opener along whorls of gastropods.

1 Common box crab — *Calappa hepatica*
To 10 cm. Tan to olive; carapace and chelipeds covered with tubercles. Carapace with flanges at rear. **Biology:** sandy areas, intertidal to 100 m. Active by day and night, usually buried. **Range:** Red Sea to Polynesia.

2 Edge-barred box crab — *Calappa* cf. *lophos*
To 10 cm. Tan with dark lines on rear edges of carapace. **Biology:** sandy seagrass beds and sand expanses to 100 m. Nocturnal, feeds on molluscs. Indentification tentative. **Range:** Red Sea to New Caledonia.

SALLY-LIGHT-FOOT AND SHORE CRABS — GRAPSIDAE

Thin-shelled somewhat flattened crabs primarily in rocky intertidal areas. Most species feed on algae just above the waterline. Wary and fast.

3 Mottled sally-light-foot — *Grapsus albolineatus*
To 5 cm. Dark greenish-brown with wavy white bands on legs. **Biology:** rocky intertidal shore of coral reefs. Common. Hides in cracks. **Range:** Red Sea to Polynesia.

4 Common sally-light-foot — *Grapsus tenuicristatus*
To 8 cm. Dark greenish-brown with fine white dashes. **Biology:** rocky intertidal shore of coral reefs. Dashes to water when approached. **Range:** Red Sea to Polynesia.

5 Purple-clawed shore crab — *Metopograpsus thukuhar*
To *ca.* 3 cm. Mottled brown to olive; claws purple with white spots. **Biology:** rocky intertidal shore of coral reefs. **Range:** Red Sea to Polynesia.

6 Large-clawed shore crab — *Geograpsus crinipes*
To *ca.* 5 cm. Tan to purplish-brown; claws large. **Biology:** nocturnal, may travel well away from shore at night to feed on insects. **Range:** Red Sea to Fr. Polynesia.

GHOST AND FIDDLER CRABS — OCYPODIDAE

Ghost crabs inhabit sandy shores above the waterline where they remain in deep burrows during the day. They forage at night on shoreline debris.

7 Common ghost crab — *Ocypode cordimanus*
To 4.5 cm. Pale with dark reticulations; eyes without horn. **Biology:** sandy areas above the high water mark. Common. **Range:** Red Sea to Fr. Polynesia.

8 Red Sea ghost crab — *Ocypode saratan*
To 6 cm. Light tan; eyes with horn. **Biology:** sandy areas above high water mark. Common. Male builds a special copulation burrow with a 15 cm high sand pyramid on beaches to attract females. **Range:** Red Sea only.

LAND HERMIT CRABS — COENOBITIDAE

9 Pale land hermit crab — *Coenobita scaevola*
To 3 cm. Live in empty gastropod shells. Exoskeleton heavy and knobby; pale. **Biology:** on land near shore, enter water on lunar cycle to spawn. Primarily nocturnal, roam beach strand for food. **Range:** Red Sea only?

4. Indonesia, RM

5. Abu Galum, RM

6. Marsa Bareka, RM

7. Sharm el Naga, HG

8. Lahami, KF

9. Marsa Shagra, RM

1. Guam, RM

2. Egypt, AK

3. Marsa Bareka, PM

HERMIT CRABS   DIOGENIDAE

Abdomen soft and curved with last pair of legs upturned for holding mollusc shell used as home.

1 Cone shell hermit crab   *Ciliopagurus strigatus*
To 6 cm. Carapace white and depressed; appendages red with narrow yellow bars. **Biology:** on rubble or algae-covered coralline rock, reef flats to at least 20 m. Prefers cone shells which it may kill to obtain a new shell. **Range:** Red Sea to Polynesia.

2 Reef hermit crab   *Dardanus lagopodes*
To 3 cm. Eyestalks white, antennae yellow, bristles white-tipped. **Biology:** reef flats to outer reef slopes over 30 m. Lives in shells with wide aperture. Common. **Range:** Red Sea to Fr. Polynesia.

3 Red Sea anemone hermit crab   *Dardanus tinctor*
To 10 cm. Eyes green, stalks with red and white bands. **Biology:** on sand or rubble of coral reefs, 1 to 30 m. Carries several *Calliactis polypus* anemones on its shell. A nocturnal carnivore and scavenger. **Range:** Red Sea.

4 Zebra-legged hermit crab   *Pylopaguropsis zebra* (?)
To *ca.* 1 cm. Right claw enlarged; carapace white, legs red with white stripes. **Biology:** deep outer reef slopes to at least 60 m. **Range:** Red Sea only?

SQUAT LOBSTERS   GALATHEIDAE

Carapace oval with sharp rostum and tail bent under; chelipeds long and held forward. Most species cryptic and commensal on sponges and cnidarians.

5 Crinoid squat lobster   *Allogathea elegans*
To 2 cm. Dark brown to black with white to yellow stripes. **Biology:** lives on crinoids (p. 342). Mimics the colour of its host and feeds on plankton stolen from it. **Range:** Red Sea to Marshall Is.

6 *Galathea* sp.
To 2 cm. Creamy with long brown-banded chelipeds. **Biology:** on fire corals or alcyonarians of protected coral reefs. Copulates belly-to-belly (?). **Range:** Red Sea only.

PORCELAIN CRABS   PORCELLANIDAE

Body and appendages flattened with bristly mouth parts specialized for filtering plankton or detritus.

7 Anemone porcelain crab *Neopetrolisthes maculata*
To 3 cm. Creamy with red to brown dots. **Biology:** live on large anemones, primarily *Stichodactyla* spp. Usually among the tentacles or under the mantle, may enter the mouth. **Range:** Red Sea to W. Pacific.

8 Tidepool porcelain crab   *Petrolisthes rufescens*
To 2 cm. Dark greenish-brown with pale reticulations. **Biology:** reef flats and shallow protected reefs. Abundant under rocks and rubble. Feeds on detritus stirred into the water by maxillipeds. **Range:** Red Sea only.

9 Sea pen porcelain crab   *Porcellanella triloba*
To 1 cm. Creamy with ochre bands. **Biology:** solitary or in groups only on large sea pens (p. 252). Nocturnal. Feeds on plankton. **Range:** Red Sea to W. Pacific.

4. Jeddah, JK

5. Maldives, WP

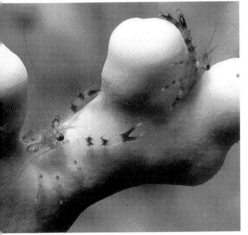

6. Sinai, MH

7. Thailand, WP

8. Marsa Shagra, RM

9. Sinai, MH

ECHINODERMS

Echinoderms are among the most conspicuous of coral reef animals. They are an ancient group that first appeared 500 million years ago. Today there are 6,500 species in six classes. Echinoderms are exclusively marine and except for a few deep-water pelagic sea cucumbers, live on the sea floor. They are either scavengers or predators of primarily sessile organisms, sifters of sediment, or filter feeders.

Echinoderms are unique among animals by having a radially symmetric body plan and water vascular system. They lack a head, brain and heart and have skeletal, circulatory and nervous systems that radiate in five directions from a central axis. Their unique water vascular (**ambulacral**) system consists of a complex of reservoirs and canals that draw water through a sieve plate and ends in numerous tiny hollow tube feet. These are powered by water pressure and used for locomotion, feeding and, in some cases, respiration. Some forms lack tubed feet but all have the ambulacral system. The digestive system is not radial and consists of a mouth and anus at opposite ends of a digestive tract running along the central axis. Some forms lack an anus. Although named for their often spiny surface, echinoderms have an internal calcareous skeleton located just beneath the skin which consists of plates or disconnected ossicles arranged radially within the skin or flesh. Spines, if present, are covered with a layer of skin. Many forms are able to survive catastrophic injury or disease and regenerate lost body parts or new individuals from body parts. Most echinoderms have separate sexes and reproduce by releasing sperm and eggs into the water. The fertilized eggs hatch into bilaterally symmetric pelagic larvae.

Crinoids (class Crinoidea) were the first echinoderms and were once the dominant animals of ancient sea floors. They have 5 to 200 narrow flexible arms usually lined with sticky cirri which capture plankton that is shunted to a central mouth. Some primitive forms are permanently attached by a long stalk. Most coral reef species have several short claw-like arms used to cling to objects exposed to currents.

Brittle stars (class Ophiuroidea) typically have 5 narrow arms connected to a discrete central disc. These may be armed with hair-like spines that can break off and irritate the skin. Most are detritivores that hide within holes or under rocks. Others are commensals of gorgonians and sponges.

Sea stars (class Asteroidea) typically have 5 arms, each with an ambulacral groove on the underside which is lined with tube-feet and converges on a central mouth. They are typically predators or detritivores that crawl on their tubed feet and evert their stomach to digest prey externally.

Sea cucumbers (class Holothuroidea) typically lie on one 'side' modified with tube feet for attachment or mobility. They are soft-bodied with skeletons reduced to spicules. Some have branched oral tentacles used for filter feeding or combing the bottom, while others feed on sediment which is gleaned of digestible matter. In defence, some sea cucumbers emit sticky, sometimes toxic, white threads from the anus (Cuvierian tubules) or eviscerate themselves, then regenerate their lost organs. The body wall of many large species is boiled and dried for human consumption. This product, known as Trepang or 'Beche-de-mer', is the subject of a valuable fishery which has decimated populations in many areas.

Sea urchins (class Echinoidea) have a globular or flattened skeleton (test) of fused plates bristling with moveable spines, tubed feet and pincer-like pedicillariae. The mouth is located on the underside and equipped with a 5-part radial set of jaws and teeth used for scraping algae or other food.

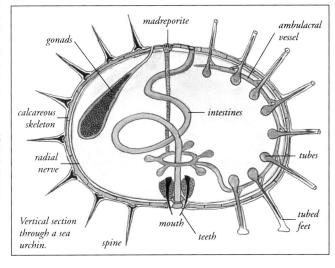

Vertical section through a sea urchin.

madreporite · gonads · calcareous skeleton · radial nerve · spine · mouth · teeth · intestines · ambulacral vessel · tubes · tubed feet

◀ **Crown-of-thorns starfish**
Acanthaster planci
(Class ASTEROIDEA:
ACANTHASTERIDAE) To 50 cm.
Covered with long venomous spines which can cause extremely painful festering wounds; olive to dark green or rust-red. **Biology:** on coral reefs, 0.5 to 30 m, usually in protected areas. A major predator of stony corals. Normally uncommon, populations periodically explode and cover entire reefs which leads to heavy destruction of corals (recently in Egypt from Hurghada to Ras Banas; p. 11). **Range:** Red Sea to Mexico. *(Photo: Hurghada, KF.)*

1. Nabq, EL

2. Marsa Alam, KH

3. Aqaba, JN

CRINOIDS, FEATHER STARS
Class CRINOIDEA

COLOBOMETRIDAE

1 Sawtoothed feather star   *Oligometra serripinna*
To 20 cm. 10 arms. Black to brown with creamy bands.
Biology: on current-swept seaward slopes, 15 to 40 m.
Occupies semi-permanent perches on sponges and the like,
with up to 100 per m². **Range:** Red Sea to New Caledonia.

2 Chadwick's feather star   *Decametra chadwicki*
To 12 cm. Cirri sparse; 5 pairs of arms joined at base.
Biology: on seaward reef slopes, common below 30 m.
Range: Red Sea only?

HETEROMETRIDAE

3 Savigny's feather star   *Heterometra savignii*
To 40 cm. Juv. with 10 arms, increasing to 20 with growth;
cirri beige to black with pale tips. **Biology:** on shallow
fore-reef slopes to 15 m. Often on fire corals. Nocturnal
above 6 m, continuously active below. Hosts up to 18
species of commensals. **Range:** Red Sea to Arab. Gulf.

MARIAMETRIDAE

4 Klunzinger's feather star *Lamprometra klunzingeri*
To 30 cm. 20 arms. Creamy with brown to red bands.
Biology: on shallow reef margins and slopes. Common on
fire corals, also on sponges and gorgonians. Hides in
crevices by day, emerges to exposed perches at dusk. Up to
70 per m². **Range:** Red Sea to Maldives, s. to S. Africa.

BRITTLE STARS, BASKET STARS
Class OPHIUROIDEA

BASKET STARS GORGONOCEPHALIDAE

5 Basket star, Gorgon head   *Astroboa nuda*
To 1 m armspread. Highly branched rays. **Biology:** reef slopes
exposed to current, 1 to 120 m. Coiled into tight ball during
day, spreads arms to form basket on exposed perches to feed
on plankton by night. **Range:** Red Sea to New Caledonia.

LONG-SPINED BRITTLE STARS
OPHIOTRICHIDAE

6 Savigny's brittle star   *Ophiothrix savignyi*
To 15 cm. Arms with needle-like spines and dark
longitudinal stripe. **Biology:** on lagoon and seaward reef
slopes, 10 to at least 40 m. On sponges, soft corals and fire
corals, hidden by day. **Range:** Red Sea to New Caledonia.

REEF BRITTLE STARS OPHIOCOMIDAE

7 Reef flat brittle star   *Ophiocoma scolopendrina*
To 15 cm. Brown to black, underside pale. **Biology:** on
reef flats and upper slopes, intertidal to 5 m. Abundant.
Territorial, defends crack from other brittle stars. Feeds on
detritus and surface film during low tide. **Range:** Red Sea.

8 Black brittle star   *Ophiocoma erinaceus*
To 15 cm. Black. **Biology:** reef flats, usually beneath
rubble by day. **Range:** Red Sea only?

OPHIURIDAE

9 Superb brittle star   *Ophiolepis superba*
To 20 cm. Spines short; beige with bold dark bands.
Biology: on deep coral reefs, 10 to 50 m. Feeds on
detritus. **Range:** Red Sea to New Caledonia.

4. Nuweiba, RM

5. Manado, RM

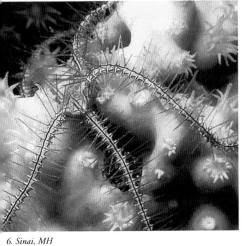

6. Sinai, MH

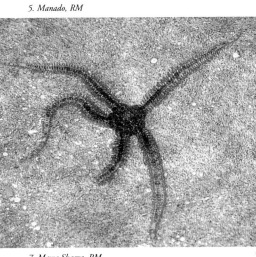

7. Marsa Shagra, RM

8. Aqaba, EL

9. Guam, RM

1. Maldives, RM

2. Oman, KF

3. Marsa Bareka, PM

SEA STARS Class ASTEROIDEA

CUSHION STARS OREASTERIDAE

1 Grainy cushion star *Choriaster granulatus*
To 25 cm. Pale pinkish-orange. **Biology:** seaward reef
slopes, 5 to 40 m, usually below 20 m. On hard bottoms.
Feeds on carrion and detritus. **Range:** Red Sea to Fiji.

2 Arabian cushion star *Culcita coriacea*
To 25 cm. Arms greatly reduced; beige to orange with brown
reticulations. **Biology:** on sand, rubble and coralline rocks
of reef flats, lagoons and slopes, to 30 m. Uncommon.
Feeds on a variety of invertebrates including corals as well
as algae. **Range:** Red Sea to Oman.

COMB SEA STARS ASTROPECTINIDAE

3 Comb sea star *Astropecten polyacanthus*
To 15 cm. Margin with prominent white spines. **Biology:** on
sandy bottoms near coral reefs, 1 to 60 m. Buried by day,
active on surface by night. Can bury itself quickly. Feeds on
bivalves. **Range:** Red Sea to New Caledonia and Hawaii.

OPHIDIASTERIDAE

4 Ghardaqa sea star *Fromia ghardaqana*
To 8 cm. Red with widely spaced pale blue spots on arms.
Biology: on rubble and sand of shallow coral reefs, to
20 m. Nocturnal, hidden under rocks by day. **Range:** Red
Sea to Mauritius.

5 Pebbled sea star *Celerina heffermani* (?)
To 10 cm. Red with white to cream plates along margin
and tops of arms. **Biology:** on sand, rubble and coralline
rocks of lagoons and protected reef slopes, 1 to 40 m.
Feeds on small invertebrates and detritus. **Range:** Red Sea
to W. Polynesia. Possibly a species of *Fromia* close to *F.
monilis*, not yet reported from the Red Sea.

6 Thorny sea star *Fromia nodosa*
To 15 cm. Small round plates with conical thorns.
Biology: on rubble and dead coral of lagoon and seaward
reef slopes, 5 to 50 m. Nocturnal, feeds on ascidians and
sponges. **Range:** Red Sea to Maldives.

7 Watson's sea star *Gomophia watsoni*
To 12 cm. Some conical thorns with blue tips; broad dark
crossbands on arms. **Biology:** on rubble and dead coral of
lagoon and seaward reef slopes, 10 to 55 m. **Range:** First
record from Red Sea? Otherwise known from Palau,
NE. Australia and New Caledonia.

8 Leach's sea star *Leiaster leachi*
To 20 cm. Orange with red spots. **Biology:** reef flats,
lagoons and slopes, 5 to 35 m. In shady areas by day,
active and exposed by night. **Range:** Red Sea to Pacific.

ECHINASTERIDAE

9 Nodular sea star *Echinaster callosus*
To 26 cm. Completely covered with cream to reddish-brown
hard round nodules. **Biology:** on lagoon and seaward reefs,
2 to over 20 m. Nocturnal. **Range:** Red Sea to New
Caledonia.

4. Lahami, KF

5. Nuweiba, PM

6. Nuweiba, EL

7. Marsa Bareka, RM

8. Marsa Bareka, PM

9. Marsa Bareka, PM

1. Nuweiba, HM

2. Sinai, PM

SEA CUCUMBERS
Class HOLOTHUROIDEA

Order ASPIDOCHIROTIDA
HOLOTHURIDAE

1 Surge sea cucumber *Actinopyga mauritiana*
To 35 cm. Rigid with small papillae; brown with white patches on sides and pale underside. **Biology:** on outer reef flats, channels and slopes exposed to surge or current to 20 m. **Range:** Red Sea to Polynesia.

2 Blackmouth sea cucumber *Pearsonothuria graeffei*
To 45 cm. Pale with brown blotches, white-tipped papillae and black tentacles. **Biology:** on algae-encrusted hard bottoms of clear lagoon and seaward slopes, 3 to 30 m. Juvenile mimics the toxic nudibranch *Phyllidia varicosa* (p. 310). **Range:** Red Sea to Fr. Polynesia.

3 Pink sea cucumber *Holothuria edulis*
To 35 cm. Pink becoming dark brown to black above. **Biology:** on sand and hard bottoms of lagoon and seaward reefs, 0.3 to 30 m. **Range:** Red Sea to Polynesia.

4 Black sea cucumber *Holothuria atra*
To 45 cm. Black with smooth skin with sand stuck to surface. **Biology:** on sand patches of shallow reef flats and seagrass beds, rarely to 30 m. Exposed, but often covered with sand or algae. Emits a toxic clear red fluid *holothurin* when handled and may eject viscera when severely disturbed. Reproduces by fission as well as sexually. **Range:** Red Sea to Polynesia.

5 Creamy sea cucumber *Holothuria fuscogilva*
To 30 cm. Creamy with dark dots around papillae. **Biology:** in lagoons and on protected reef slopes, 3 to 25 m. On open sand with *Caulerpa* and *Halimeda* algae, often buried. **Range:** Red Sea to Fr. Polynesia.

6 White-teated cucumber *Holothuria nobilis*
To 50 cm. Black with large white spots and teat-like projections along base; sand stuck to surface. **Biology:** on sand and rocks of open areas of reef terraces, 0.3 to 40 m. Feeds on detritus and tiny invertebrates. Highly esteemed for trepang markets. **Range:** Red Sea to W. Indian Ocean. **Similar:** *H. whitmaei* is black.

STICHOPODIDAE

7 Tubercle cucumber *Stichopus pseudohorrens*
To 40 cm. Olive-brown with long sharply pointed papillae. **Biology:** on open sand and rubble of lagoons and seaward slopes, 3 to 30 m. **Range:** Red Sea to W. Indian Ocean. **Similar:** *H. whitmaei* is black.

8 Prickly cucumber *Thelenota ananas*
To 70 cm. Reddish-brown with large and mostly branched papillae. **Biology:** deep lagoon and seaward reefs, 2 to 30 m. On open sand near reefs. **Range:** Red Sea to New Caledonia.

Order APODIDA: SYNAPTIDAE
Thin-skinned with hooked spicules and extremely extensible bodies that contract when touched. Feed on detritus by inserting sticky oral tentacles in mouth.

9 Worm cucumber *Synapta maculata*
To 2.5 m. **Biology:** on sand, mud, seagrass and rubble of shallow reefs. **Range:** Red Sea to Fr. Polynesia.

3. Mövenpick Bay, EL

4. Siyul Reefs, Egypt, EL

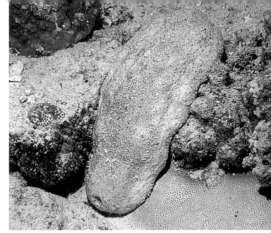

5. Canyon, Dahab, EL

6. Quseir, EL

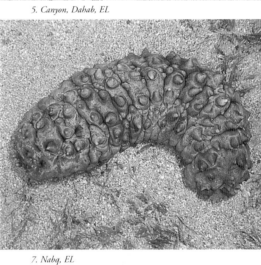

7. Nabq, EL

8. Caroline Is., RM

9. Mangrove Bay, EL

1. Nuweiba, EL

2. Aqaba, EL

3. Abu Galum, RM

SEA URCHINS Class ECHINOIDEA
CIDARIDAE
Blunt primary spines lack skin, often fouled.

1 Imperial urchin *Phyllacanthus imperialis*
To 8 cm. Spines thick and blunt with rough texture; test
wine-red. **Biology:** on shallow sheltered reefs, to 15 m.
Nocturnal, in crevices and holes by day. Omnivorous.
Range: Red Sea to Tonga.

FIRE URCHINS ECHINOTHURIDAE
2 Red Sea fire urchin *Asthenosoma marisrubri*
To 15 cm. Short spines with venomous globular swelling at
tips. May cause painful injuries. **Biology:** on sand, rubble,
dead coral or among seagrasses of lagoon and seaward reef
slopes, 3 to 30 m. Hosts commensal shrimps and crabs.
Nocturnal. **Range:** Red Sea only.

ECHINOMETRIDAE
3 Rock-boring urchin *Ehinometra* cf *mathaei*
To 9 cm. Stout brown to wine-red spines with white basal
rings. **Biology:** on exposed reef flats and upper reef slopes, to
8 m. Erodes rounded holes and furrows in dead coral rock as
it feeds on algae. Common. **Range:** Red Sea to Polynesia.
4 Slate-pencil urchin *Heterocentrotus mammillatus*
To 30 cm. Red primary spines thick, blunt and widely
spaced, these once used for writing. **Biology:** on exposed
reef crests and shallow fore-reef slopes. Wedged in holes or
crevices by day. **Range:** Red Sea to Polynesia.

NEEDLE-SPINED URCHINS
DIADEMATIDAE
5 Common long-spined urchin *Diadema paucispinum*
To 40 cm. Spines extremely long; may inflict painful
injuries. Anal opening with red ring. **Biology:** on subtidal
reef flats and protected reef slopes to 25 m. May form tight
clusters on open areas as a protection from predators. Feeds
on algae. **Range:** Red Sea to Fr. Polynesia.
6 Banded urchin *Echinothrix calamaris*
To 30 cm. Spines banded; anal sac dark with white dots.
Biology: on shallow reefs to 40 m. Nocturnal. Spines
venomous. **Range:** Red Sea to Polynesia.
7 Blue-black urchin *Echinothrix diadema*
To 30 cm. Spines banded when small; white to black when
large; anal sac with dark spots. **Biology:** on hard surfaces
of reef flats and slopes to 20 m. Hosts shrimp *Stegopontonia
commensalis*. **Range:** Red Sea to Polynesia.

TOXOPNEUSTIDAE
8 Collector urchin *Tripneustes gratilla*
To 12 cm. Spines short; five double bars of pedicellariae;
creamy to red or black. **Biology:** shallow reefs to 25 m.
On rubble, sand, or among seagrasses. Often covered with
debris. **Range:** Red Sea to Polynesia.

TEMNOPLEURIDAE
9 Bald-patch urchin *Microcyphus rousseaui*
To 5 cm. Spines short; sides with 5 pale spineless zigzag-
shaped bars. **Biology:** among rubble or seagrasses and on
rocks of shallow reefs, 10 to 30 m. Rare. **Range:** Red Sea and
Oman to E. Africa.

4. Marsa Shagra, RM

5. Nuweiba, RM

6. Sewul, EL

7. Marsa Shagra, RM

8. Marsa Shagra, RM

9. Sharm el Naga, HG

TUNICATES Phylum CHORDATA, Subphyllum UROCHORDATA

Tunicates are in the same phylum as the vertebrates and as larvae share with them many characteristics such as gill slits, a muscular tail, and a notocord, a primitive 'backbone' consisting of a stiff rod and associated nerve cord. As adults they lose all of these traits except the notocord and are either permanently attached to the bottom or drift in the plankton as solitary or colonial filter feeders. They have barrel-like bodies covered with a protective tunic consisting of living cells embedded in a non-cellular material containing cellulose. Most of the interior is occupied by a sac-like pharynx perforated by pharyngeal slits. Cilia maintain a steady flow of water through the top opening which passes through the slits into a surrounding atrium then leaves through an excurrent opening on the side, taking metabolic wastes and gametes with it. A special gland continuously supplies the gill slits with mucus to trap and transport food particles to the deepest point of the pharynx where it is digested in an intestine. A heart with two pacemakers forces blood containing vanadium upwards to the pharynx for several strokes to be oxygenated, then reverses the flow downward to supply the visceral sac with oxygen. A brain and neural gland produce hormones to coordinate biorythms.

Adult **Ascidians** (class Ascidiacea) are permanently attached sessile animals that may be either solitary ('sea squirts') or colonial. They may be transparent or opaque, either vividly coloured or cryptic. Some may be covered with encrusting algae or animals. Some solitary forms range up to the size of a fist while individuals of colonial forms (zooids) are smaller than 3 cm in length, but may form colonies of up to a metre across. In colonial forms, individuals may be attached to their neighbours by a stolon or a continuous mat of common tissue. Mat-forming ascidians may resemble sponges and share a common excurrent (cloacal) opening. Ascidians are hermaphrodites with a 'tadpole' pelagic larval stage. Gametes of solitary forms are released into the sea for fertilization. Those of colonial forms may be retained for internal brooding of eggs which hatch into a brief larval stage lasting from a few minutes to an hour. Some colonial forms may also reproduce by budding.

Salps (class Thalacea) are transparent pelagic tunicates with oral and excurrent siphons at opposite ends of a tubular body. They drift in the current or move by slow jet propulsion, feeding as they move. They may be solitary or connected to form chains or tubes which in some species may attain a length of over 20 m. Many are luminescent. Salps have primitive organs for detecting light and movement in order to migrate vertically with the zooplankton. By being entirely pelagic they have lost their larval stage.

◄ **Pink clinging synascidia** *Didemnum* sp. (Class Ascidacea: DIDEMNIDAE) Colonies to at least 25 cm. Encrusting, may 'drip' from one point of attachment to another; purplish-pink with tiny pale tubercles, edges of excurrent siphons pale. **Biology:** on gorgonians and corals of shallow protected lagoon and seaward reefs, 2 to 20 m. **Range:** Red Sea to W. Pacific. Many **similar** species. *(Photo: Oman, HE)*

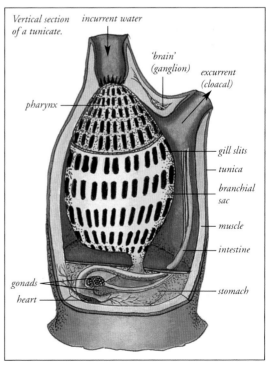

Vertical section of a tunicate.

incurrent water

'brain' (ganglion)

excurrent (cloacal)

pharynx

gill slits

tunica

branchial sac

muscle

intestine

gonads

heart

stomach

1. Hurghada, JH

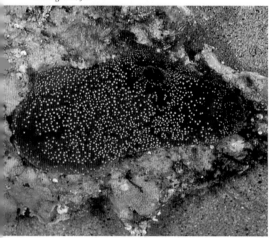

2. Hurghada, JH

3. Oman, KF

ASCIDIANS   Class ASCIDIACEA
SOLITARY ASCIDIANS   ASCIDIIDAE

1 Dark-mouthed ascidian   *Ascidia melanostoma*
To 12 cm. Tranlucent purplish-brown, darker at top;
siphon entrance with 8 flaps. **Biology:** under overhangs or
rocks. Colour due to vanadium which may deter predators.
Range: Red Sea to Indonesia.

2 Golden-dotted ascidian   *Phallusia julinea*
To 30 cm. Brown to yellow with tiny yellow dots; without
epibionts. **Biology:** in crevices or under rocks below 5 m.
Yellow dots due to pigmented blood cells which become
green in air. **Range:** Red Sea to New Caledonia.

3 Blue phallusia   *Phallusia* sp.
To 18 cm. Blue to apricot; barrel-shaped to globular.
Biology: shallow rocky and coral reefs, 3 to 15 m. Yellow
dots due to pigmented blood cells which become green in
air. **Range:** Gulf of Aden and Oman.

4 Spiny ascidian   *Halocynthia spinosa*
To 30 cm. Tunic leathery with small warts and spines,
siphon entrance with 8 flaps. **Biology:** hard bottoms of
poorly developed coral reefs. Solitary or in small groups.
Range: Red Sea to S. Africa.

5 Herdman's ascidian   *Herdmania momus*
To 8 cm. Reddish-brown; siphons trumpet-shaped. **Biology:**
in crevices or under overhangs of lagoon and sheltered
seaward reefs, 5 to at least 100 m. Stores iron in tissues. Gill
sac hosts small commensal shrimps. Large adults usually
covered by sediments and algae. **Range:** circumtropical.

CLAVELINIDAE
6 Red Sea glassy ascidian   *Clavelina* sp.
To 4 cm. Solitary or in groups; barrel-shaped; transparent,
edge of siphon with distinctive white stripes. **Biology:** on
shady hard bottoms of coral reefs, 3 to at least 20 m.
Range: Red Sea only?

7 Golden-gut ascidian   *Clavelina auracea*
Clusters to 4 cm. In colonies attached by stolons. Zooids
tadpole-shaped; clear with green brachial basket and yellow
U-shaped gut and stolon. **Biology:** on reef flats and
protected slopes to at least 25 m. **Range:** Red Sea to New
Caledonia.

DIAZONIDAE
8 Arabian ascidian   *Rhopalaea* sp.
To 5 cm. Solitary; transparent with white sperm duct.
Biology: hard bottoms of reef flats and slopes. **Range:** Red
Sea only.

STYELIDAE
9 Pigmented ascidian   *Polycarpa pigmentata*
To 10 cm. Solitary; interior of siphon with white dots;
often covered with epizoic growth. **Biology:** on dead corals
and rubble of lagoon and sheltered seaward slopes. **Range:**
Red Sea to Micronesia.

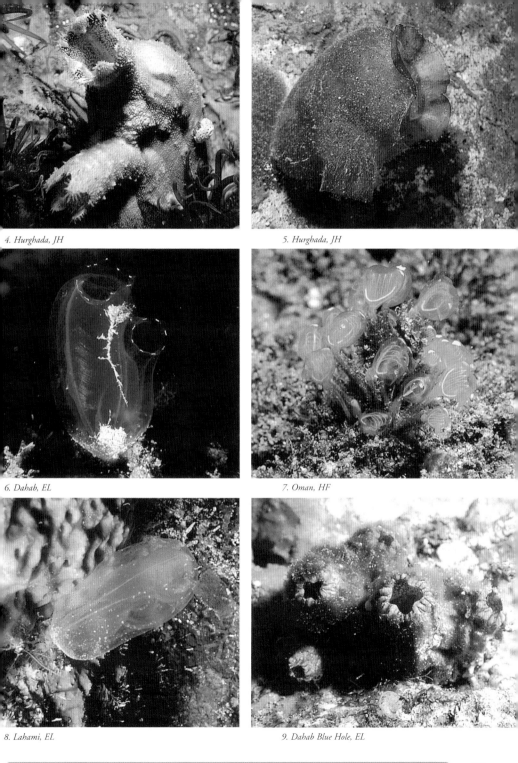

4. Hurghada, JH

5. Hurghada, JH

6. Dahab, EL

7. Oman, HF

8. Lahami, EL

9. Dahab Blue Hole, EL

1. Nuweiba, EL

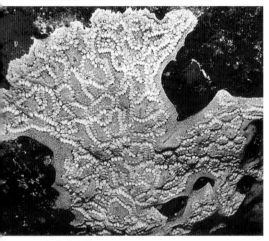

2. Hurghada, HE

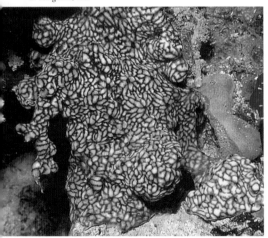

3. Hurghada, HE

STYELIDAE (*cont.*)

1 Black synascidia   *Botryllus niger*
Zooids to 2 cm. Colonial; yellowish-brown, dead ones black. **Biology:** on sand, coral heads, or among seagrasses of reef flats and protected reef slopes to 20 m. **Range:** circumtropical.

2 Labyrinth synascidia   *Botryllus* cf. *firmus*
Colonies to 10 cm. Encrusting; creamy yellow with pale worm-like markings. **Biology:** on dead coral or rocks, 5 to 20 m. **Range:** Red Sea to Maldives.

3 Carpet synascidia   *Botryllus* sp.
Colonies to 20 cm. Encrusting; white to blue with distinctive black markings. **Biology:** on hard bottoms of rocky and coral reefs. **Range:** Red Sea only?

4 Apricot synascidia   *Eusynstyela misakiensis*
Zooids to 1 cm, colonies to 50 cm. Zooids apricot and cone-like, connecting mat white. **Biology:** on rocks, dead corals and shells of coral reefs below 5 m. **Range:** Red Sea to New Caledonia.

5 Cherry synascidia   *Metandrocarpa manina*
Zooids to 1 cm, clusters to 30 cm. Zooids barrel-like, red. **Biology:** on hard bottoms under overhangs and in shady areas of reef slopes below 5 m. Reproduces asexually by stolons. **Range:** Red Sea to W. Pacific.

DIDEMNIDAE

6 Moseley's ascidian   *Didemnum moseleyi*
Colonies to 25 cm. Encrusting, sponge-like; siphon openings crater-like with white edges and fine pores, tunic with spicules; colour variable. **Biology:** on hard bottoms under overhangs and in shady areas, 4 to over 10 m. **Range:** Red Sea to Micronesia.

7 Grey synascidia   *Didemnum* sp.
Colonies to 15 cm. Encrusting, sponge-like; grey with creamy pores and dark siphon edges. **Biology:** on gorgonians and corals of shallow protected lagoon and seaward reefs, 2 to 20 m. **Range:** Red Sea only.

8 Lapping synascidia   *Polysyncratos* cf. *thallomorpha*
Colonies to 15 cm. Encrusting, sponge-like; colour variable from rosy to pale grey, due to embedded sclerites (2–3 mm). **Biology:** on dead corals and gorgonians. **Range:** Red Sea to W. Pacific.

Class THALACEA

Salps   **Order SALPIDA**
SALPIDAE
Pelagic tunicates connected into chains or hollow tubes that drift or swim and feed on zooplankton. Some (*Pyrosoma atlanticum*) are spectacularly bioluminescent. About 20 species worldwide.

9 Vagina salp   *Tethys vagina*
Chains to 50 cm, zooids to 5 cm. In short chains; transparent except for internal organs. **Biology:** drifting with current, usually near surface. Common after storms. **Range:** circumtropical.

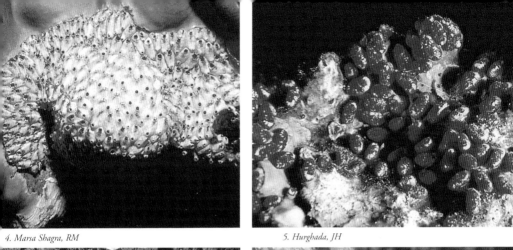

4. Marsa Shagra, RM

5. Hurghada, JH

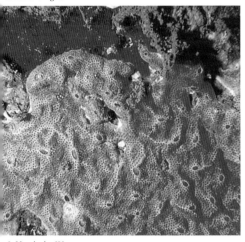

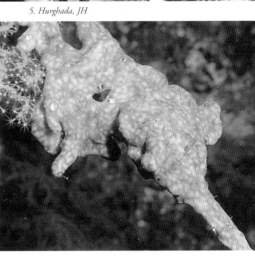

6. Hurghada, JH

7. Nabq, EL

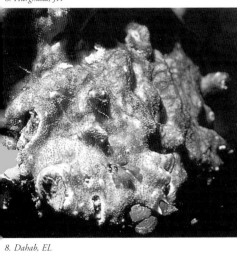

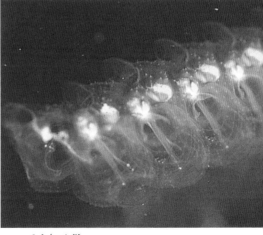

8. Dahab, EL

9. Lahami, EL

MARINE PLANTS

Plants form the base of the world's food chain. They are the primary producers that, through the process of photosynthesis, convert sunlight and nutrients into the biological materials and oxygen that sustain the vast majority of all life. Unlike many marine animals, marine plants are limited to the shallow sunlit depths, the photic zone. In the clearest tropical seas they may occur as deep as 268 m, but their greatest abundance and diversity is in the upper 30 m. Marine plants serve as food either directly for all the sea's herbivorous animals or indirectly for most other marine animals. They are the food of the reef's grazing animals including snails, crustaceans, sea urchins and fishes. Single-celled phytoplankton are the primary food of many zooplankton, particularly copepods, and the early larval stages of nearly all marine animals. **Zooxanthellae** live symbiotically within the tissues of reef-building corals (p. 258) and many other animals. They feed their hosts from within by converting the host's metabolic wastes into nutrients and oxygen.

There are two major groups of marine plants, algae ('seaweeds') and seagrasses (p. 364). A third group, the mangroves (p. 365), are land plants adapted to take root in shallow marine waters. Algae are by far the most abundant and widespread marine plants, occurring in all the world's natural bodies of water. They are simple primitive plants that lack true leaves, stems and roots. However, their cells organize to form the corresponding structures of blades, stalks, rhizoids and gas-filled bladders. Seagrasses and mangroves are true flowering plants that have flowers, fruits and setting seeds as well as vascular tissue and roots.

Plants, like animals, are classified hierarchically into phyla, classes, orders (ending in 'ales'), families (ending in 'acaea'), genera and species. Marine algae are classified into six major phyla. The **diatoms** (Chrysophyta) and **dinoflagellates** (Pyrrhophyta) are microscopic single-celled forms. Diatoms may form conspicuous gelatinous mats. Dinoflagellates include the zooxanthellae whose pigments give their host animals their colour. The remaining algae are large enough to be seen directly and are classified according to the colour of their dominant photosynthetic pigments: the **Blue-green** algae (Cyanophyta), **Green** algae (Chlorophyta), **Brown** algae (Phaeophyta) and **Red** algae (Rhodophyta).

Algae reproduce in diverse ways, either asexually by spores or fragmentation, or sexually by gametes. Many algae alternate generations between spore- and gamete-producing phases. In some, the appearance of each phase may be different enough to resemble entirely different species. Others may have only one phase or the other. Unlike most marine animals, algae may be quite ephemeral in their occurrence. Many may grow rapidly and appear seasonally or in short-lived blooms when conditions are favourable. Others such as the coralline Red algae grow slowly and may form long-lived or semi-permanent structures and may help cement the reef together.

Algae lack the complex structures of higher plants. Blue-greens form mats or simple chains of cells. Greens, brown and reds form a variety of structures which may be many cells thick, but are not differentiated at the cellular level as in higher plants. Their primary structural integrity comes from the cell walls or from calcium carbonate incorporated into the structure.

◄ A field of seagrasses (p. 364) on a protected sandy slope. Seagrass meadows serve as important nurseries for many species of fishes and crustaceans and are vital food for turtles and dugongs. *(Photo: Abu Dabab, BE.)*

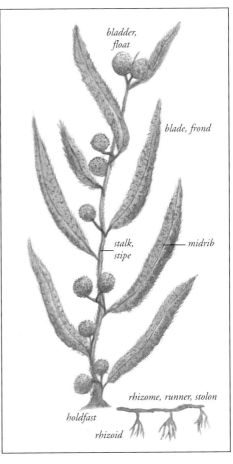

bladder, float

blade, frond

stalk, stipe

midrib

rhizome, runner, stolon

holdfast

rhizoid

1. Nabq, RM

2. Abu Galum, RM

3. Mangrove Bay, KF

BLUE-GREEN ALGAE CYANOPHYTA

The simplest and oldest multicellular algae. Formed 3 billion-year-old stromatolites that oxygenated the atmosphere allowing evolution of all higher life forms. Able to self-fertilize by converting atmospheric nitrogen into organic nitrogen. Form slimy gelatinous clusters, mats, or hair-like chains. Cover most shallow exposed surfaces and splash zones, some can withstand prolonged desiccation, some become calcified with age. Causes black-band disease in corals. Identification difficult.

1 Brown slime unid. sp. A (*Schizothrix calcicola?*)
Delicate slimy brown mats and tufts. **Biology:** on surfaces of sand, rubble, rock, or seagrasses, intertidal to over 30 m. Forms oxygen bubbles during the day. Easily breaks, pieces often float to surface. **Range:** all seas. **Similar:** *Hormothamnion enteromorphoides* is green.
2 Green-glove mat unid. sp. B
Bright blue-green mat. **Biology:** symbiotic within tissue of ascidian *Diplosoma viridis*, 0 to 30m. Note juv. frogfish, *Antennarius pictus* in photo. **Range:** Red Sea.
3 Jelly tuft unid. sp. C
Gelatinous brown tuft. **Biology:** on exposed hard surfaces. **Range:** circumtropical.

GREEN ALGAE CHLOROPHYTA

Predominately green, contain chlorophyll a and b. Wide variety of shapes. Heavily calcified *Halimeda* skeletons are a major component of sand on most coral reefs. Occur as deep as 185 m.

SIMPLE GREENS ULVACEAE
4 Green guts *Enteromorpha clathrata*
Flattened hollow tubes, to 20 cm. **Biology:** primarily in nutrient-rich areas of estuaries and bays, in intertidal zone and on shallow flats. **Range:** circumtropical.
5 Sea lettuce *Ulva lactuca*
Flat twisted blades, to 10 cm wide, to 1 m. **Biology:** on rocks and rubble of intertidal zone and shallows in nutrient-rich areas. Edible. **Range:** circumtropical.

STALKED GREENS CAULERPACEAE
6 Sea grapes *Caulerpa racemosa*
Clusters of lentil-sized spheres or flattened plates connected to prostrate branches. **Biology:** most habitats from tidepools to seaward slopes over 30 m. In strands or dense stands to 20 cm thick. Edible. **Range:** circumtropical.
7 Feather caulerpa *Caulerpa sertularioides*
Upright feather-like branches, to 20 cm high connected to prostrate branches. **Biology:** on sand or hard bottoms, tidepools to 10 m. **Range:** circumtropical.

SPONGE GREENS CODIACEAE
8 Branching sponge weed *Codium* sp.
Fuzzy rounded pencil-thick branches in round clumps. **Biology:** on shallow reefs. **Range:** Red Sea to W. Pacific.

CALCAREOUS CHAIN GREENS HALIMEDACEAE
9 Chain halimeda *Halimeda copiosa*
Discs to 1.5 cm; in chains to 40 cm. **Biology:** in shady areas of coral slopes to over 40 m. **Range:** circumtropical.

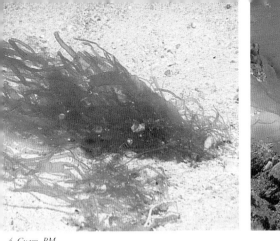
4. Guam, RM

5. Marsa Shagra, EL

6. Guam, RM

7. Bali, RM

8. Abu Galum, RM

9. Mangrove Bay, KF

1. Ras Mohammed, EL

2. Ras Mohammed, EL

3. Mangrove Bay, KF

1 Cluster halimeda *Halimeda opuntia*
Discs to 1 cm, in dense randomly branched clumps to
25 cm. **Biology:** in seagrass beds or on rubble or hard
bottoms of reef slopes to over 25 m. **Range:** circumtropical.

2 Green spongeball chain *Tydemania expeditionis*
Grape-sized balls of fine filaments in chains to 15 cm.
Biology: attached to hard surfaces of lagoon and seaward
reef slopes, 1 to over 30 m. **Range:** Red Sea to Fiji.

BUBBLE GREENS VALONACEAE
3 Sailor's eyeball *Valonia ventricosa*
Smooth shiny dark green ball to 6 cm, solitary or in
clumps, sometimes encrusted with epiphytes. **Biology:**
among algae or in small crevices, reef flats to 80 m. A single
huge cell with a giant vacuole. **Range:** circumtropical.

BROWN ALGAE PHAEOPHYTA
Predominately brown seaweeds that contain chlorophylls a
and c and xanthophylls. Many have gas-filled bladders.
Reach their greatest diversity and size in cold waters.

KELPS ALARIACEAE
4 Southern kelp *Ecklonia radiata*
Thick broad blades with spiny surface attached to central
stipe, strong holdfast, to 1.5 m. **Biology:** on exposed rocky
shores and slopes in areas of upwelling (16–23°C). Part of
warm-temperate community. **Range:** Arabian Sea from
E. Yemen to S. Oman; S. Africa to New Zealand.

SIMPLE BROWNS DICTYOTACEAE
5 Irdescent ribbonweed *Dictyota bartayrsii*
Delicate dichotomously branched blades, tips rounded, to 10
cm; may have blue iridescence. **Biology:** attached to rubble
in protected areas, 1 to 40 m. **Range:** circumtropical.

6 Variegated fan weed *Lobophora variegata*
Encrusting yellow to reddish-brown fans in 3 forms: ruffles
with lobed margins; shelf as flat blades; crust as tightly
adhering patches. **Biology:** on lagoon and seaward slopes
from mangrove roots to over 100 m. Ruffles form in
shallow areas of low herbivory, 1 to 8 m; crust form in
areas of intense grazing, subtidal to 30 m; shelf form from
shallow shaded areas to deep reef slopes below 100 m.
Range: circumtropical.

7 Funnelweed *Padina* cf. *gymnaspora*
Clusters of curled fan-shaped blades with concentric lines
of tiny hairs and light calcification, to 15 cm tall; margin
rolled inward. **Biology:** sheltered to moderately exposed
reefs, intertidal to at least 14 m. Attached to rubble and
other hard surfaces. **Range:** circumtropical.

8 Fragile Funnelweed *Padina* sp.
Blades smaller than in other *Padina*. **Biology:** on rubble
and other hard surfaces of moderately exposed reefs,
intertidal to at least 14 m. **Range:** Red Sea.

BLADDER BROWNS
SCYTOSIPHONACEAE
9 Glob weed *Colpomenia sinuosa*
Smooth, slick, highly convoluted hollow globules to 30 cm
across; golden-brown. **Biology:** on gravel or rubble, intertidal
to 30 m. **Range:** circumtropical.

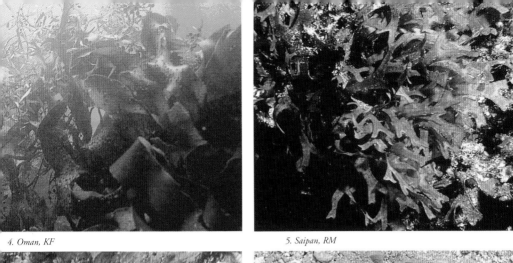

4. Oman, KF

5. Saipan, RM

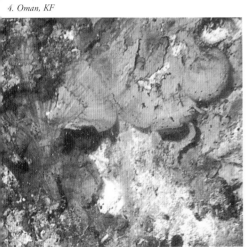

6. Marsa Shagra, RM

7. Nuweiba, EL

8. Abu Galum, RM

9. Abu Galum, EL

1. Nuweiba, RM

2. Mövenpick Bay, RM

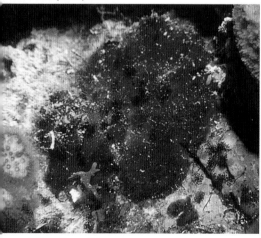

3. Marsa Shagra, RM

SARGASSUM WEEDS   SARGASSACEAE
1 Sargassum weed   *Sargassum* sp.
Buoyant fronds of small oblong blades and spherical floats around tough stipe; to over 1 m. **Biology:** attached to rubble and hard surfaces of shallow protected areas to about 5 m. **Range:** Red Sea.

2 Triangle turbinweed   *Turbinaria decurrens*
Extremely tough; blades around central stipe, tips with flattened triangular swelling containing internal bladder; in clusters or solid carpet. **Biology:** outer reef flats and margins exposed to surge. **Range:** Red Sea.

RED ALGAE   RHODOPHYTA
Predominately red seaweeds that contain chlorophylls a and d (in some), and red biliproteins. Many finely branched species form a dense turf on hard surge-swept surfaces. Coralline reds (Corallinaceae) are highly calcified branched tufts or crusts that cement rubble and sand into solid structures and build up exposed reef margins.

SQUAMARIACEAE
3 Blue crust weed   *Peyssonnelia* cf. *capensis* (?)
Small blue rounded encrustations with free edges, to 10 cm. Colour possibly highly variable. **Biology:** on hard surfaces, coral slopes from shady shallow areas to depths of 200 m. **Range:** Red Sea to W. Pacific?

SLIMY REDS HELMINTHOCLADIACEAE
4 Slimy noodle weed   *Trichogloea* (*Liagora* (?)) sp.
Slimy noodle-like tufts of white branches with red gelatinous halo. **Biology:** on hard surfaces of lagoon and seaward reefs, 1 to 15 m. **Range:** Red Sea only (?).

FLUFFY REDS   BONNEMAISONIACEAE
5 Asparagus weed   *Asparagopsis taxiformis*
Soft finely branched reddish- to bluish-brown tufts to 15 cm. **Biology:** in clusters on shallow subtidal slopes exposed to wave action. Edible. **Range:** circumtropical.

CORALLINE REDS   CORALLINACEAE
6 Fragile coral weed   *Amphiroa fragilissima*
Narrow (1 mm) brittle dichotomously jointed calcified branches in tufts, to 8 cm. **Biology:** lightly attached to hard surfaces and small crevices of lagoon and seaward reefs to at least 10 m. **Range:** circumtropical.

7 Laminar coral weed   *Mesophyllium mesomorphum*
Dark red to pink brittle overlapping plates. **Biology:** on clear lagoon and seaward reefs to over 35 m. In shady pockets of shallows, more exposed or epiphytic on other algae deeper. **Range:** Red Sea to Polynesia; W. Atlantic?

8 Crustose coralline red   *Sporolithon* (?) sp.
Reddish-grey calcareous crust, white undersurface. **Biology:** on hard surfaces of most zones, particularly clear exposed areas to over 20 m location. **Range:** circumtropical. Many **similar** species.

BRANCHING REDS RHODOMELIACEAE
9 Papillose laurencia   *Laurencia papillosa*?
Yellowish-tan; stubby side branches with knobby tips. **Biology:** on rubble and shell fragments of outer lagoon and seaward reefs. **Range:** Red Sea.

4. Abu Galum, EL

5. Guam, RM

6. Marsa Shagra, RM

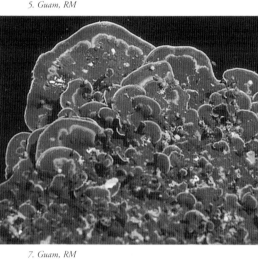

7. Guam, RM

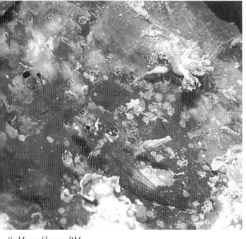

8. Marsa Shagra, RM

9. Marsa Bareka, EL

1. Mangrove Bay, RM

SEAGRASSES
ANGIOSPERMAE

Seagrasses are flowering plants with roots, rhizomes (underground stems) and leaves interconnected by a vascular system. Some have distinct erect stems. Most species have linear to slightly curved ribbon-like leaves with longitudinal veins. *Halophila* has tiny oval leaves (under 2.5 cm) and *Syringodium* has string-like cylindrical leaves. All seagrasses propagate horizontally and reproduce sexually by pollination with seed production. Seagrass meadows are a vitally important coastal habitat that is more productive than agricultural land and is closely linked to coral reefs. They stabilize sediments, trap nutrients for recycling and serve as nursery grounds for many species of crustaceans and fishes. Turtles, dugongs, sea urchins and a few molluscs and fishes feed directly on seagrasses. Many more species feed on the epiphytic algae that grow on them. At least 9 species of seagrasses occur in the Red Sea.

CYMODOCEACEAE
1 Clump seagrass　　　　　　*Cymodocea serrulata*
Leaves broad (4 to 9 mm), curved, 6 to 15 cm long, in fan-like clumps connected to short shoot; rhizome smooth. **Biology:** forms vast meadows in lagoons and bays, 1 to 25 m. **Range:** Red Sea to W. Pacific. **Similar:** *C. rotundata* has narrow leaves (Red Sea to New Caledonia); *Thalassia hemprichii* has longer leaves with short black bars of tannin pigment, shredded remnants of old leaves at base of each clump and rhizomes with a scale at each scar (Red Sea to Gilbert Is.).

POTAMOGETONACEAE
2 Canopy seagrass　　　　*Thalassodendron ciliatum*
Leaves broad (to 1.4 cm), curved, to 15 cm long, in cluster at end of long shoot (to 65 cm); rhizome thick and woody. **Biology:** in semi-protected areas, along lower bases of coral slopes and on sand, 1 to over 12 m. Rooted to hard bottoms as well as sand. Forms deep dense stands with closed canopy. Important shelter for many fishes and favoured habitat of the parrotfish *Leptoscarus vaigiensis*. **Range:** Red Sea to Vanuatu.
3a Ribbon seagrass　　　　　　*Halodule uninervis*
Leaves narrow (5 mm), flat, tip with 3 points, to 20 cm long; attached directly to thin rhizome with many leaf scars. **Biology:** in mud to coarse sand of protected and semi-exposed bays and lagoons, shoreline to 6 m, rarely to 30 m. In pure meadows or mixed with *Cymodocea*, *Halophila*, *Syringodium* (also in photo) or *Thalassia*. **Range:** Red Sea to Fiji. **Similar:** *Cymodocea rotundata* has smooth rhizomes (Red Sea to New Caledonia).
3b Dugong grass　　　　　　*Syringodium isoetifolium*
Leaves narrow (1–2 mm), cylindrical, pointed, 7–30 cm long, 1 to 3 per shoot; rhizome smooth. **Biology:** in protected areas of lagoons and bays, shoreline to 15 m. In pure meadows or mixed with *Cymodocea*, *Halodule*, *Halophila* or *Thalassia*. **Range:** Red Sea to Samoa.

2. Nabq, RM

3. Marsa Abu Dabab, RM

▲ White mangove
Avicennia marina (Order
Tubiflorae: VERBENACEAE)
Small tree with many pencil roots,
prop and stilt roots, small yellow
flowers and heart-shaped fruit with
point at base. **Biology:** in groves
along intertidal zone of protected
shores of bays, wadis and estuaries.
The only species in N. Egypt. About
2,700 km² of mangroves occur in the
Red Sea and Gulf of Aden. **Range:**
Red Sea and Arabian Gulf to W.
Pacific, s. to S. Africa. **Similar:** Red
mangrove *Rhizophora mucronata*
(order Myrtales: Rhizophoraceae) has
extensive prop-roots, no pencil roots
and spear-shaped fruit (S. Red Sea to
S. Africa). *(Photo: Mangrove Bay, RM.)*

MANGROVES ANGIOSPERMAE

Mangroves form a distinct ecosystem closely linked to coral reefs, which
bridges the marine and terrestrial worlds. Mangroves are a group of trees
from several families that grow in the intertidal zone with roots and
trunks submerged in brackish or marine waters. They share distinct
adaptations for removing excess salt and obtaining oxygen while rooted
in mud. Excess salt may be excluded from the water taken up by the roots
by an ultra-fine filtration system, filtered through special glands in the
leaves, or stored in older leaves before they are shed. Mangrove roots
obtain oxygen through modifications that expose them to air including
aerial roots, stilt roots, and pencil and knee roots. The distinctive buoyant fruits have a spear-like or pointed embryonic root-stem that hangs
down and lodges into soft sediment to start a new plant.

Mangroves stabilize sediments, remove pollutants, trap nutrients for
recycling and provide habitat for many terrestrial and marine organisms.
Nutrients from leaf litter, insects and bird droppings enter the marine
food chain through mangroves. By providing food and shelter, mangroves are important nurseries for juveniles of many species of fishes that
inhabit coral reefs as adults. This may occur only during high tide in
small groves or continuously in large groves riddled with permanent
creeks. Mudflats associated with mangroves are important habitat for
mudskippers (*Periophthalmus*) and crabs, particularly fiddler crabs and
the large edible mangrove crab (*Scylla serrata*). The roots of mangrove
may be encrusted with oysters and other bivalves, while the canopy is a
favoured nesting habitat for herons, ibises, ospreys and other birds.

BIBLIOGRAPHY

Allen, G.R. 1991. *Damselfishes of the World*. Mergus, Melle. 271 p.

Allen, G.R., R. Steene, and M. Allen. 1998. *A Guide to Angelfishes and Butterflyfishes*. Tropical Reef Research, Perth. 250 p.

Baranes A. and D. Golani. 1993. An annotated list of deep-sea fishes in the northern Red Sea, Gulf of Aqaba. *Israel J. Zool.* **39**: 299–336.

Bergbauer, M. 1997. *Giftige und Gefährliche Meerestiere*. Müller Rüschlikon, Cham.

Colin, P.L. and C. Arneson. 1995. *Tropical Pacific Invertebrates*. Coral Reef Press, Beverly Hills. 296 p.

Debelius, H. 1996. *Nudibranchs and Sea Snails*. Ikan, Frankfurt.

_____. 1998. *Reef Guide Red Sea*. Ikan, Frankfurt. 321 p.

_____. 1999. *Crustacea Guide of the World*. Ikan, Frankfurt. 321 p.

Dekker, H. and Z. Orlin. 2000. Checklist of Red Sea Mollusca. *Spirula* 47 Suppl. 11.

Erhardt, H. and H. Moosleitner. 1995. *Meerwasser-Atlas, Band 2–3*. Mergus Verlag, Melle. 1322 p.

Eichler, D. and E. Lieske. 1996. *Korallenfische-Indischer Ozean*. Jahr Verlag, Hamburg. 368 p.

Eichler, D. and R.F. Myers. 1997. *Korallenfische-Zentraler Indopazifik*. Jahr Verlag, Hamburg. 489 p.

Eichler, D. 1997. *Tropische Meerestiere*. BLV, Munich. 271 p.

Fabricius, K. and P. Alderslade. 2001. *Soft Corals and Sea Fans*. Australian Inst. of Marine Science, Townsville, Qld. 264 p.

Fautin, D. and G.R. Allen. 1992. *Field Guide to Anemonefishes and their Host Anemones*. W. Australian Museum, Perth. 157 p.

Field, R. and M. Field. 1998. *Reef Fishes of the Red Sea*. Kegan Paul Publ., London.

Goren, M. and M. Dor. 1994. *An Updated Checklist of the Fishes of the Red Sea*. Israel Academy of Sciences and Humanities, Elat. 120 p.

Gosliner, T.M., D.W. Behrens, and G.C. Williams. 1996. *Coral Reef Animals of the Indo-Pacific*. Sea Challengers, Monterey, CA. 314 p.

Göthel, H. 1994. *Farbatlas Meeresfauna Fische*. Ulmer Verlag, Stuttgart.

Hoover, J.P. 1998. *Hawaii's Sea Creatures*. Mutual Publ., Honolulu. 366 p.

Khalaf, A.M. and A.M. Disi. 1997. Fishes of the Gulf of Aqaba. *Marine Station Aqaba*, 8.

Kemp, M.J. 1998. Zoogeography of coral reef fishes of the Socotra Archipelago. *Journal of Biography* 25.

_____. 2000. Zoogeography of coral reef fishes of the north-eastern Gulf of Aden. *Fauna of Arabia* 18.

Klausewitz, W. 1989. Evolutionary history and zoogeography of the Red Sea ichthyofauna. *Fauna of Arabia* 10.

Kuiter, R. 1998. *Fishes of the Maldives*. Atoll Editions, Apollo Bay, Australia. 267 p.

_____. 2000. *Seahorses, Pipefishes and Their Relatives*. TMC Publ., UK. 240 p.

Kuiter, R. and T. Tonozuka. 2001. *Indonesian Reef Fishes*. Zoonetics, Seaford, Australia. 893 p.

Lieske, E. and R.F. Myers. 2002. *Coral Reef Fishes*. HarperCollins (UK), Princeton (USA). 400 p.

Lourie, S., A.C.J. Vincent, and H.J. Hall. 1999. *Seahorses, an Identification Guide to the World's Secies and their Conservation*. Project Seahorse, Surrey, UK. 214 p.

Mebs, D. 1989. *Gifte im Riff, Wissenschaftl*. Verlagsgesellschaft, Stuttgart.

Mergner, H. and H. Schuhmacher. 1974. Morphologie, Ökologie, und Zonierung von Korallenriffen bei Aqaba, Helgoländer wiss. *Meeresunters.*, 26.

Michael, S.W. 1993. *Reef Sharks and Rays of the World*. Sea Challengers, Monterey, CA. 107 p.

_____. 1998. *Reef Fishes Vol. 1*. Microcosm, Shelburne, Vermont. 624 p.

Myers, R.F. 1999. *Micronesian Reef Fishes, a Comprehensive Guide to the Fishes of Micronesia*. Coral Graphics, Guam. 330 p.+192 pls.

Randall, J.E. 1983. *Red Sea Reef Fishes*. Immel Publ., London.

_____. 1986. *Sharks of Arabia*. Immel Publ., London.

_____. 1995. *Coastal Fishes of Oman*. Univ. Hawaii Press, Honolulu. 439 p.

Randall, J.E., G.R. Allen, and R. Steene. 1997. *Fishes of the Great Barrier Reef*. Univ. Hawaii Press, Honolulu. 557 p.

Ritter, E. 2002. *Uber die Körpersprache von Haien*. Verlag Dr. Steinert, Witten.

Schuhmacher, H. 1991. *Korallenriffe*. BLV, Munich.

Schuhmacher, H. and J. Hinterkircher. 1996. *Niedere Meerestiere*. BLV, München.

Sharabati, S. 1984. *Red Sea Shells*. KPI Ltd, London.

Smith, M.M. and P.C. Heemstra (eds). 1991. *Smith's Sea Fishes*. Southern Book Publ., Johannesburg. 1048 p. +144 pls.

Spalding, M.D., C. Ravilious, and E.P. Green. 2001. *World Atlas of Coral Reefs*. Univ. of California Press, Berkley. 424 p.

Veron, J.E.N. 2000. *Corals of the World*. Australian Inst. of Marine Science, Townsville, Qld. 3 vols, 1382 p.

Vine, P. 1996. *Red Sea Invertebrates*. Immel Publ., London. 224 p.

Weinberg, S. 1996 *Rotes Meer – Indischer Ozean*. Delius Klasing, Stuttgart.

Yonow, N. 2000. Red Sea Opisthobranchia 4. *Fauna of Arabia* 18.

Yonow, N. 1996. Une breve revue du regime alimentaire des Nudibranches et Sacoglosses. *Rev. fr. Aquariol.* 23.

INDEX TO SCIENTIFIC NAMES

INDEX TO ENGLISH NAMES

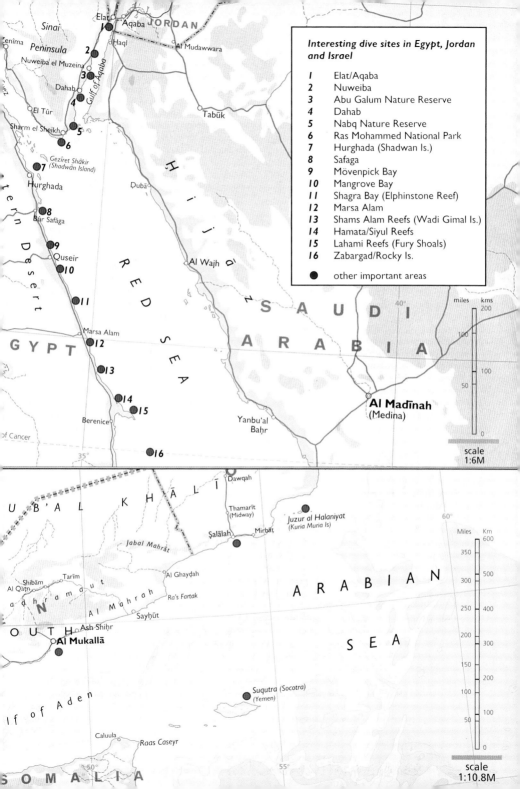

Interesting dive sites in Egypt, Jordan and Israel

1	Elat/Aqaba
2	Nuweiba
3	Abu Galum Nature Reserve
4	Dahab
5	Nabq Nature Reserve
6	Ras Mohammed National Park
7	Hurghada (Shadwan Is.)
8	Safaga
9	Mövenpick Bay
10	Mangrove Bay
11	Shagra Bay (Elphinstone Reef)
12	Marsa Alam
13	Shams Alam Reefs (Wadi Gimal Is.)
14	Hamata/Siyul Reefs
15	Lahami Reefs (Fury Shoals)
16	Zabargad/Rocky Is.
●	other important areas

scale 1:6M

scale 1:10.8M

ACKNOWLEDGEMENTS

This book would not have been possible without the generous support of the following persons and institutions: Myles Archibald, Johann Vifian (Subex, Mövenpick Hotel), Pierre De La Motte (LTU), Wolfgang Thaler and Wolfgang Clausen (Orca, Hamata), Robert Fellermeier (Mövenpick Hotel, Quseir), Hossam Helmy and Heshem Mostafa Kamel (Marsa Shagra, Red Sea Diving Safaries), Essam Hassan and Robert Furtner (Mangrove Bay Resort), Dr A. Grissac and Aymann Maybrouk (directors of Ras Mohammed National Park), Fares Abdel Aziz Hassan (Abu Galum Nature Reserve), Kerstin Ehlert and Medhat El Dakory (Shams Alam Hotel), Tarek Mansour (Nabaa Camp), Rolf Schmidt (Sinai Divers), Ingrid El Kabany and Mohammed El Kabany (Inmo Diver's Lodge, Dahab), Lagona Divers (Dahab), René Hugenschmidt (Hugy underwater housings), Thomas Ritscheck (Tauchen Magazine), R. Boll (Fuji-Film, Germany), Susanne Rossbach (Daniella Village, Dahab), Dr A. Baranes (Marine Station, Elat) and Dr M.A. Khalaf (Marine Station, Aqaba). Special thanks to Dr Michael Hackenberg and Stefanie Moehrle who helped us during our diving expeditions to the Sinai.

The following scientists lent their expertise in species identification: John E. Randall, Ofer Gon, David Greenfield, and Thomas Fraser (Apogonidae); Sara Lourie (*Hippocampus*); Prof. Daniel Robineau, Natural History Museum, Paris (mammal distributions); Gustav Paulay, Florida Natural History Museum (many Cnidaria and Crustacea); Dr Nathalie Yonow, University of Wales, Swansea (Opisthobranchia); Dr Leslie Newmann, Southern Cross University, NSW (Flatworms); Dr C.H.J.M. Fransen Natural History Museum, Amsterdam (Crustacea); Dr Robert van Soest, University of Leiden (Porifera); Dr Vollrath Wiese, Cismar, Germany (Mollusca). Patrice Marker assisted with the layout and various computer problems. Thanks to Helen Brocklehurst, our editor at HarperCollins.

Finally, it is the photographs that make this book beautiful and useful. Special thanks go to Klaus E. Fiedler who provided more photographs than any other collaborator and co-authored introductions to several invertebrate groups. The late Jürgen Kuchinke and Dagmar Kuchinke located and photographed many small creatures including nearly all of the flatworms and nudibranchs during 14 years in Saudi Arabia.

We thank the highly talented individuals who generously contributed their photographs. Photographic credit is given in the form of their initials in the caption of each photograph. They are:

TA Tim Allen
MB Mathias Bergbauer
WB Waltraut Binanzer
JB Jerry Borrell
BE Birgit Eichenseher
DE Dieter Eichler
TE Tyrid Engstler
HE Harry Erhardt
KF Klaus E. Fiedler
HF Herbert Frei
BF Borut Furlan
GG Gottfried Genser
HG Helmut Göthel
HGr Hilmar Graulich
MH Michael Hackenberg
SH Simon Hackett
HHH Hans Hermann Harms

PH Pit Hausmann
KH Klaus Hilgert
JH Johann Hinterkircher
JPH John P. Hoover
LI Linda Ianniello
RK Ralf Kiefner
MK Manuela Kirschner
AK Andreas Koffka
JK Jürgen Kuchinke
RK Rudie H. Kuiter
EL Ewald Lieske
HL Holger Lucas
MM Minas Mavrikakis
PM Patrice Marker
KM Kevin Metcalfe
SM Scott Michael
SMo Stefanie Moehrle

HM Horst Moosleitner
RM Robert F. Myers
PN Peter Nahke
JN John Neuschwander
WP Winfried Persinger
JR John E. Randall
CR Carl Roessler
GS Georg Scheer
HS Hagen Schmid
LS Larry Sharron
DS David Shen
HSj Hans Sjöholm
MT Markus Thiele
WT Werner Thiele
FV Fred Voigt
HV Herwarth Voigtmann
VW Vollrath Wiese